Sustainable Textiles: Production, Processing, Manufacturing & Chemistry

Series Editor

Subramanian Senthilkannan Muthu, Head of Sustainability, SgT and API, Kowloon, Hong Kong

This series aims to address all issues related to sustainability through the lifecycles of textiles from manufacturing to consumer behavior through sustainable disposal. Potential topics include but are not limited to: Environmental Footprints of Textile manufacturing; Environmental Life Cycle Assessment of Textile production; Environmental impact models of Textiles and Clothing Supply Chain; Clothing Supply Chain Sustainability; Carbon, energy and water footprints of textile products and in the clothing manufacturing chain; Functional life and reusability of textile products; Biodegradable textile products and the assessment of biodegradability; Waste management in textile industry; Pollution abatement in textile sector; Recycled textile materials and the evaluation of recycling; Consumer behavior in Sustainable Textiles; Eco-design in Clothing & Apparels; Sustainable polymers & fibers in Textiles; Sustainable waste water treatments in Textile manufacturing; Sustainable Textile Chemicals in Textile manufacturing. Innovative fibres, processes, methods and technologies for Sustainable textiles; Development of sustainable, eco-friendly textile products and processes; Environmental standards for textile industry; Modelling of environmental impacts of textile products; Green Chemistry, clean technology and their applications to textiles and clothing sector; Eco-production of Apparels, Energy and Water Efficient textiles. Sustainable Smart textiles & polymers, Sustainable Nano fibers and Textiles; Sustainable Innovations in Textile Chemistry & Manufacturing; Circular Economy, Advances in Sustainable Textiles Manufacturing; Sustainable Luxury & Craftsmanship; Zero Waste Textiles.

More information about this series at http://www.springer.com/series/16490

Subramanian Senthilkannan Muthu · Ali Khadir
Editors

Dye Biodegradation, Mechanisms and Techniques

Recent Advances

Editors
Subramanian Senthilkannan Muthu
SgT Group and API
Hong Kong, Kowloon, Hong Kong

Ali Khadir
Islamic Azad University of Shahre Rey
Branch
Tehran, Iran

ISSN 2662-7108 ISSN 2662-7116 (electronic)
Sustainable Textiles: Production, Processing, Manufacturing & Chemistry
ISBN 978-981-16-5934-8 ISBN 978-981-16-5932-4 (eBook)
https://doi.org/10.1007/978-981-16-5932-4

This Springer imprint is published by the registered company Springer Nature Singapore Pte Ltd.
The registered company address is: 152 Beach Road, #21-01/04 Gateway East, Singapore 189721, Singapore

Contents

About the Editors

Dr. Subramanian Senthilkannan Muthu currently works for SgT Group as Head of Sustainability, and is based out of Hong Kong. He earned his Ph.D. from The Hong Kong Polytechnic University, and is a renowned expert in the areas of Environmental Sustainability in Textiles & Clothing Supply Chain, Product Life Cycle Assessment (LCA) and Product Carbon Footprint Assessment (PCF) in various industrial sectors. He has 5 years of industrial experience in textile manufacturing, research and development and textile testing, and over a decade's experience in life cycle assessment (LCA), and carbon and ecological footprints assessment of various consumer products. He has published more than 100 research publications, and written numerous book chapters and authored/edited over 100 books in the areas of Carbon Footprint, Recycling, Environmental Assessment and Environmental Sustainability.

Ali Khadir is an environmental engineer and a member of the Young Researcher and Elite Club, Islamic Azad University of Shahr-e-Rey Branch, Tehran, Iran. He has published/prepared several articles and book chapters under reputed international publishers, including Elsevier, Springer, Taylor & Francis and Wiley. His articles have been published in journals with IF of greater than 4, including Journal of Environmental Chemical Engineering and International Journal of Biological Macromolecules. He also has been the reviewer of journals and international conferences. His research interests center on emerging pollutants, dyes, and pharmaceuticals in aquatic media, advanced water and wastewater remediation techniques and technology. At present, he is editing other books in the field of nanocomposites, advanced materials and the remediation of dye-containing wastewaters.

Dyes: Classification, Pollution, and Environmental Effects

Said Benkhaya, Souad M'rabet, Hassane Lgaz, Abderrahim El Bachiri, and Ahmed El Harfi

Abstract There are a wide variety of textile dyes including reactive, direct, vat, sulfur, disperse, basic, and acid dyes. Therefore, the classification of dyes has become mandatory due to the increase in the annual global production of these compounds. They can be classified according to their chemical compositions (azo, anthraquinone, nitroso, nitro, indigoïde, cyanine, phtalocyanine, and triphenylmethane) or according to their field of application to different substrates such as textile fibers, paper, leathers, and plastics. In our new investigation, we have been able to describe the different families of textile dyes according to their chemical structures (chromophoric/auxochromic), application methods in the textile industry, and their Color Index (**C.I**). The presence of these dyes in the liquid effluents from washing textiles cause a negative impact on the balance of the aquatic environment, which requires prior treatment of these effluents.

Keywords Textile dyes · Azo dyes · Reactive dyes · Color index · Chromophoric/Auxochromic · Aquatic environment

---

S. Benkhaya (✉) · A. El Harfi
Laboratory of Advanced Materials and Process Engineering, Faculty of Sciences, University Ibn Tofail, Kénitra, Morocco
e-mail: said.benkhaya@uit.ac.ma

S. M'rabet
Laboratory of Geosciences and Environment, Team Development Geo Resource and Renovation Territories, Department of Geology, Faculty of Science, University Ibn Tofail, Kenitra, Morocco

H. Lgaz
Department of Crop Science, College of Sanghur Life Science, Konkuk University, Seoul 05029, South Korea

A. El Bachiri
Royal Naval School, University Department, Boulevard Sour-Jdid, Casablanca, Morocco

S. S. Muthu and A. Khadir (eds.), *Dye Biodegradation, Mechanisms and Techniques*, Sustainable Textiles: Production, Processing, Manufacturing & Chemistry,
https://doi.org/10.1007/978-981-16-5932-4_1

1 Introduction

The mauveine, the first synthetic dye, was discovered by chance by William Henry Perkin in 1856, then 18 years old [16]. It was obtained from aniline (obtained from coal tar) by the action of sulfuric acid in the presence of potassium bicarbonate and made it possible to dye the silk purple. The first so-called "Azo" dyes were discovered in Great Britain in 1860. They quickly ousted aniline-based dyes, which had poor light resistance. But it was the German industry (Badische Anilin und Soda Fabrik: BASF) that made the biggest contribution to the rise of the dye industry [101].

A dye is a colored substance, natural or synthetic, which interacts with the medium into which it is introduced and colors it by dissolving and dispersing therein. These dyes constitute a real industry and a capital of modern chemistry, thanks to their chemical stability, ease of synthesis, and variety of colors. They have two specific properties: color and the ability to be fixed to solid substrates such as textiles, by dyeing or printing techniques. In general, dyes are widely used in printing, food products, cosmetics, and in particular in the dyeing of textile substrates (fibers, leathers, furs, woods, plastics, elastomers, etc.). In addition, they can be used in the field of research, in order to reveal small transparent structures by microscopy [19].

An enormous amount of synthetic dyes is used annually in the textile, leather, plastics, paper, and dye industries due to their coloring properties [29]. This results in a large amount of colored wastewater, unsuitable for recycling without proper treatment. Dyes are ubiquitous and persistent environmental contaminants due to their large-scale production and numerous areas of application [16]. Azo dyes are among the most widely used dyes in several industrial applications such as cosmetics, the paper industry, and the textile industry, all of which generate effluents loaded with residual coloring substances [16, 15]. Like most dyes, azo dyes have been shown to have an adverse effect on aquatic organisms and humans, and lead to brain, central nervous system, carcinogenicity, and mutagenicity dysfunction in humans [17]. This is because they contain many toxic substances such as aromatics and heavy metals and could reduce the transmission of sunlight. Wastewater containing dyes from textile industries is a serious problem due to the complex chemical structures of the dyes, their high pH, high chemical oxygen demand, and high temperature which make the degradation of dyes very difficult when they occur. They are present in any type of complex matrix [111]. Therefore, the emergence of true textile wastewater treatment is of great ecological interest around the world.

However, the objective of this review is to describe the different classes of textile dyes, including their chemical structures, their chromophoric, auxochromic, color index (C.I), and their application in the textile industry. Finally, we pointed out the impact of these dyes on aquatic environments.

2 Chemical Structure of Textile Dyes

Dyes have a unique chemical structure for each color. These structures are classified in the ChemSpider database under the data collection "NCSU Max Weaver Dye Library" [89]. The chemical structure of a textile dye is made up of three components: skeleton, chromophore groups (responsible for its color), auxochrome (promotes fixation on the substrate), and solubilizing groups (responsible for the solubility in water or organic solvents), Fig. 1 [16, 15].

An auxochrome is a functional group that contains isolated electron pairs that increase the intensity of the color, and are also important parts of dyes [3, 15, 39]. Table 1 shows some examples of auxochrome and chromophore groups [5, 16, 39, 118, 150, 151].

Table 2 shows the classification of textile dyes according to their chromogen [16, 15, 93, 129, 133].

Therefore, textile dyes have a unique chemical structure for each dye. This structure is made up of two main parts: The chromophore responsible for its color and the auxochrome which improves the capacity of this chromophore to absorb light. They

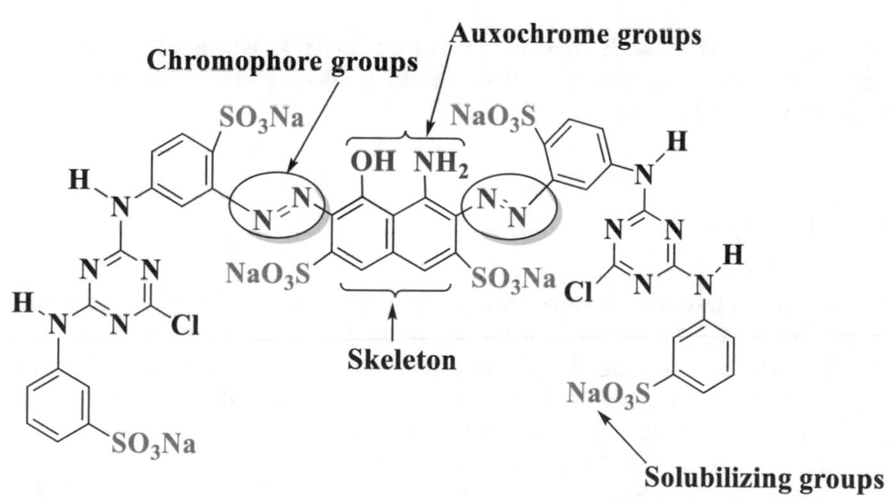

Name : C.I.Reactive Blue 171

Molecular Formula : $C_{40}H_{23}Cl_2N_{15}Na_6O_{19}S_6$

CAS Registry Number : 77907-32-5

Molecular Weight : 1418.93(g/mol)

Molecular Structure : Double azo class

Fig. 1 Chemical structure of a textile dye (S. Benkhaya)

Table 1 Chromophore and auxochrome examples (S. Benkhaya)

Chromophore group		Auxochrome group	
Azo	– N=N–	Hydroxyl	– OH
– Carbonyl – Carboxylate	– C=O – R′CO2R (R and R′ ≠ H)	– Amino – Ammonia	– NH2 – NH3
– Nitro – Sulfonate	– N=O – RSO2OH	– Aldehyde and/or – Carboxylic acid – Sulfonic acid	– CHO – COOH – SO3H
Quinoid	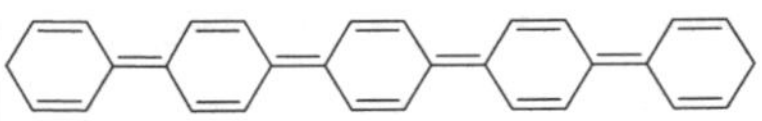	Methyl mercaptan	– SCH3

themselves fail to introduce color, but when it is present with the chromophores in an organic compound, it intensifies the color of the chromogen.

3 Classification of Textile Dyes

Dyes are organic chemicals containing aryl rings to delocalize electrons. They are mainly classified based on their structure, source, color, and method of application in the color index [16, 64].

3.1 Azo Dyes

Azo dyes have one or more than one azo groups (–N=N–) in their chemical structures (Ahmed et al. 2020) [158]. More than 70% of the dyes used in the world belong to the azo group, and they are also the most common dyes (60–70%) used for dying textiles (Alexander and Joseph Thatheyus 2021). They are widely used in a variety of industries, textile, pharmaceutical, paper, non-linear optical systems, medical and biomedical fields, etc. [62, 136, 142]. They contain an amino or dialkylamino group, and undergo a pronounced color change in different solvents and pH [77]. Azo dyes based on heterocyclic moiety are known for their excellent coloring properties, tinctorial strength, fastness, and thermal stability [120]. These dyes are generally characterized by chemical groups capable of forming covalent bonds with textile substrates [16, 15]. They represent more than 50% of synthetic dyes used worldwide [159]. They cover the whole spectrum, but mainly yellow, orange, and red dyes [1]. The azo group is responsible for the color of the dye [87]. Cleavage of the azo bond can lead to products containing aromatics, carcinogenic to humans and other organisms [11]. These dyes represent the largest class of industrially synthesized organic dyes [142]. They are one of the most widely used synthetic dyes due to their

Table 2 Classification of textile dyes according to the chromogen (S. Benkhaya)

Chromogen	Colour Index Generic Name	ColourIndex Constitution Number	CAS No	Structural formula of dye
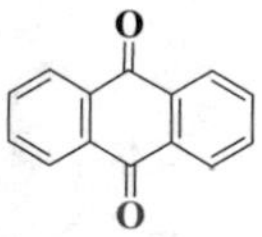 Anthraquinone	C.I. mordant red 4	C.I.58240	82-29-1	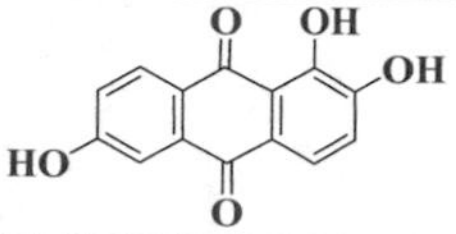
	C.I.Pigment Violet 5	C.I.58055	22297-70-7	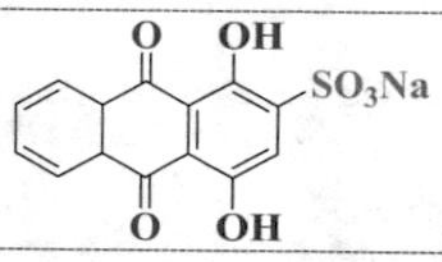
	C.I.MordantRed 11	C.I.58000	72-48-0	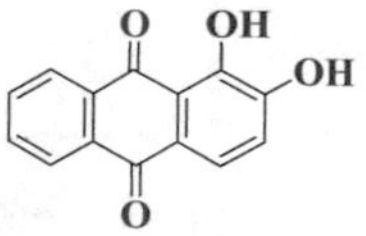
	C.I.Pigment Blue 60	C.I.69800	81-77-6	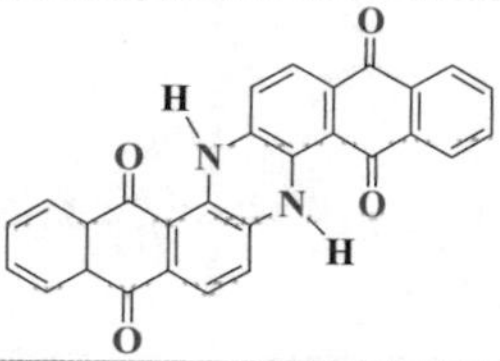
	C.I. Pigment Red 83	C.I. 58000:1	22.6.12	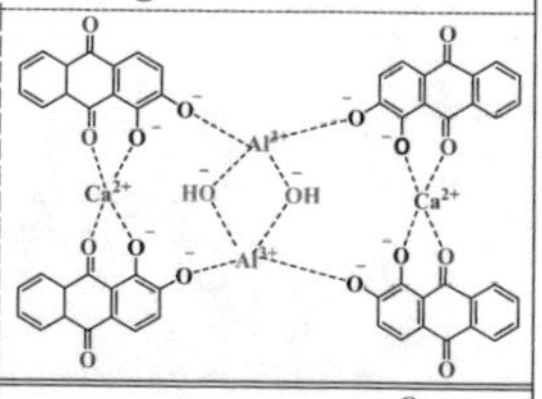
—N=O Nitroso	C.I. acid green 1	C.I. 10020	19381-50-1	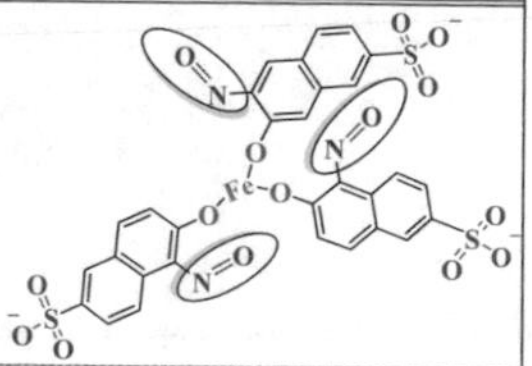

(continued)

cost-effectiveness and stability under a wide range of pH, temperature, and light conditions [139]. Figure 2 shows some chemical structures of azo dyes.

The synthesis of an azo compound is based on the appropriate oxidation/reduction reaction or diazotization/coupling reaction in the presence of a diazonium salt and a coupling component. Figure 3 shows the process for the synthesis of an azo compound [16, 15]. Penthala et al. [115] synthesized azo and anthraquinone dyes according to the reactions shown in Figs. 4, 5, and 6.

Table 2 (continued)

	C.I. Pigment Green 8	C.I. 10006	15.8.2	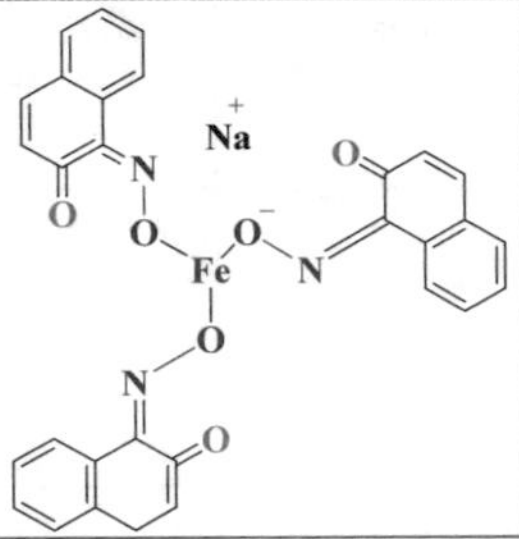
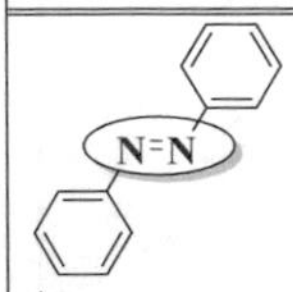 Azo	C.I.Solvent Yellow 10	C.I.11840	621-66-9	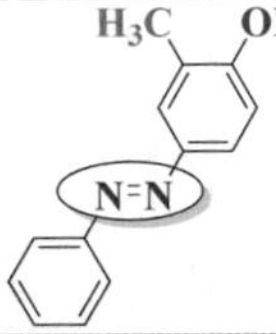
	C.I.Pigment Orange 1	C.I. 11725	11.1.1	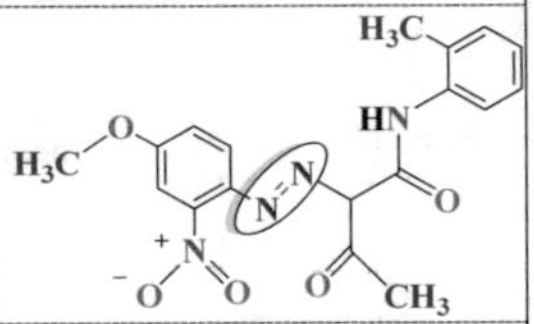
	C.I. Acid Red 112.	C.I. 27195	12.300.10	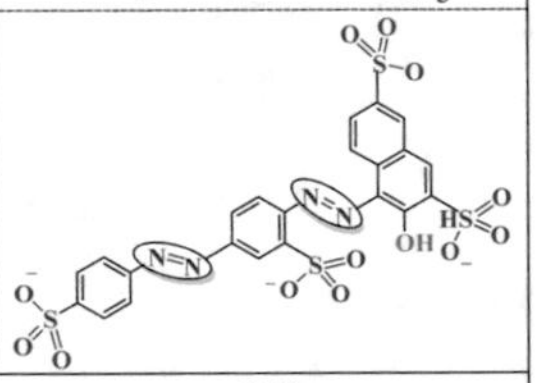
	C.I. Mordant Black 11.	C.I.14645	17.300.2	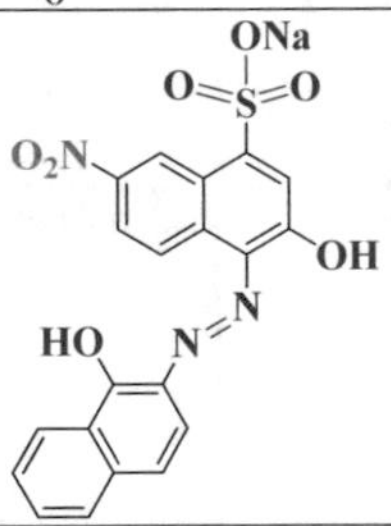

(continued)

For their part [121] synthesized azo dyes containing a thiazole group. The dye synthesis routes are illustrated schematically in Fig. 7. Concerning Ghanavatkar et al. [59], they synthesized heterocyclic azo dyes according to the synthesis scheme mentioned in Fig. 8.

Table 2 (continued)

	C.I.Acid Blue 74	C.I.73015	860-22-0	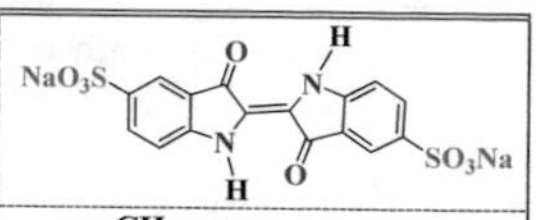
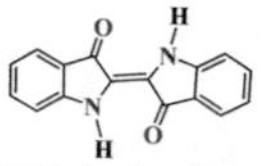 Thioindigoid	C.I.Pigment Red 181	C.I. 73360	12.181.1	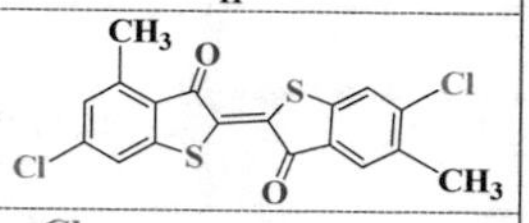
	C.I.Pigment Red 88.	C.I. 73312	12.88.1	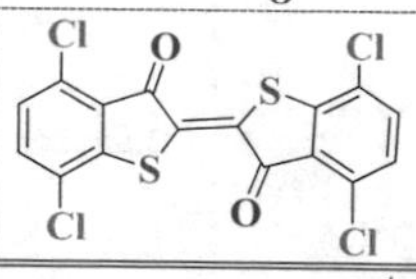
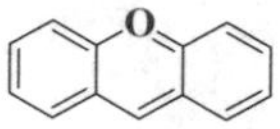 Xanthene	C.I. acid red 52	C.I. 45100	3520-42-1	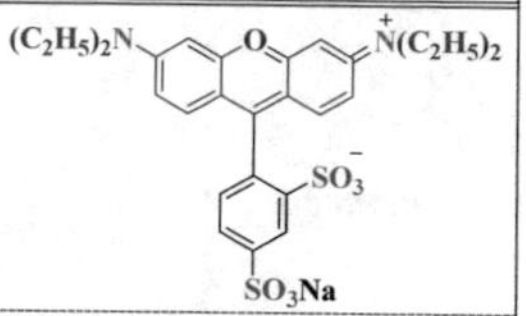
	C.I.Acid red 91.	C.I.45400	548-24-3	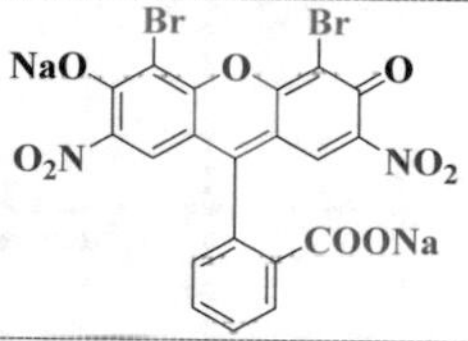
	C.I.Basic Red 1.	C.I. 45160	70.1.28	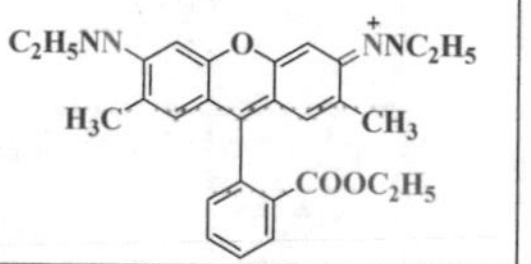
	C.I.Pigment Red 90	C.I.45380:1	29.3.1	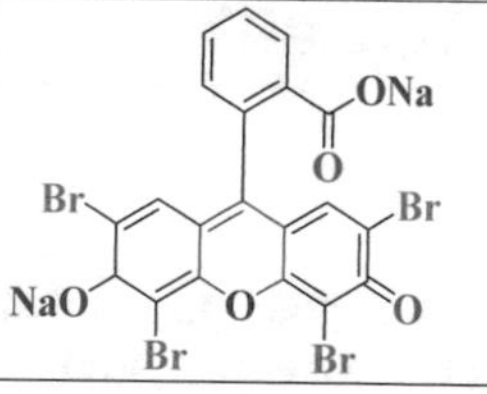

(continued)

3.2 *Reactive Dyes*

Reactive dyes are soluble anionic dyes which, in solution, are repelled by the negatively charged surface of cotton fiber [85]. The first commercial reactive dyes for cotton were based on the dichloro-s-triazine reactive group [85]. Reactive dyes are

Table 2 (continued)

	C.I. Acid red 91	C.I. 45400	-	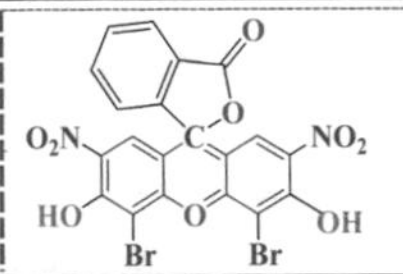
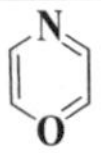 Oxazine	C.I. basic blue 12	C. I. 51180	3625-57-8	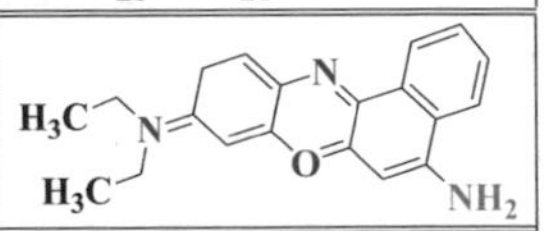
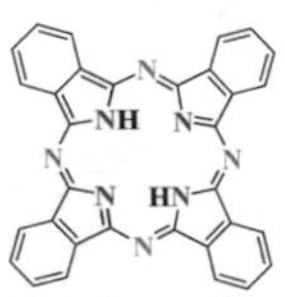 Phthalocyanine	C.I.direct blue 199	C.I. 74190	12222-04-7	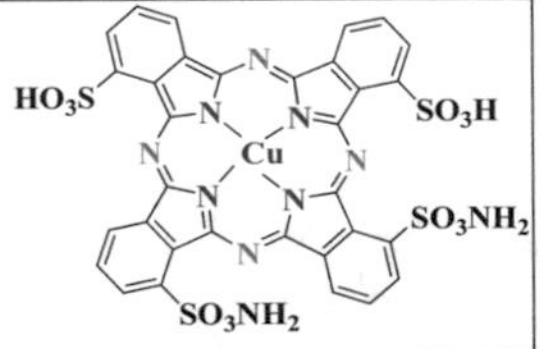
	C.I.Direct Blue 86	C.I.74180	1330-38-7	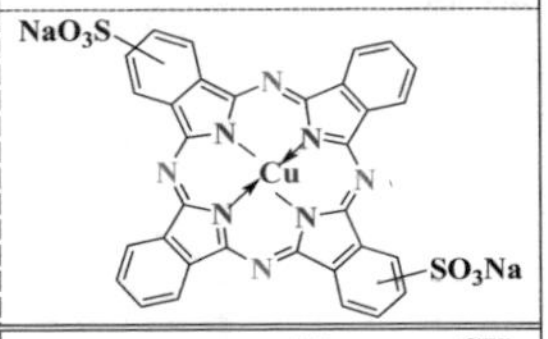
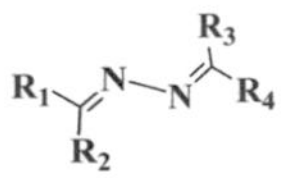 Azine	C.I. basic red 5	C. I. 50040	553-24-2	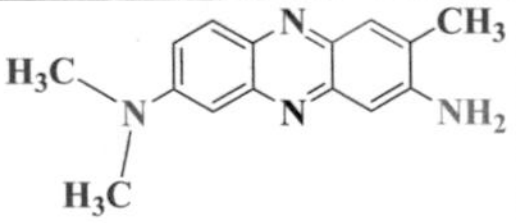
Perylene	C.I. Pigment Red 179	C.I. 71130	12.179.1	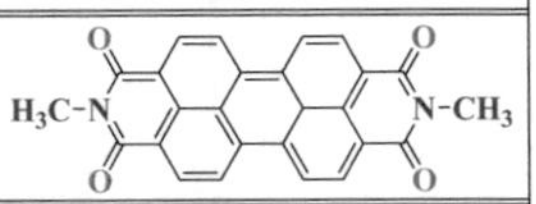
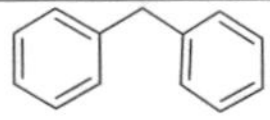 Diphenylmethane	C.I. basic yellow 2	C.I. 41000	2465-27-2	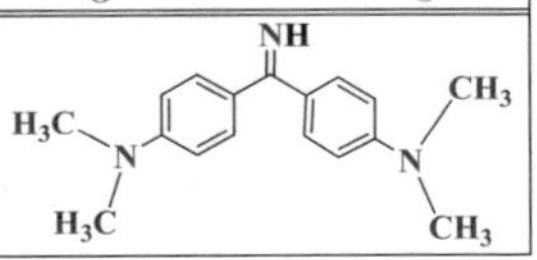

(continued)

widely applied due to their brilliant colors, excellent long-term colorfastness properties, vivid colors, high photolytic stability, and a variety of profile shades [28, 114, 135]. These dyes are also known for their high solubility in water and low degradability [65]. Reactive dye is the most important class of dyes for cellulosic fibers and is also used today for protein fibers such as wool and silk [130]. The application of reactive dyes to the cellulosic fiber requires large amounts of salt like common salt (NaCl) or Glauber's salt (Na_2SO_4) to reduce the electrostatic repulsion between the anionic dye and the anionic cellulosic material [6]. These compounds are classified as azo and anthraquinone dyes based on their complex aromatic molecular structure [7]. These are the second largest classes of dyes [104, 143]. They are capable of forming a covalent bond with the amine or sulfhydryl groups of proteins in textile fibers [16].

Table 2 (continued)

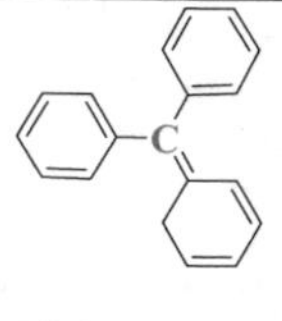 Triphenylmethane	C.I. basic green 4	C.I. 42000	2437-29-8	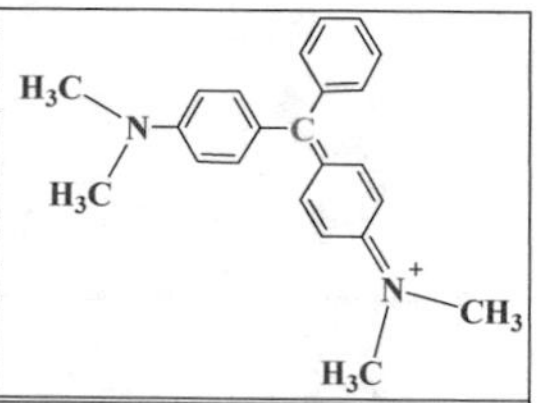
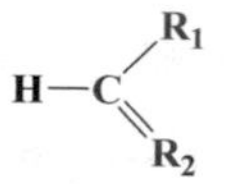 Methine	C.I.Basic Red 22	C.I.11055	12221-52-2	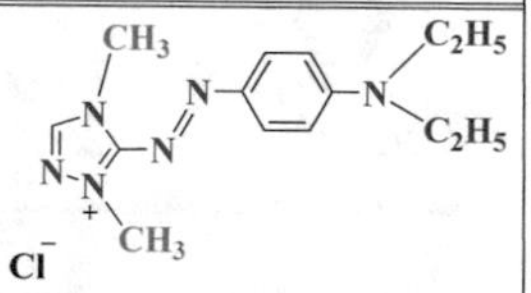
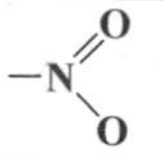 Nitro	C.I.Sulphur Blue 11	C.I.53235	1326-98-3	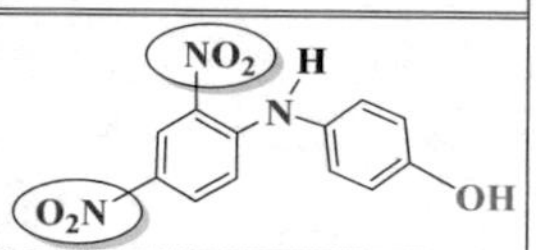
	C.I.Acid Yellow 24	C.I. 10135	10142-54-8	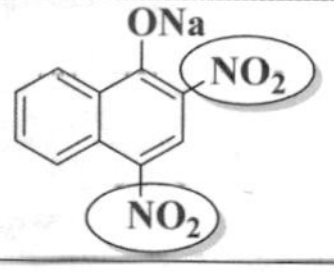

We can list three groups of reactive dyes: Azo, Anthraquinone, and Phthalocyanine [86].

This class of dyes were developed on the basis of direct dye; the structural difference between them is that reactive groups in reactive dye molecules form covalent bonds with cellulose fibers, which overcomes the disadvantage of the weak wet fastness of direct dye [131]. The rate of fixation is 50–70% for monofunctional reactive dyes and 70–85% for bifunctional reactive dyes, even under optimized dyeing conditions [47]. The dyeing of reactive dyes when dyeing cellulosic substrates leaves between 20 and 60% of the dyes unbound [37]. Reactive dyes left in the effluent cannot be reused as they become unreactive due to hydrolysis [68]. Reactive dyes which carry the azo group are present in wastewater from textile dyers at concentrations ranging from 5 to 1500 mg l^{-1} due to their poor attachment to fabrics (Gottlieb et al. 2003). These dyes are major threats to the environment because they have mutagenic properties and also affect the discoloration kinetics [4, 16]. Some chemical structure reactive dyes and the color index (**CI**) number are represented in Table 3 [44].

Zhang et al. [156] have synthesized five green anthraquinone reactive dyes according to Fig. 9. A new reactive azo dye containing heterofunctional reactive groups was synthesized by Siddiqua et al. [134]. The synthetic steps include condensation reaction, diazotization, and coupling. Figure 10 shows the overall reaction sequences.

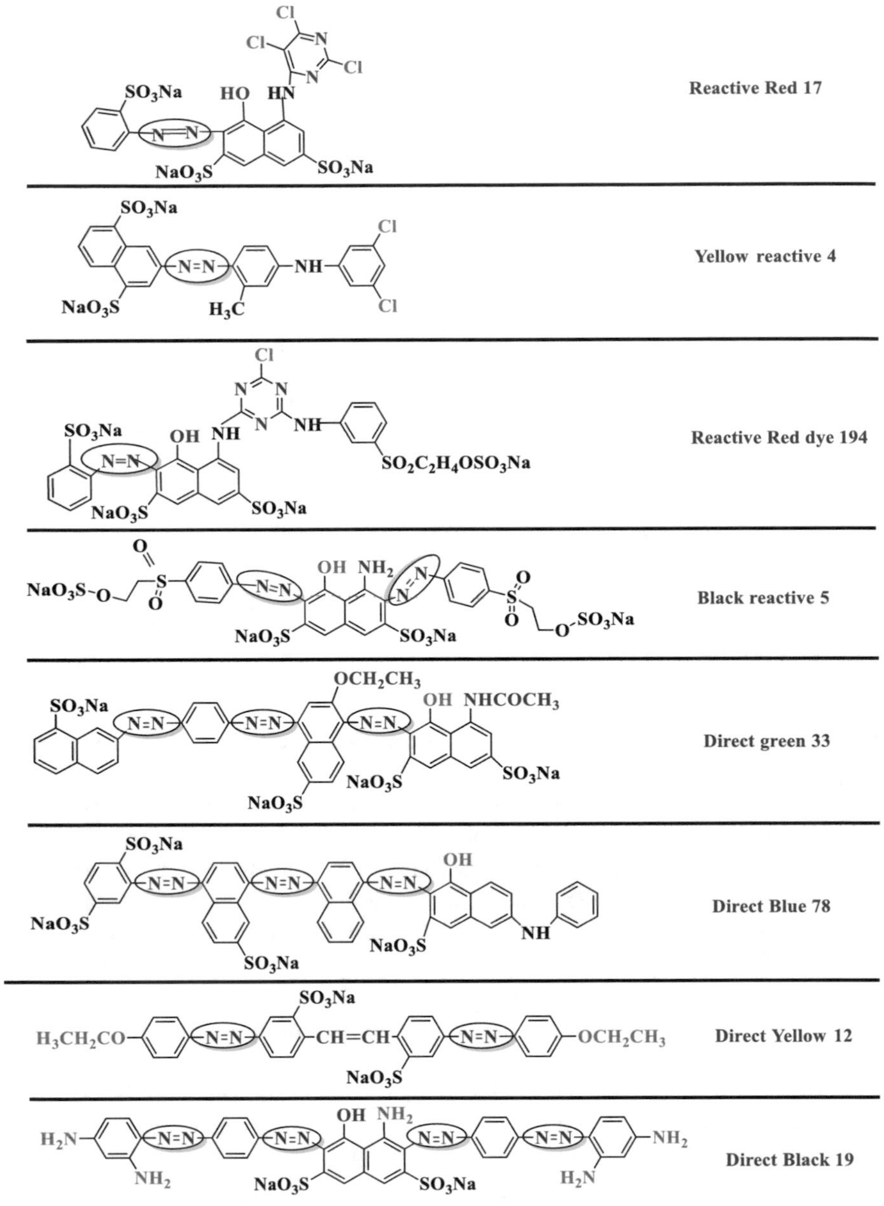

Reactive Red 17

Yellow reactive 4

Reactive Red dye 194

Black reactive 5

Direct green 33

Direct Blue 78

Direct Yellow 12

Direct Black 19

Fig. 2 Structures of some azo dyes and names (S. Benkhaya)

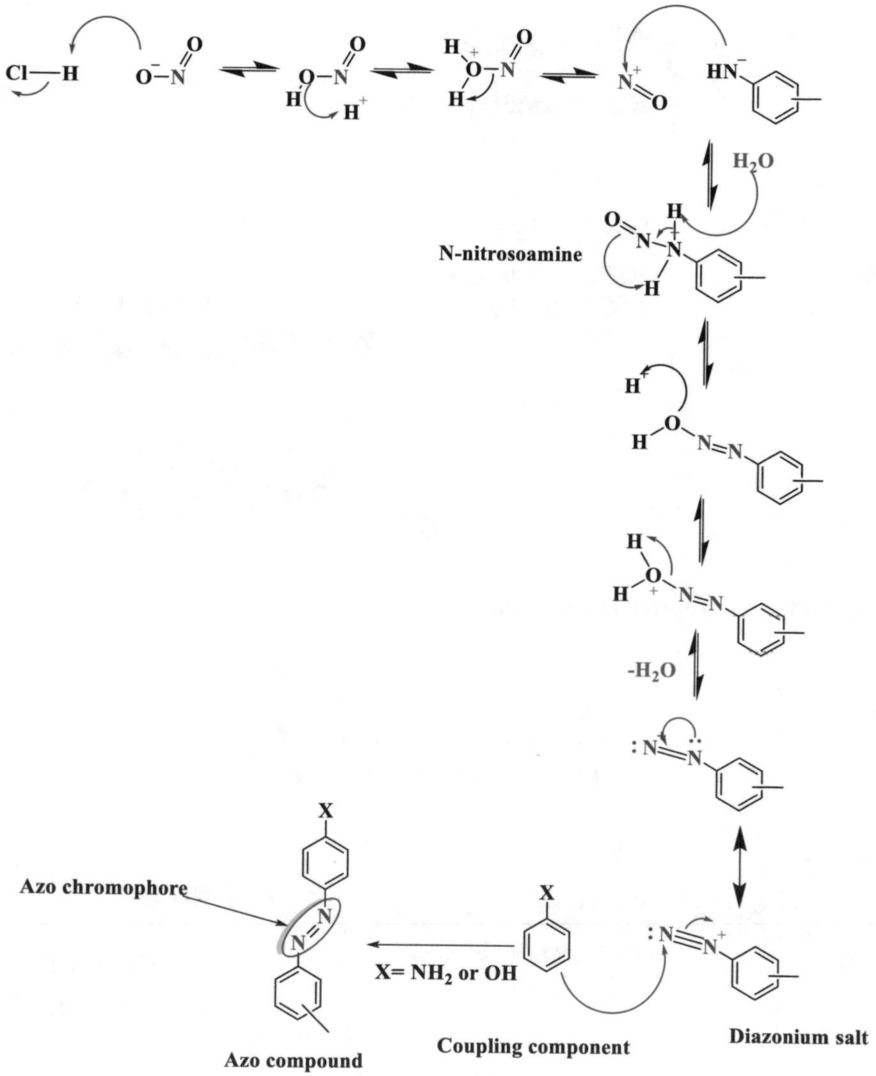

Fig. 3 Synthesis process of an azo compound (S. Benkhaya)

3.3 Vat Dyes

The annual consumption of vat dyes has been around 33,000 tonnes since 1992 with 15% of the total consumption of textile dyes [81, 124]. Vat dye is the most popular among dye classes used for the coloration of cotton [90]. They represent 24% of the cellulose fiber dyestuff market by value [90], and their application on nanofibers has not been investigated to date [84]. In addition, around 120,000 tonnes of vat dyes

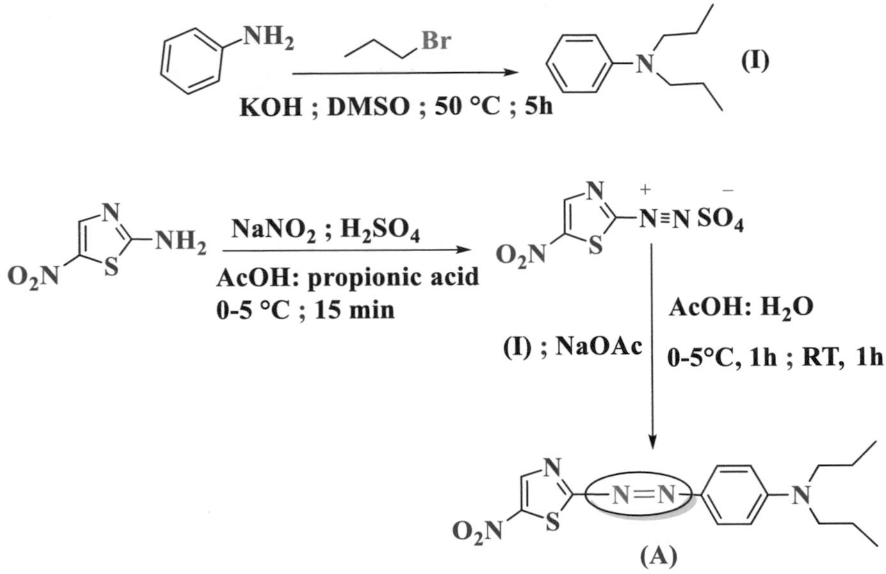

NH_2 Br

KOH ; DMSO ; 50 °C ; 5h

N (I)

N NH_2 $NaNO_2$; H_2SO_4 N + −

O_2N S AcOH: propionic acid O_2N S $N{\equiv}N$ SO_4

0-5 °C ; 15 min

AcOH: H_2O

(I) ; NaOAc 0-5°C, 1h ; RT, 1h

N

O_2N S N=N N

(A)

Fig. 4 Synthesis of azo dye (**A**) (S. Benkhaya)

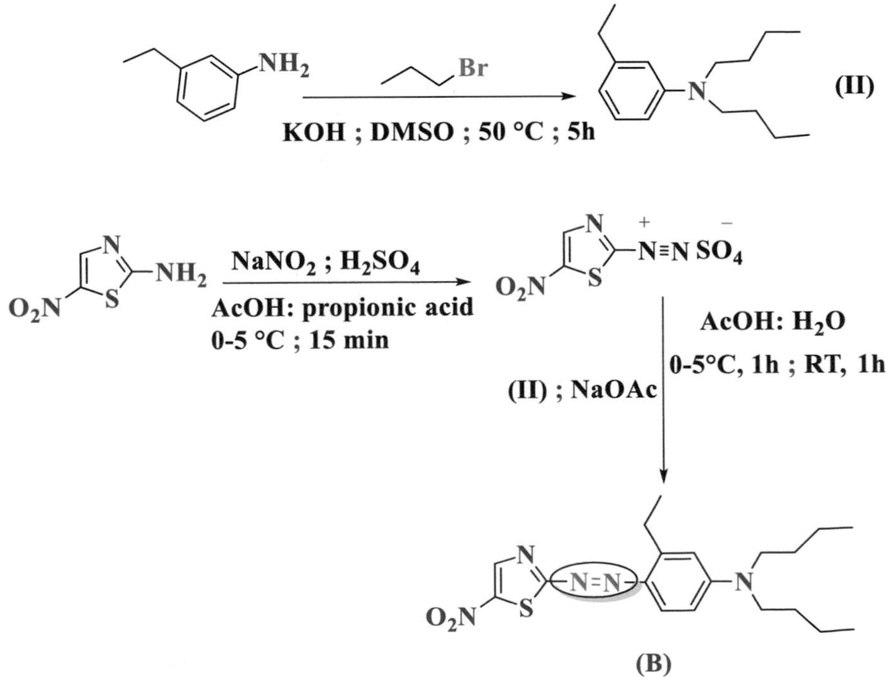

NH_2 Br

N (II)

KOH ; DMSO ; 50 °C ; 5h

N NH_2 $NaNO_2$; H_2SO_4 N + −

O_2N S AcOH: propionic acid O_2N S $N{\equiv}N$ SO_4

0-5 °C ; 15 min

AcOH: H_2O

0-5°C, 1h ; RT, 1h

(II) ; NaOAc

N

O_2N S N=N N

(B)

Fig. 5 Synthesis of azo dye (**B**) (S. Benkhaya)

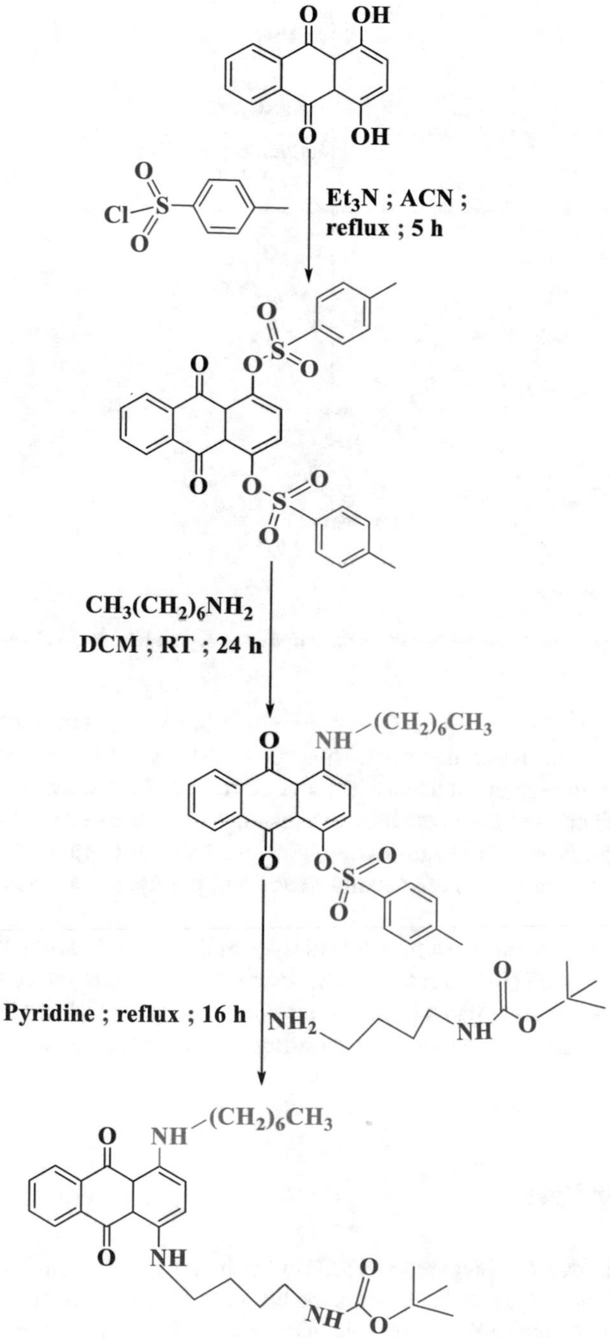

Fig. 6 Synthesis of anthraquinone dye (S. Benkhaya)

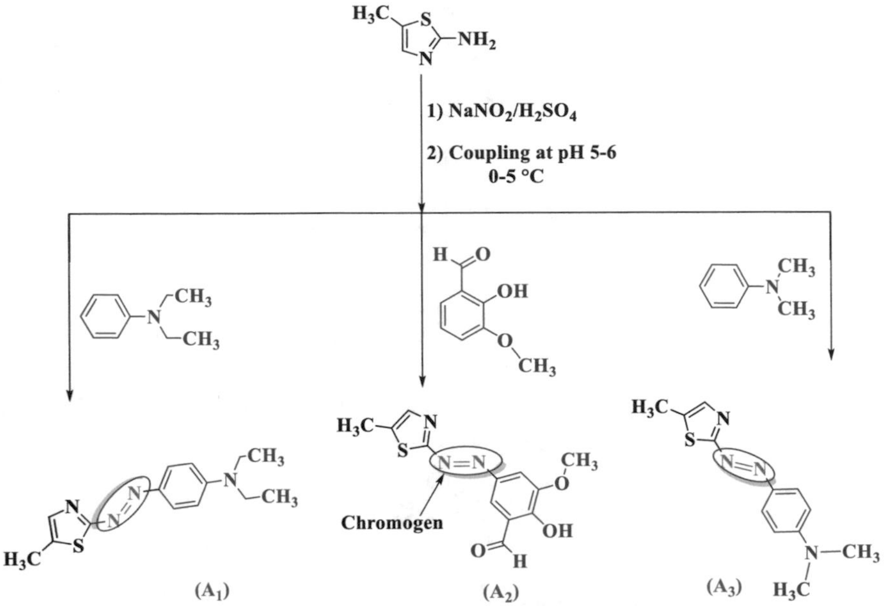

1) $NaNO_2/H_2SO_4$

2) Coupling at pH 5-6
0-5 °C

Chromogen

(A₁) (A₂) (A₃)

Fig. 7 Synthesis route for the preparation of azo dyes $\mathbf{A_1}$, $\mathbf{A_2}$, and $\mathbf{A_3}$ (S. Benkhaya)

are used each year [20]. Most vat dyes require a reducing agent to solubilize and are soluble only in their reduced (oxygen-free) form [8] Fig. 11 illustrates the reaction of reduction/oxidation of vat dyes. The reduced form of the vat dye exhibits affinity to cellulose fibers and thus becomes exhausted from the dye bath [12]. Vat dyes have some significant advantages over other dyes, viz., color value, reproducibility of color, fastness properties are usually better and the dyeing is easier to wash-off [35].

Table 4 shows the two most popular vat dyes in the textile industry "Vat Yellow 1 (**a**) and Vat Black 25 (**b**)" with their color index number, color, wavelength (λ_{max}), and chemical structure. Figure 12 illustrates some vat dyes [84, 70, 123]. A new vat dye (Vat Yellow 2) was synthesized according to the reaction scheme illustrated in Fig. 13 [144].

3.4 Sulfur Dyes

The first sulfur dye was prepared in 1873 by Croissant and Bretonnière [71]. Global production of these dyes has been estimated between 110,000 and 120,000 tonnes per year [109]. They play an important role in the textile dyeing industries [107]. Sulfur dyes are one of the most popular dye classes for cellulosic fibers and their blends, being widely used to produce inexpensive, medium to high depths [22]. They are widely used to produce economical black, blue, brown, and green shades on medium

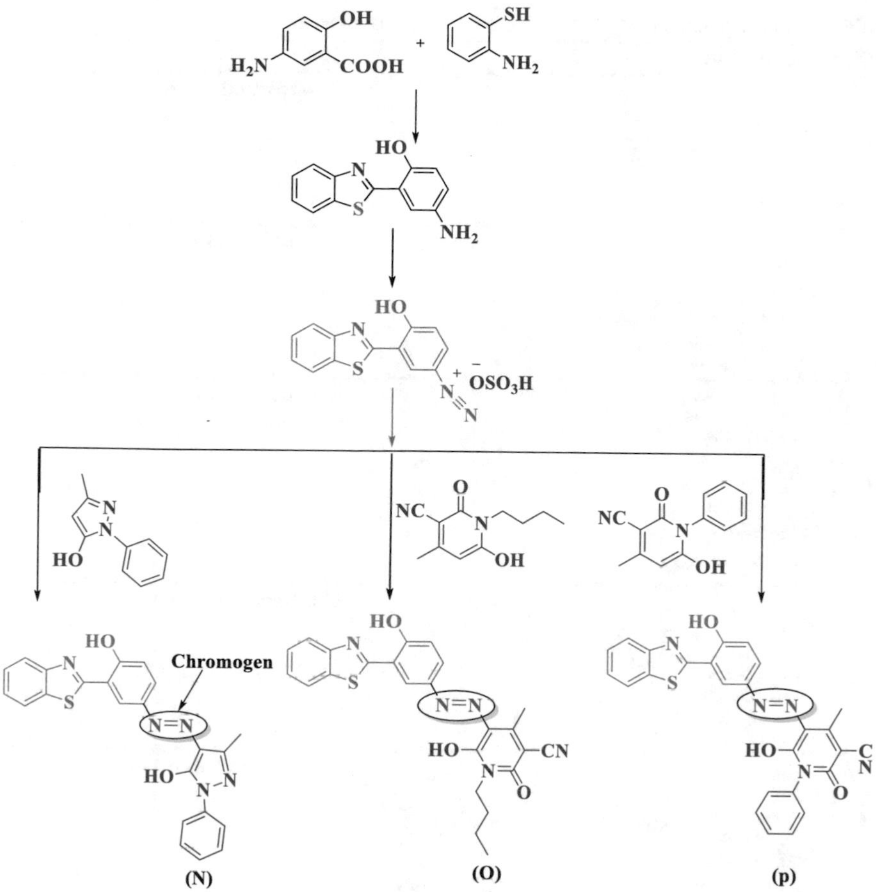

Fig. 8 Synthesis of heterocyclic azo dyes (S. Benkhaya)

to high depth cellulosic fibers [149]. Many sulfur dyes contain sulfide heterocyclics such as benzothiazole, thiazone, and thianthrene in their chemical structures [137]. Sulfurized vat dyes are produced by a sulfurization process which is used for other sulfur dyes, but are reduced with $Na_2S_2O_4$ and applied as vat dyes [34].

Stolte et al. [140] demonstrated that sulfur dyes are high molecular weight dyes obtained by the sulfurization of organic compounds. These dyes are commonly used for cellulose. The use of sulfur dyes on cellulosic fibers with sodium sulfide as an effective reducing agent is still known to be a traditional and inexpensive dyeing process [108]. Sulfur dyes on nylon 6,6 showed excellent resistance to moisture and rubbing, while lightfastness of the dyes was poor [23]. Sulfur dyes are converted to sodium-derived leuco by reduction using sodium sulfide [140]. Leuco sulfur dyes on cotton were enhanced by the application of two commercial cationic fixing agents when applied as a post-treatment using both the Pad-Dry and Pad-Flash Cure methods

Table 3 Chemical structures of some reactive dyes (S. Benkhaya)

Chemical structure	CI no
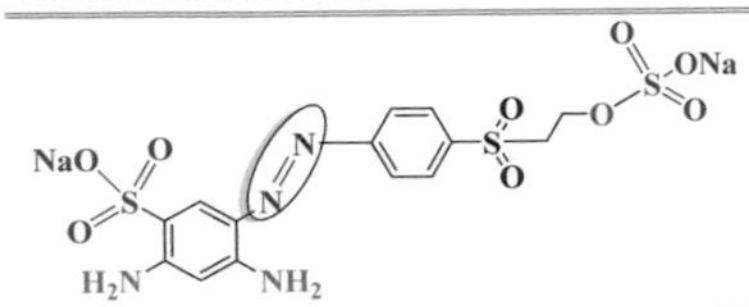	**ReactiveYellow 201**
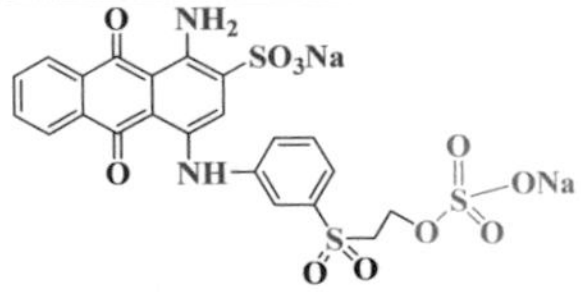	**ReactiveBlue19**
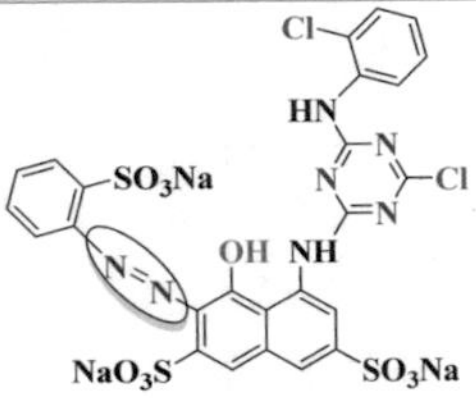	**Reactive RR24**
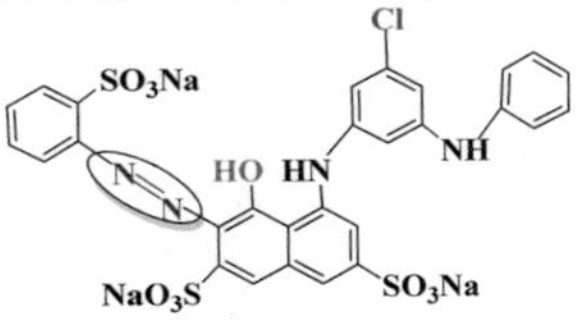	**C.I. Reactive Red 3**
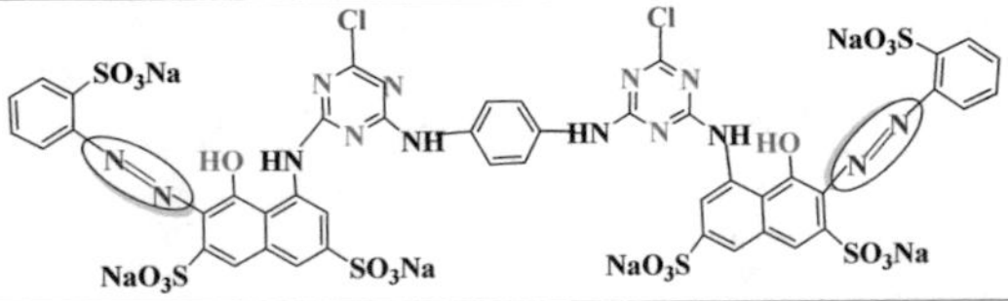	**C.I. Reactive Red 120**
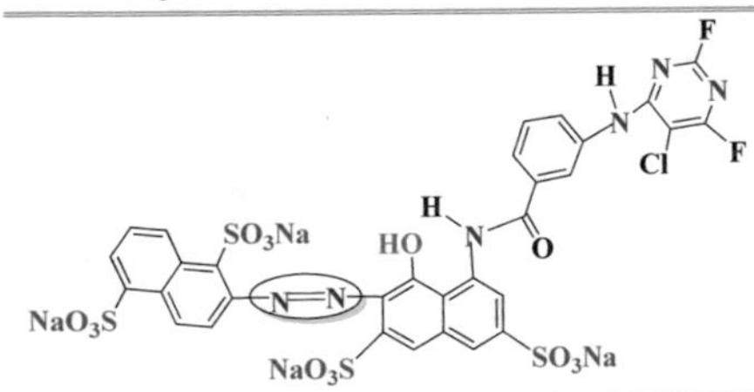	**C.I. Reactive Red 147**

(continued)

[21]. Table 5 illustrates some sulfur dyes with color index number and Cas Registry Number [41].

Table 3 (continued)

Structure	Name
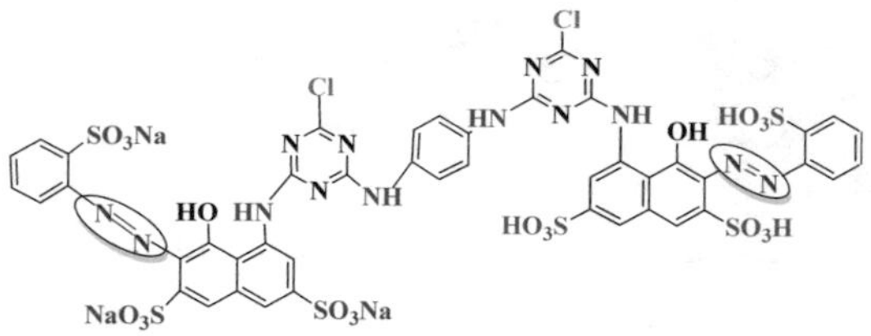	**C.I. Reactive Red 120.**
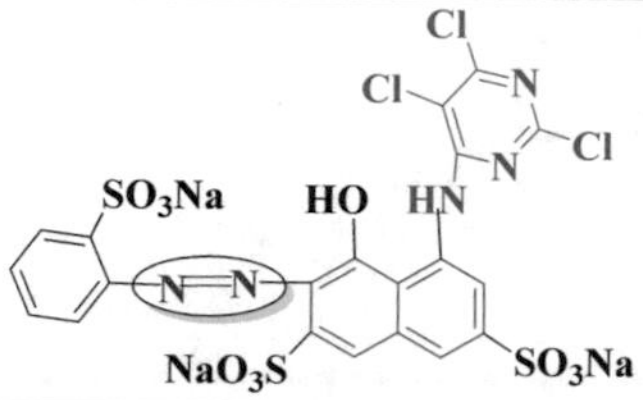	**C.I. Reactive Red 17**
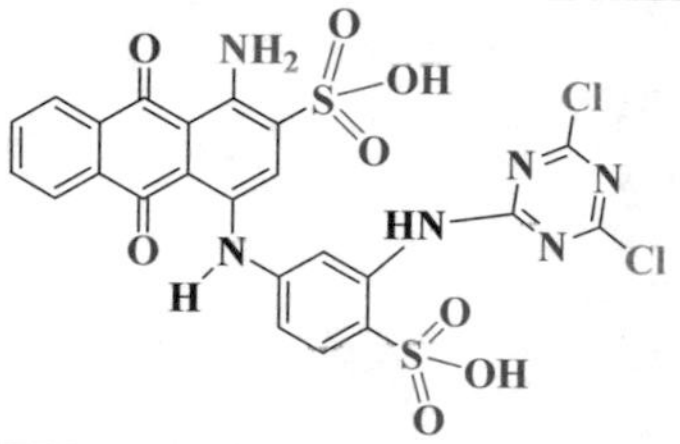	**Reactive Blue 4**

3.5 *Acid Dyes*

As the name suggests, they are "acids", and the molecule has one or more acid functions (**SO_3H^-** and **COOH**) [13]. They are anionic sulfonated dyes, and their acid nature explains their affinity for the basic functions of fibers, such as polyamides [13]. The usage of acid dyes constitutes about 30–40% of the total consumption of dyes, and they are applied extensively on nylon, cotton, wool, silk, polyamides, and leather [63, 103]. They are usually applied at acidic pH [16]. Mainly, the acid dyes are more stable in an acidic medium and hence a majority of the applications were carried out in a weakly acidic bath of pH 4.5–5 [132]. These dyes, especially sulfonic acid ones, are widely used in the textile, pharmaceutical, printing, leather, paper, and other fields, thanks to their bright colors and high solubility [153]. As a representative element of this family of dyes, mention may be made of the red congo. The acid dyes were divided into three groups because according to their application properties and variable strength, (i) the first group has little affinity under neutral or weakly acidic conditions, (ii) the second group of dyes escapes on the nylon in the pH range from 3.0 to 5.0, and (iii) the third group has a strong affinity for nylon under neutral or weakly acidic conditions (pH 5.0–7.0) [38]. Examples of these dyes are shown in Fig. 14 [43]. Sun et al. [141] have synthesized some surfactant-type acid dyes. The

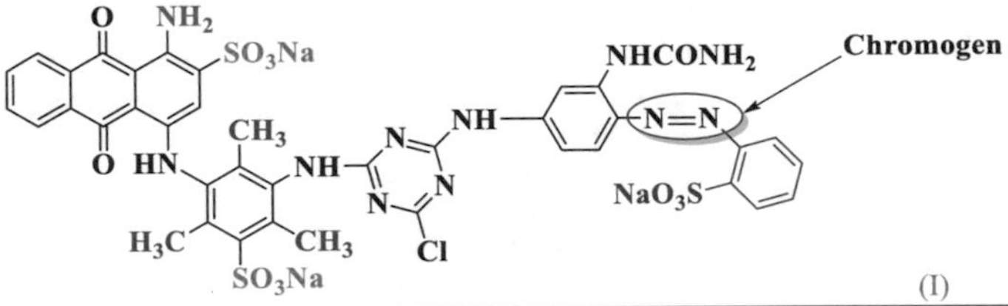

Fig. 9 Green anthraquinone reactive dyes (I–V) (S. Benkhaya)

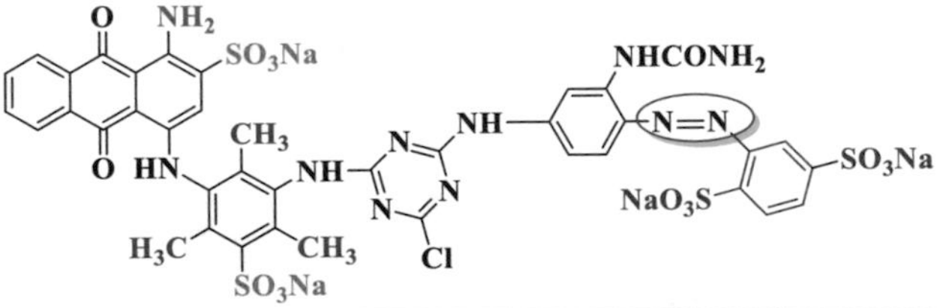

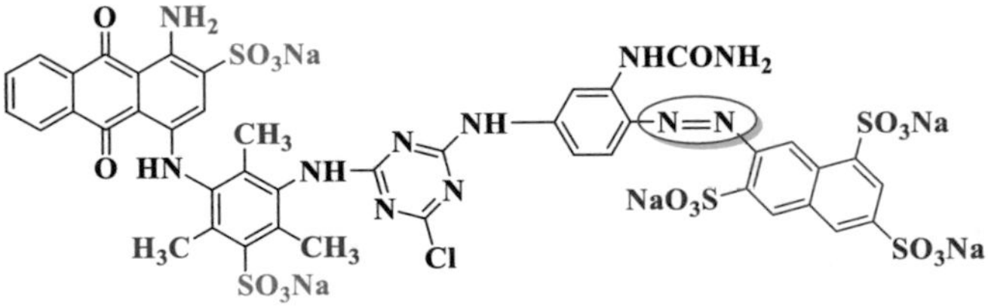

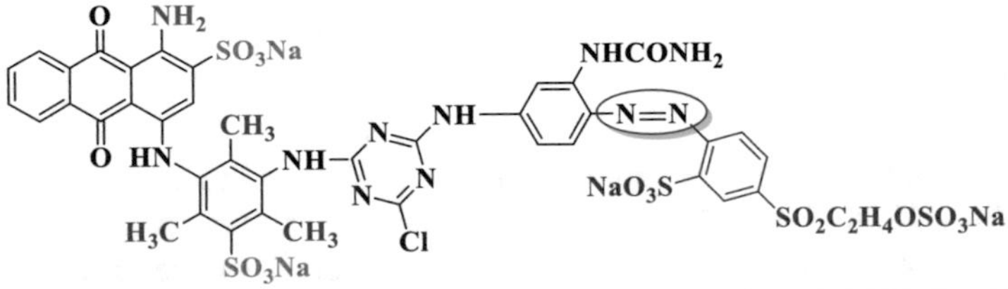

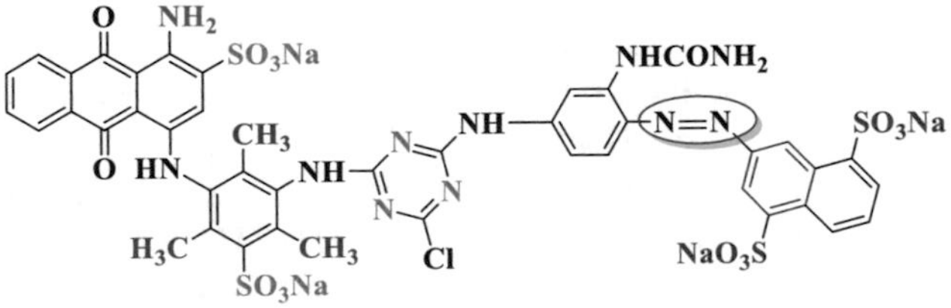

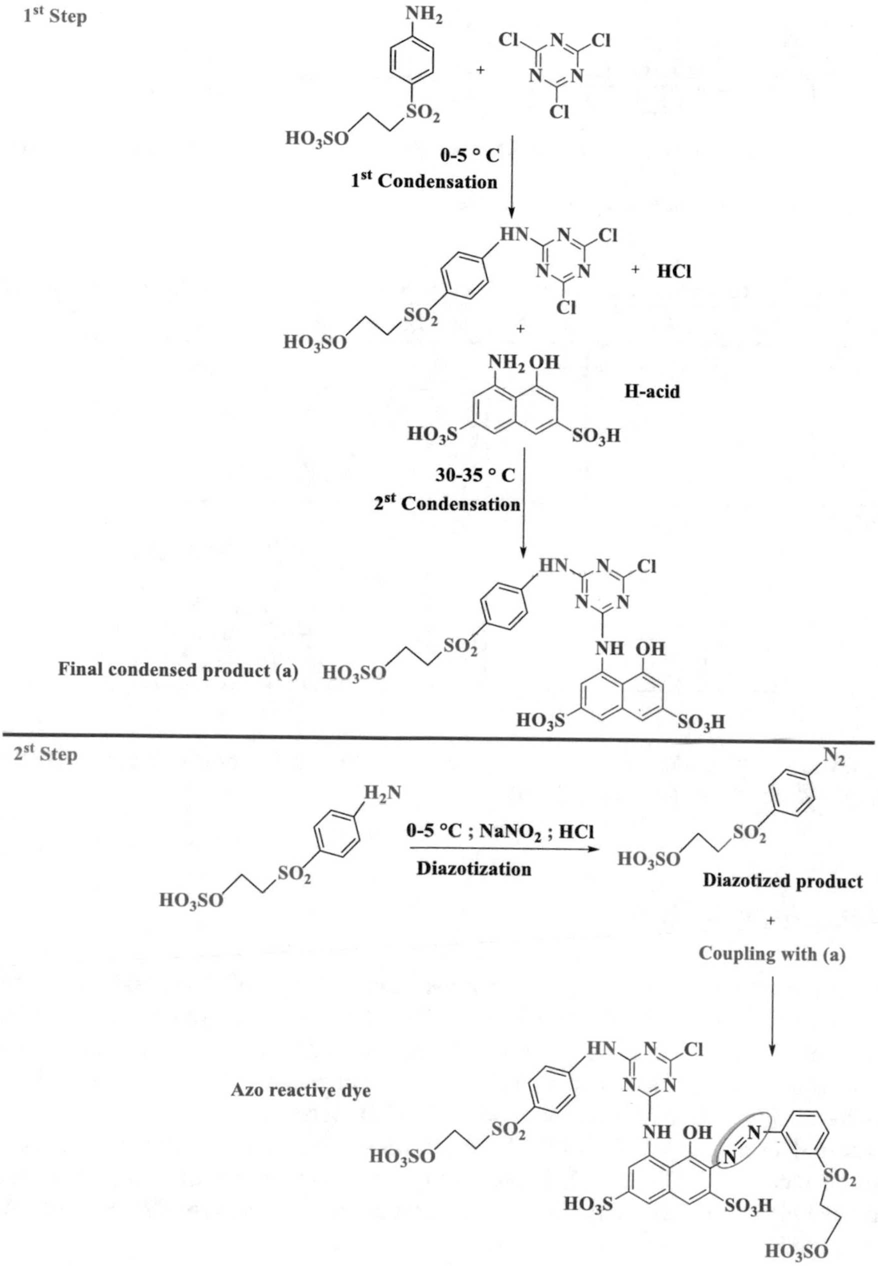

1st Step

NH_2 + Cl N Cl N N Cl

HO_3SO SO_2

0-5 ° C
1st Condensation

HN N Cl N N Cl + HCl

HO_3SO SO_2

+

NH_2 OH

H-acid

HO_3S SO_3H

30-35 ° C
2st Condensation

HN N Cl N N

Final condensed product (a) HO_3SO SO_2

NH OH

HO_3S SO_3H

2st Step

H_2N

HO_3SO SO_2

0-5 °C ; $NaNO_2$; HCl
Diazotization

N_2

HO_3SO SO_2

Diazotized product

+

Coupling with (a)

Azo reactive dye

HN N Cl N N

HO_3SO SO_2

NH OH N=N

HO_3S SO_3H SO_2

HO_3SO

Fig. 10 Synthesis scheme of azo reactive dye (S. Benkhaya)

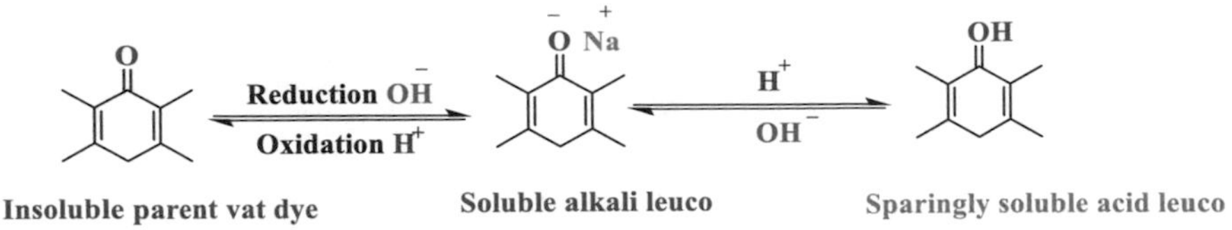

Fig. 11 Reduction/oxidation of vat dyes (S. Benkhaya)

Table 4 Physico-chemical characteristics of Vat Yellow 1 (**a**) and Vat Black 25 (**b**) (S. Benkhaya)

Color index no.	Color	(λ_{max})	Chemical structure
70600	Yellow	587	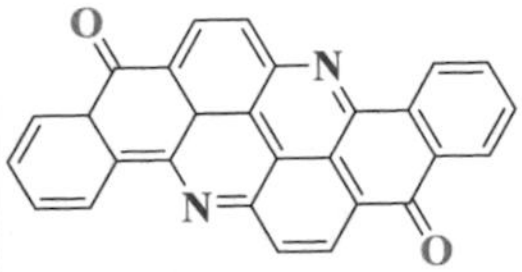 (a)
69525	Darkblue	675	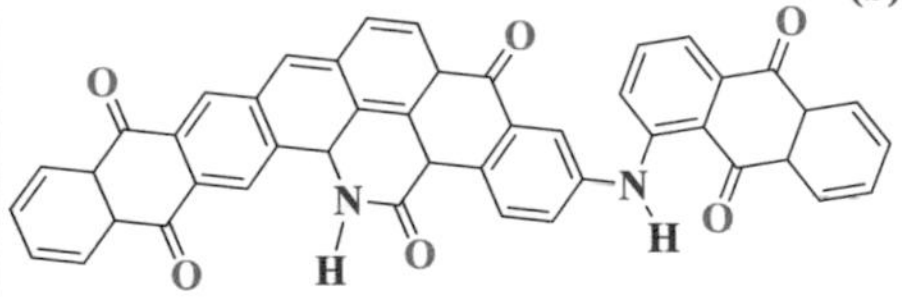 (b)

synthesis route is described in Fig. 15. Figures 16 and 17 show different possible acid dye structures according to (**n**).

3.6 *Disperse Dyes*

Disperse dyes are for their part the most common type of dyes, and their total production was around 366,500 tonnes. They are synthetic dyes and are applied to hydrophobic fibers from an aqueous dispersion. Disperse dyes are among the persisting class of dyes due to their recalcitrant nature and non-biodegradable behavior [78]. They represent around 44% of the total dye production [117, 157]. These dyes are frequently insoluble or poorly soluble in water, and are non-ionic in character [40]. Disperse dyes are colored, non-ionic aromatic compounds that commonly contain azo and nitro groups, and are widely used for dyeing synthetic fibers [53].

Among disperse dyes, azo disperse dyes have attracted particular attention and they have been widely applied for dyeing natural and synthetic fibers [92]. Heterocyclic azo dyes are as disperse azo dyes which exhibit good tinctorial strength, large molar extinction, and brighter dyeing than those obtained from aniline-based intermediates [105].

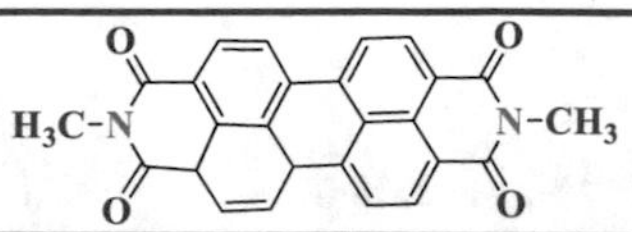	**C.I. Vat Red 23**
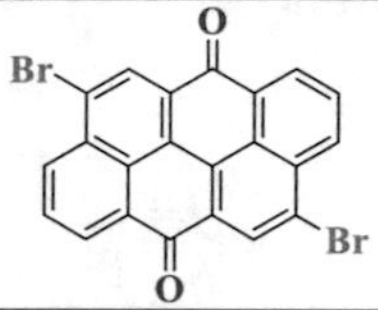	**Pigment Red 168 (Dibromoanthanthrone)**
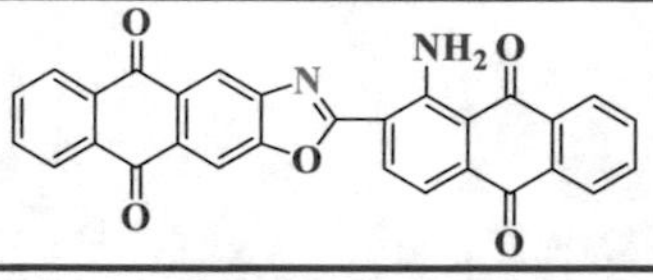	**C.I. Vat Red 10**
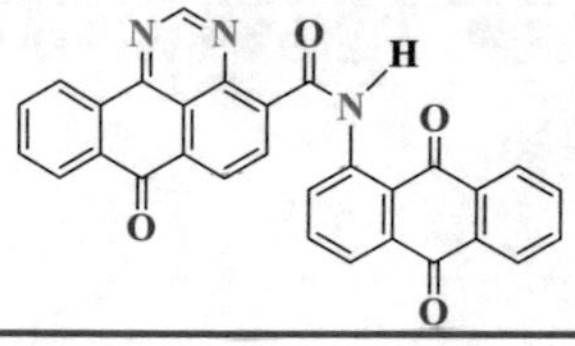	**Pigment Yellow 108**
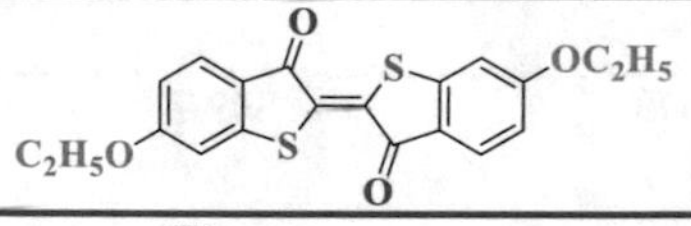	**C.I.Vat Orange 5**
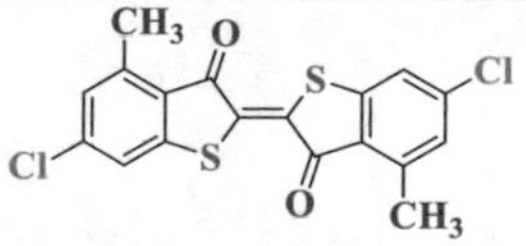	**C.I. Vat Red 1**
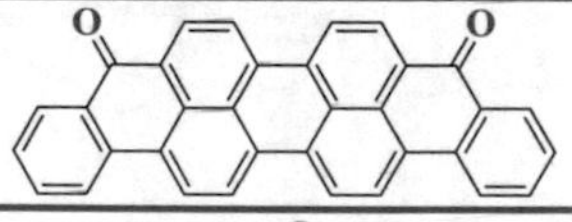	**C.I. Vat Blue 20**
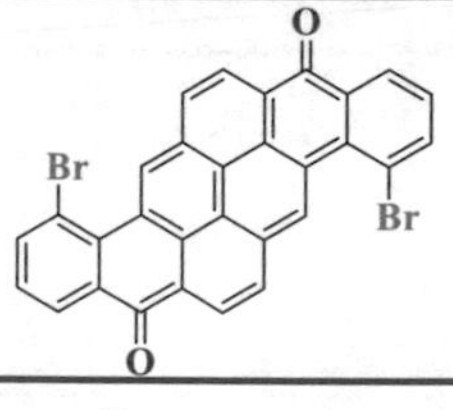	**C.I. Vat Orange 2**
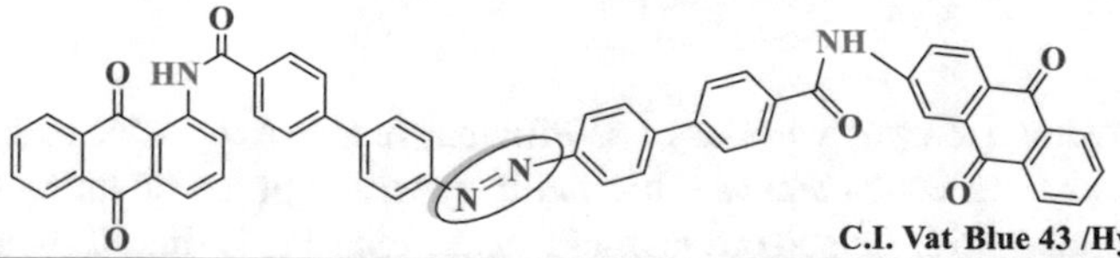	**C.I. Vat Blue 43 /Hydronblau, 9**

Fig. 12 Some examples of vat dyes (S. Benkhaya)

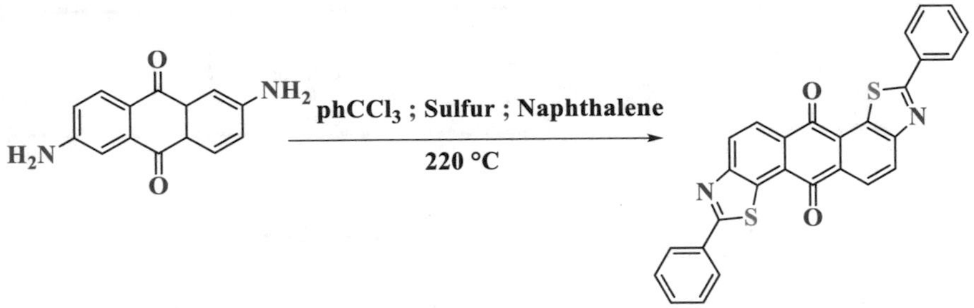

phCCl$_3$; Sulfur ; Naphthalene
220 °C

Fig. 13 Synthesis of Vat Yellow 2 (S. Benkhaya)

Table 5 Chemical structures of some sulfur dyes (S. Benkhaya)

Chemical structure	Color Index no	CAS Registry Number
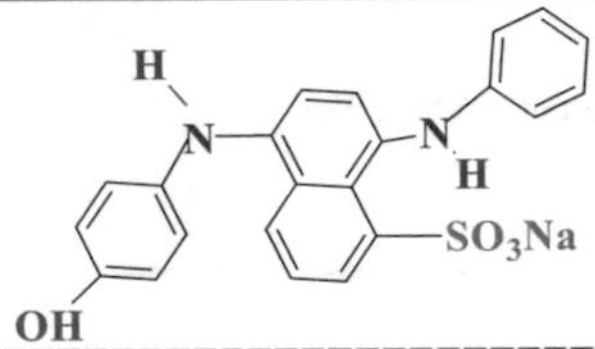	CI 53570	12262-33-8
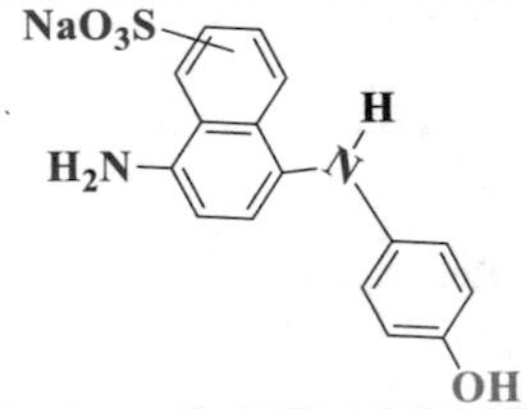	C.I.53540	1327-69-1
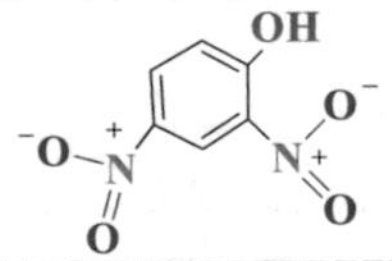	C.I 53185	1326-82-5
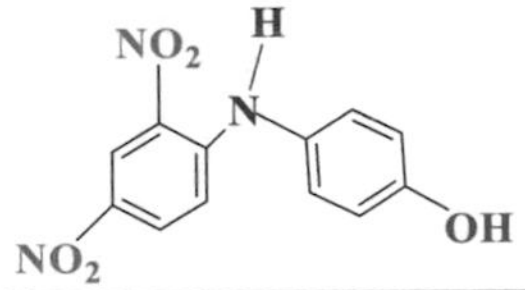	CI 53235	1326-98-3

In addition, aryl hydrazone dyes were used as efficient disperse dyes for high-temperature polyester dyeing which shows a high affinity with high color strength [10]. For polyester dyeing with disperse dyes, less water cleaner dyeing without additives is always the best and desired goal for researchers [55]. In the dyeing polyester with these dyes, ultrasonication decreased the size of dye particles [69]. Typical disperse dye structures are illustrated in Fig. 18. Phthalimide disperse dyes

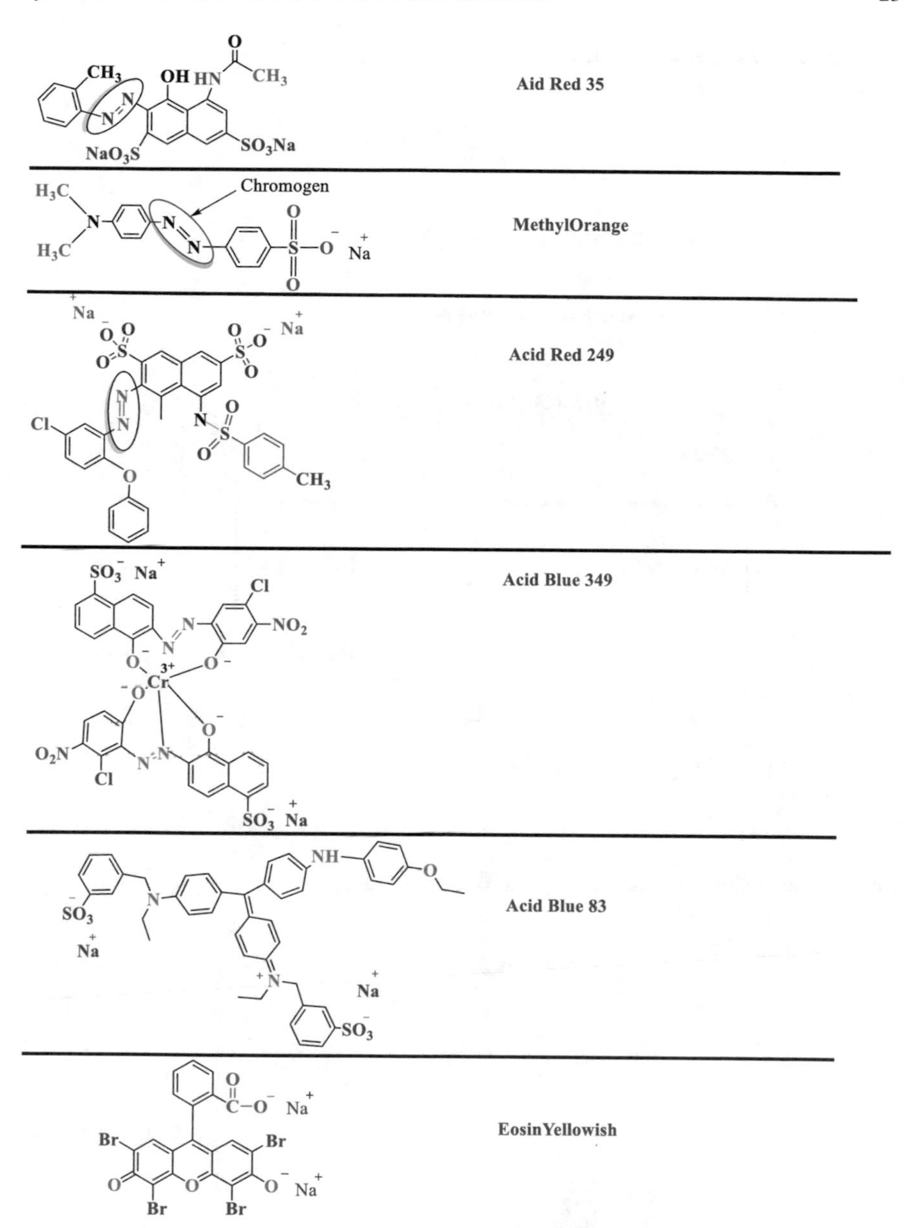

Fig. 14 Examples of the acid dyes (S. Benkhaya)

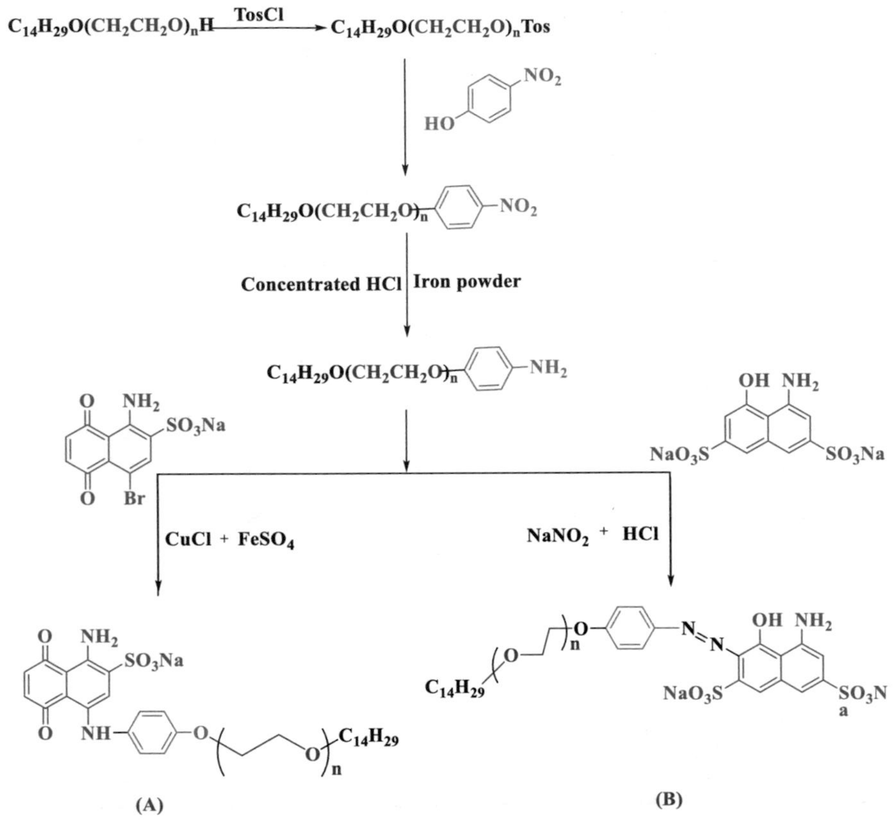

Fig. 15 The synthesis of acid dyes (**A**) and (**B**) (S. Benkhaya)

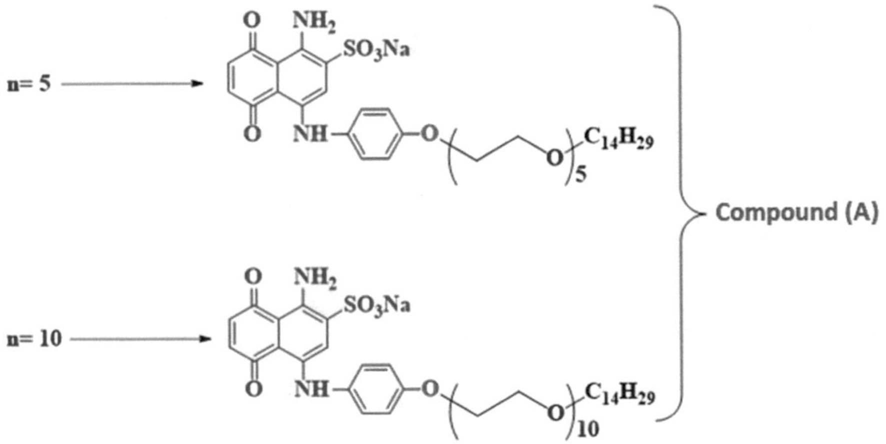

Fig. 16 Structure of compound (**A**) according to (**n**) (S. Benkhaya)

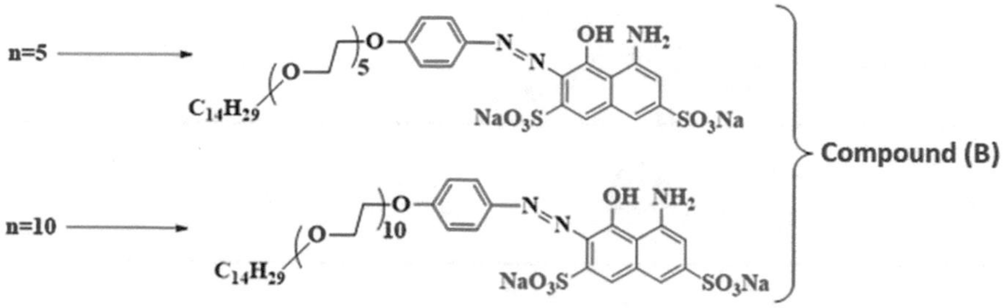

Fig. 17 Structure of compound (**B**) according to (**n**) (S. Benkhaya)

(Dye **1**, Dye **2**, and Dye **3**) were synthesized by Yizhen et al. [155] according to Figs. 19 and 20.

Karci and Bakan [80] have synthesized diazo pyrazole disperse dyes (Dye **4** and Dye **5**) from 5-amino-4-arylazo-3-methyl-1-phenylpyrazoles (aminoarylazopyrazoles), according to the reaction scheme shown in Fig. 21. Figure 22 shows different structures of Dye 4 and Dye 5 according to X substitution.

Maliyappa et al. synthesized a novel substituted aniline-based heterocyclic dispersed azo dyes coupling with 5-methyl-2-(6-methyl-1, 3-benzothiazol-2-yl)-2, 4-dihydro-3H-pyrazol-3-one [99]. The schematic representation of the synthesized dispersed azo molecules (Dye **6**) is shown in Fig. 23. Figure 24 shows different possible dispersed dye structures (Dye **7**) according to R_1 and R_2 substitutions.

Maliyappa et al. [98] have synthesized some new dispersed azo dyes. The general route for the synthesis of 6-substituted benzothiazole-based disperse azo dyes (Dye **8**) is described in Fig. 25. Figure 26 shows different possible structures (Dye **9**) according to R substitution.

3.7 *Basic Dyes*

Basic dyes are typically hydrochloride salts, so the formation of cations by loss of $Cl^-$is reasonable [138]. They include monoazoic, diazoic, and azine compounds [110]. Basic dyes have high brilliance and intensity of colors and are highly visible even in a very low concentration [73]. They are commercially available as chlorides and other water-soluble salts, and, on the other hand, they were treated with tannin or antimony potassium tartrate to yield insoluble colorants, i.e. pigments [128]. Cationic dyes commonly known as basic dyes are widely used in acrylic, nylon, silk, and wool dyeing [45, 146]. Basic dyes are used without adding any salt during the dying process [102]. They are easily decomposed when irradiated with light due to their poor photo-stability, owing to which their color intensity reduces and might even result in a color change [152]. Some basic dyes with their chemical structures and Color Index Generic Name are shown in Fig. 27 [14, 61, 95, 102, 126, 45].

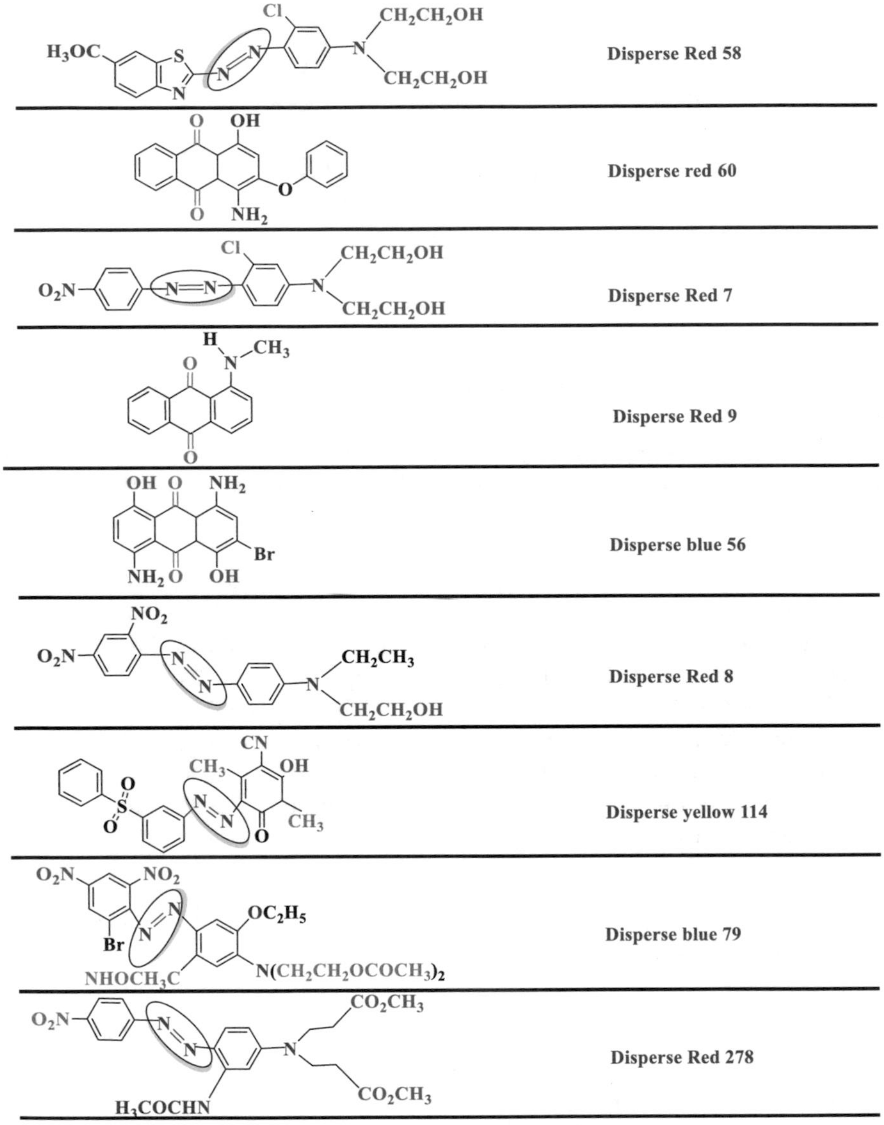

Disperse Red 58

Disperse red 60

Disperse Red 7

Disperse Red 9

Disperse blue 56

Disperse Red 8

Disperse yellow 114

Disperse blue 79

Disperse Red 278

Fig. 18 Chemical structures of some disperse dyes (S. Benkhaya)

3.8 Direct Dyes

Direct dyes are water-soluble and anionic in nature. They are widely used in dying industries due to their easy application and economic factors [57]. The usage of direct dyes increased from 53,848 tonnes in 1992 to 181,998 tonnes in 2011 (237.98%) [58].

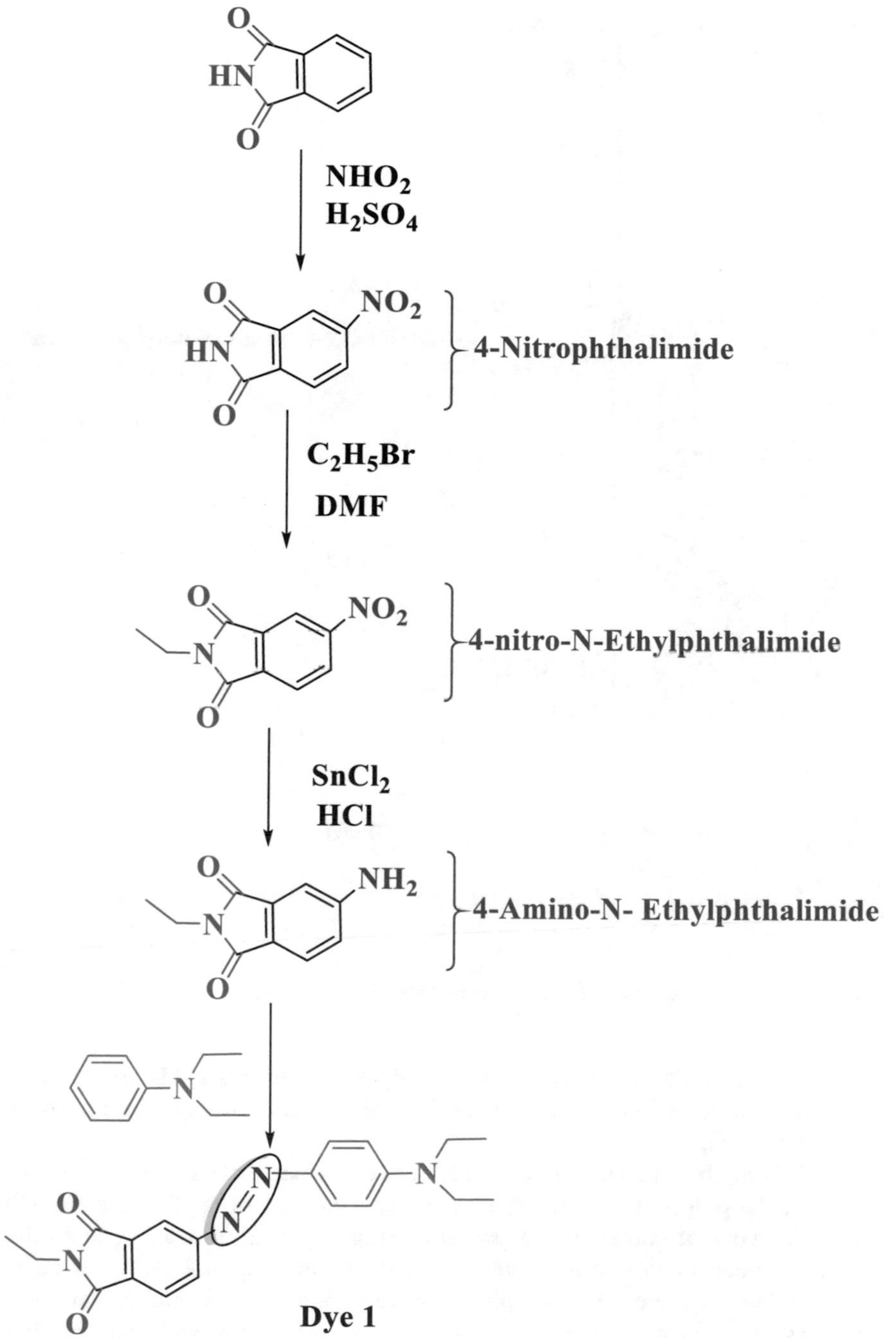
HN
O
O
NHO2
H2SO4
NO2
4-Nitrophthalimide
C2H5Br
DMF
N
4-nitro-N-Ethylphthalimide
SnCl2
HCl
NH2
4-Amino-N- Ethylphthalimide
N
N=N
Dye 1

Fig. 19 Synthesis of dispersed azo dyes (Dye **1**) (S. Benkhaya)

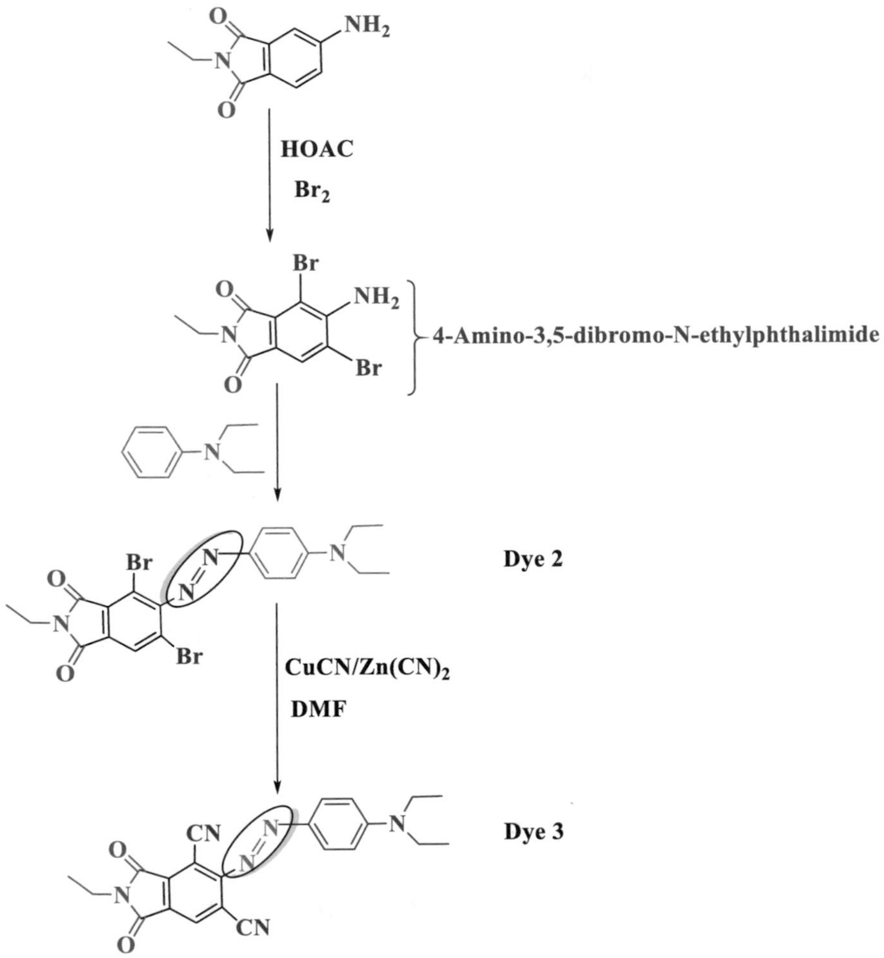

Fig. 20 Synthesis of disperse dyes (Dye **2** and Dye **3**) (S. Benkhaya)

They are structurally very similar to reactive dyes, in that both dye types are essentially long, planar, anionic molecules solubilized by one or more ionized sulfonate groups [26, p. 8].

In addition, they are classified according to many parameters such as the chromophore, the properties of solidity or the characteristics of application [91, 97]. The main types of chromophores are azo, phthalocyanine, dioxazine, and other smaller chemical classes such as formazan, anthraquinone, quinolone, and thiazole. Although these dyes are easy to apply and have a wide range of shades, their resistance to washing is only moderate; this has led to their replacement by reactive dyes which have much higher resistance to humidity and washing properties on cellulosic substrates. Figures 28 and 29 showed some structures of direct dyes [9, 24, 25, 67,

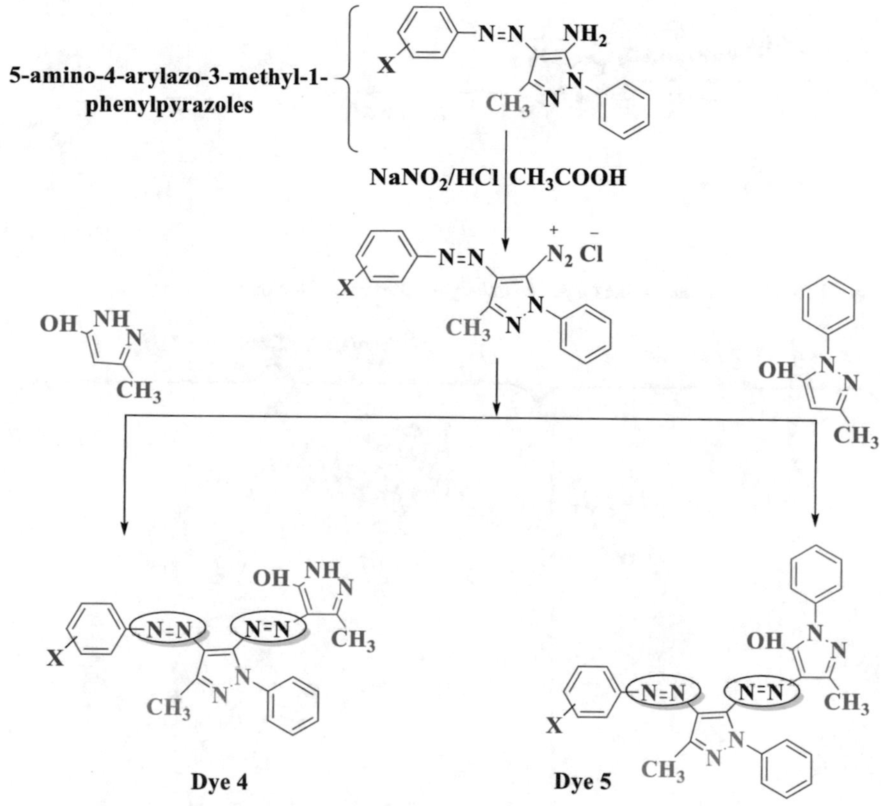

5-amino-4-arylazo-3-methyl-1-phenylpyrazoles

$NaNO_2/HCl$ | CH_3COOH

Dye 4 **Dye 5**

Fig. 21 Synthesis of disperse diazo pyrazole dyes (Dye **4** and Dye **5**) (S. Benkhaya)

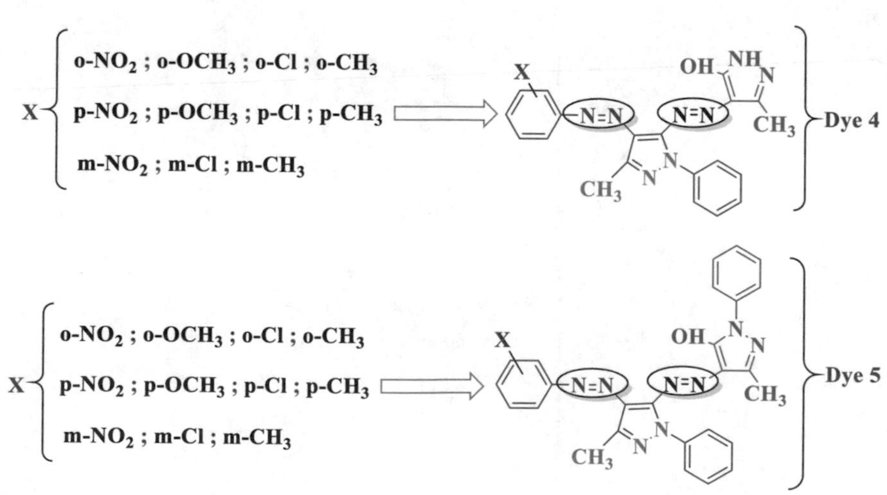

X: o-NO_2 ; o-OCH_3 ; o-Cl ; o-CH_3
p-NO_2 ; p-OCH_3 ; p-Cl ; p-CH_3
m-NO_2 ; m-Cl ; m-CH_3
→ Dye 4

X: o-NO_2 ; o-OCH_3 ; o-Cl ; o-CH_3
p-NO_2 ; p-OCH_3 ; p-Cl ; p-CH_3
m-NO_2 ; m-Cl ; m-CH_3
→ Dye 5

Fig. 22 Structure of disperse diazo pyrazole dyes Dye **4** and Dye **5** according to X substitution (S. Benkhaya)

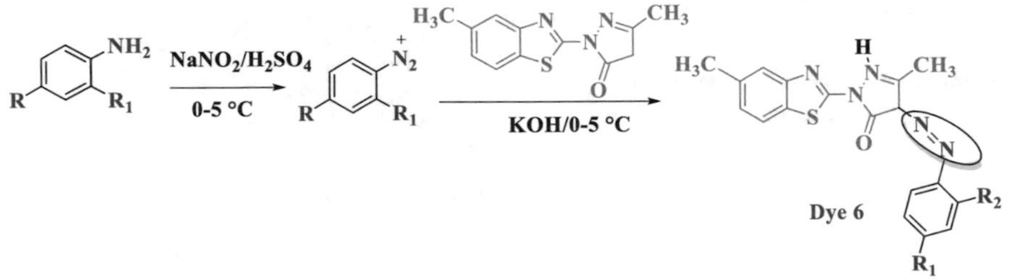

Fig. 23 Synthesis route for the dispersed azo dyes (Dye 6) (S. Benkhaya)

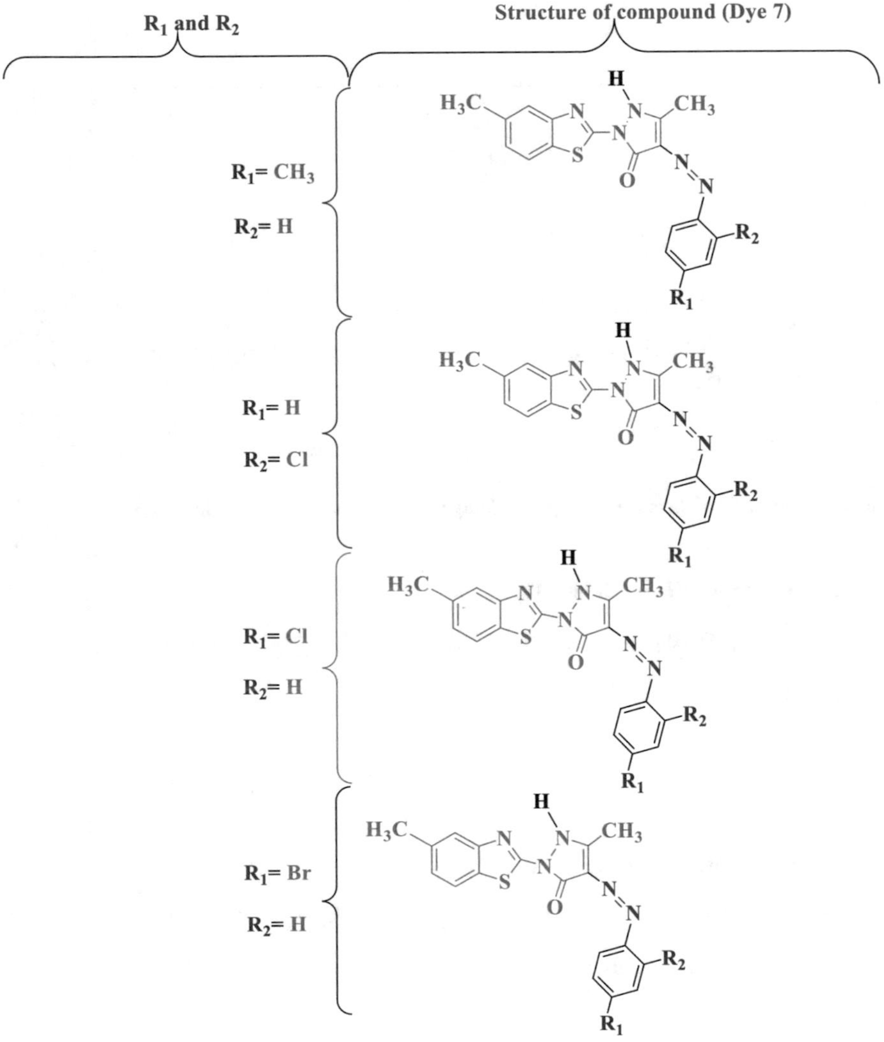

Fig. 24 Different structures of dispersed dyes (Dye 7) (S. Benkhaya)

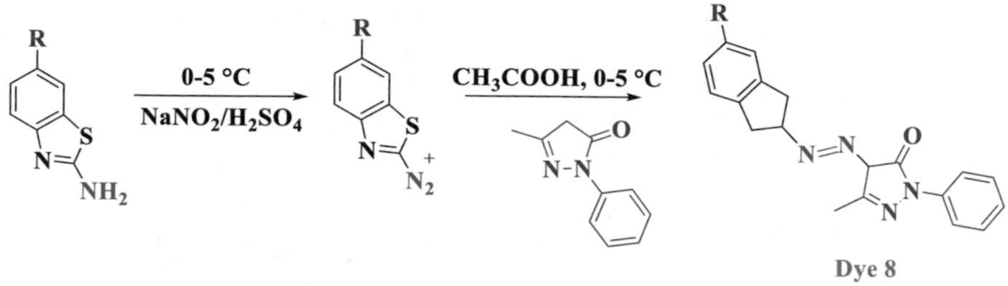

Fig. 25 The synthetic pathway of disperse azo dyes (Dye 8) (S. Benkhaya)

106, 119]. Table 6 groups the classification of textile dyes according to their various application methods and their characteristics [15].

4 Dyes and Color Index Numbers

In the Color Index, each dye or pigment is presented with two numbers referring to the basis of the coloristic and chemical classification [160]. Color Index Name and Color Index Constitution Number have been developed for identifying the dyes [16, 127, 145]. So, the most important reference work dealing with the classification of dyes and organic/inorganic pigments (Fig. 30) [64] is the Color Index, Fig. 31 [16, 39]. It provides useful information for each dye and pigment on the methods of application and on the range of fastness properties that may be expected [39]. Reactive Red 6, which is a member of mono azo dyes, has been given as a representative example to illustrate the color index classification, Fig. 32.

5 Pollution and Environmental Effects

The textile industry produces large quantities of strongly colored wastewater with a high load of inorganic salts, dyes, pigments, chemical products, heavy metals (Pb, Cr, Ar), etc. [88, 94]. It is estimated that approximately 3,106 L of wastewater is produced after the treatment of approximately 20,000 kg of textiles per day [118]. The toxicity of liquid textile waste can come from either the metallic part of the dye molecule, such as chromium in acid dyes or copper in direct dyes, or other materials used in the process of tincture, such as traces of mercury present in various chemical reagents [112]. Heavy metal contents of less than 100 mg/l have been reported in the case of non-metallic dyes. In the case of metallic dyes and metallic salt dyes, the reported metal levels are considerably higher [72]. About 72 toxic chemicals have been identified in textile effluents, and it is estimated that approximately 200 million liters of effluents are produced annually worldwide by the textile industry

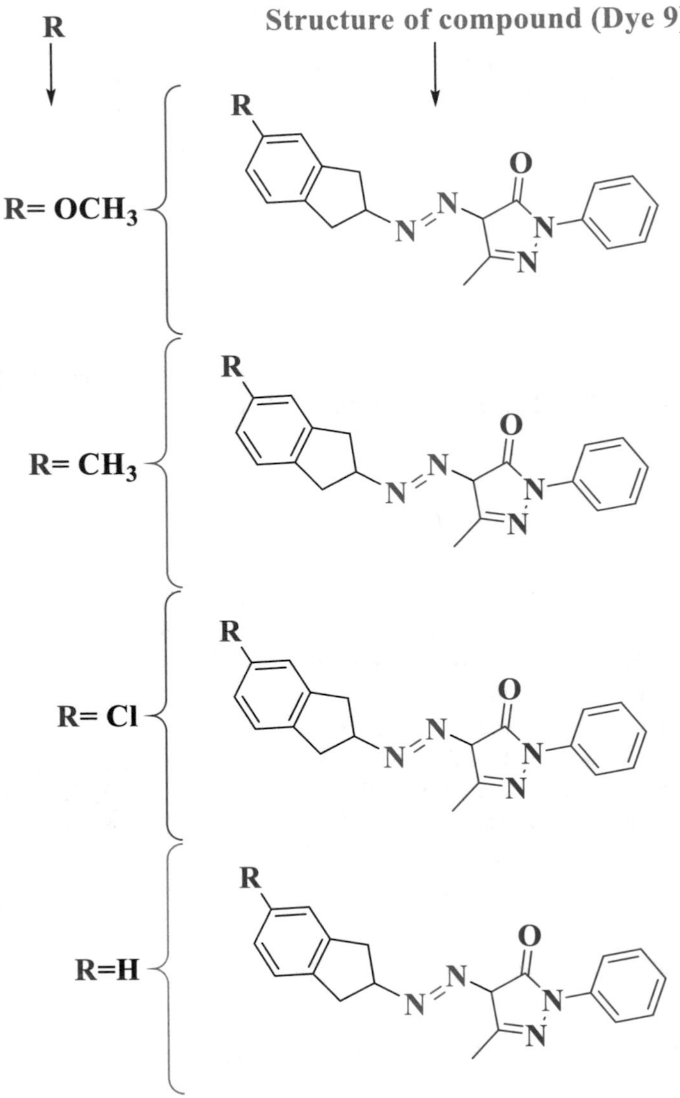

Fig. 26 Different structures of disperse azo dyes (Dye **9**) (S. Benkhaya)

[56]. More about 40% of textile dyes contain organically bound chlorine, which is a known carcinogen [60, 79, 83].

It has been reported that approximately 100 tons/year of dyes are rejected into the aquatic environment with a consumption of greater than 104 tons/year coming mainly from the textile industry [18]. During the dyeing process, not all dyes are attached to the fabrics. There is always a part of the unfixed dye which is washed away with water and which constitutes the main pollutant in textile effluents. The

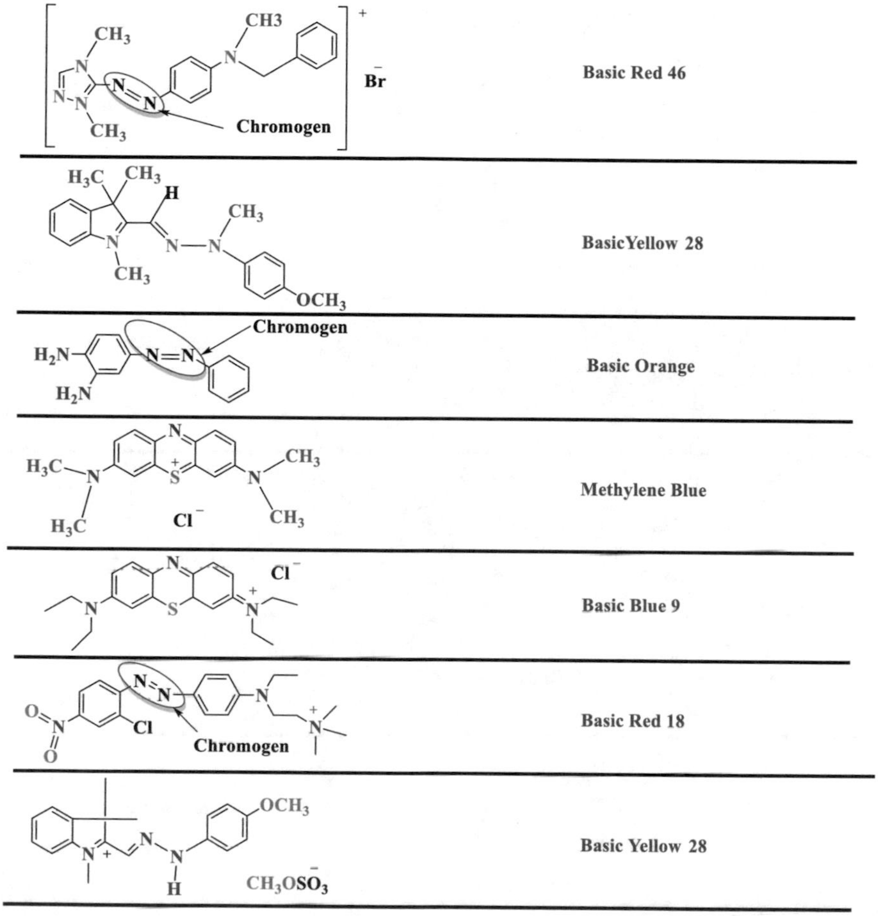

Fig. 27 Some chemical structures of basic dyes (S. Benkhaya)

reactive dye is a non-fixed water-soluble dye applied to cotton fabrics, and it spills 50–90% into the textile effluent [2]. Azo dyes have complex structures and show high resistance toward natural, biological, and physical degradation [154]. Most of the dyes are non-biodegradable and toxic, and they hinder light transmittance, which disturbs the ecosystem cycle, and dye-removal water treatment becomes more critical [100].

These dyes have harmful effects on the environment as well as on human health. Some dyes are toxic and/or carcinogenic, and the biodegradation of many dyes yields aromatic amines, which may be carcinogenic or otherwise toxic [82]. In aquatic environments, dyes affect photosynthetic activities by preventing the penetration of oxygen and light [49]. Despite this, most textile wastewater is characterized by high values of physico-chemical and biological parameters (temperature, salinity, pH, biological oxygen demand, chemical oxygen demand, biotoxicity, a large amount

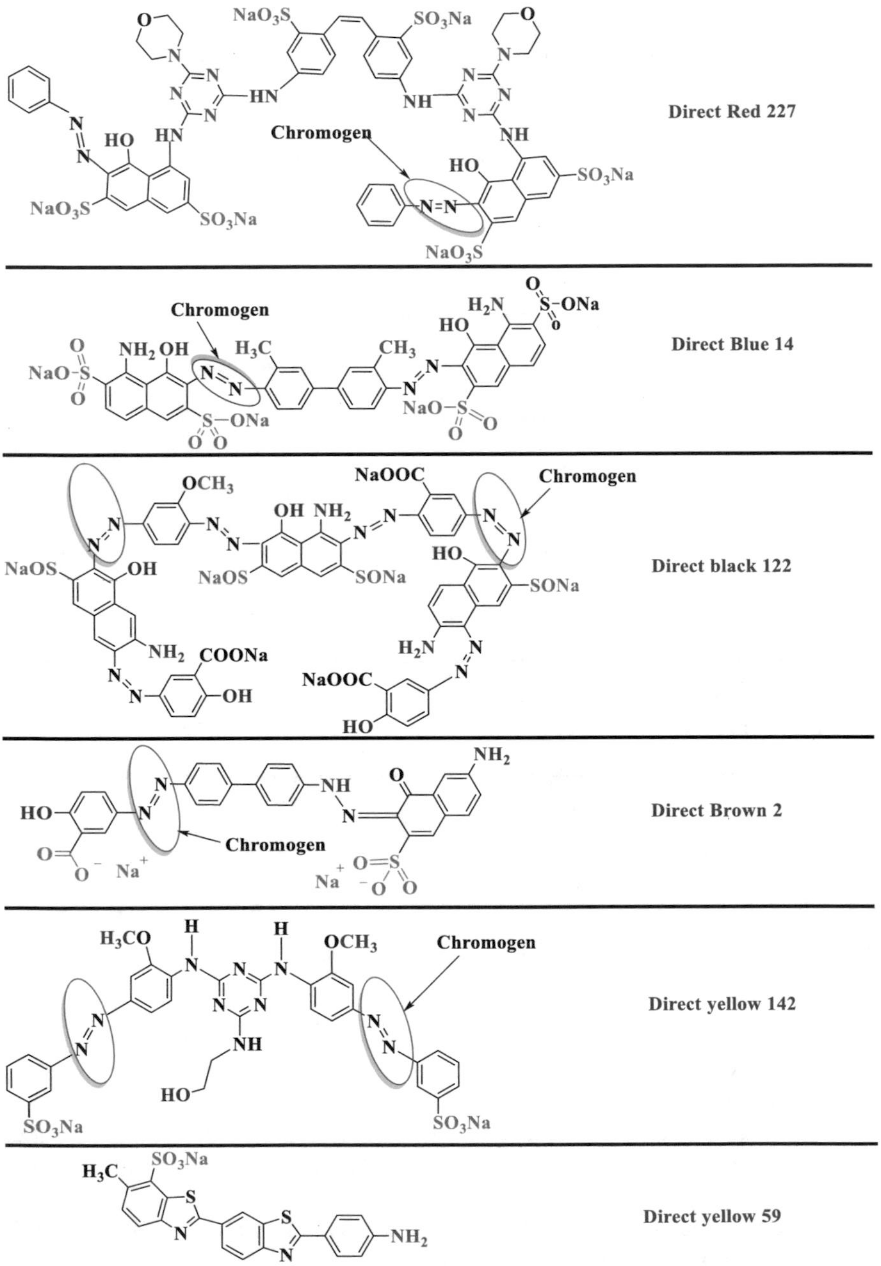

Direct Red 227

Direct Blue 14

Direct black 122

Direct Brown 2

Direct yellow 142

Direct yellow 59

Fig. 28 Some chemical structures of direct dyes (S. Benkhaya)

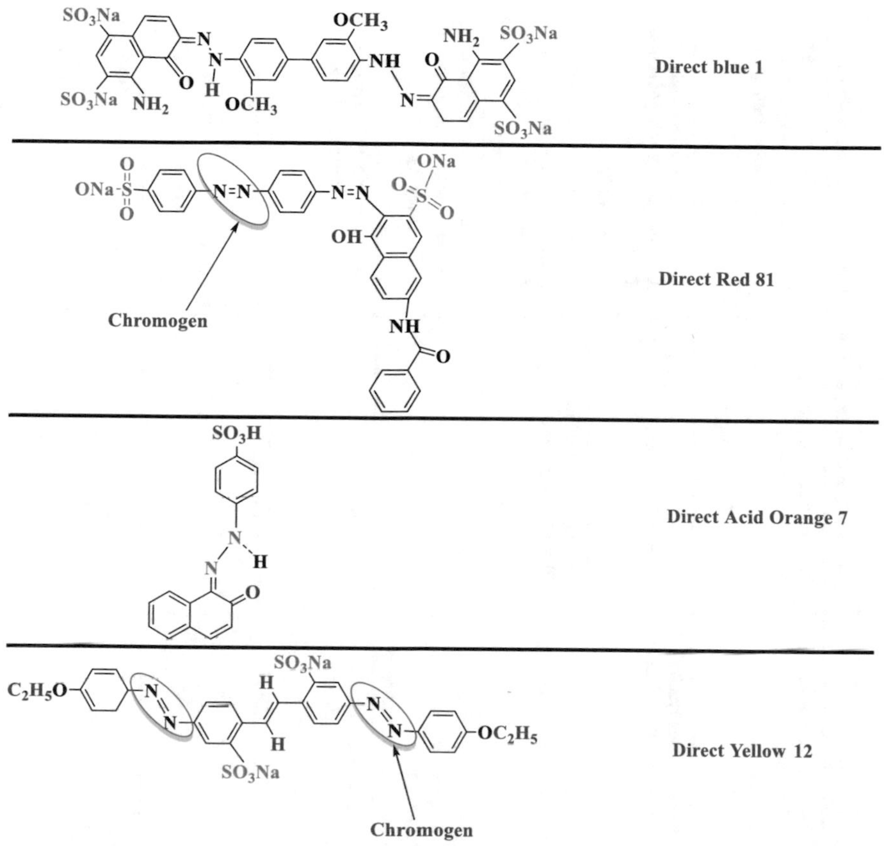

Direct blue 1

Direct Red 81

Chromogen

Direct Acid Orange 7

Direct Yellow 12

Chromogen

Fig. 29 Some chemical structures of direct dyes (S. Benkhaya)

of suspended solids, etc.) [102]. In addition, textile wastewater is characterized by high values of physico-chemical and biological parameters (temperature, salinity, pH, biological oxygen demand, and chemical oxygen demand) [50]. There are also organometallic compounds [42] which have harmful effects on the aquatic flora and fauna and, consequently, on the environment and humans in particular [48, 54]. In addition, colored water causes a scarcity of light essential to the development of aquatic organisms. Some studies have shown that certain dyes and surfactants can lead to the inhibition of biological systems and be toxic to fish [122].

Until now, a wide range of biological and physico-chemical techniques such as membrane filtration processes, ozonation, coagulation/flocculation, adsorption, and electrochemical oxidation have been used to remove dyes from their substances. Waste effluents [113]. Obviously, each of these techniques has its own advantages and limitations, and therefore, the implementation of an appropriate method is a decision-maker's choice based on cost–benefit analysis. Among the different water

Table 6 Classification of textile dyes [16, 74, 75, 125] (S. Benkhaya)

Dye class and examples	General description	Chemical structure	Properties and fixation degree (%)	Applications and loss in effluent (%)
Acid: – Acid blue 25 – Acid red 57 – Methyl orange	The degree of colorfastness for the size of the dye molecule [148]	Anthraquinone, nitroso, azine, azo, xanthene, nitro, and triphenylmethane	– Anionic, water soluble – 80–95% (Polyamide)	– Polyamide 70–75%, wool 25 to 30%, silk, wool, paper, nylon, inks, leather, inkjet printing and cosmetics, etc. [36] – 5–20% (Polyamide)
Reactive: – Reactive black 5 – Reactive yellow 2	Fixing of the reactive dye requiring a temperature above 60 °C	Anthraquinone, phthalocyanine, azo, oxazine, formazan, and basic	– Anionic, water soluble – 50–90% (Cellulose)	– Use of reactive dyes mainly in the dyeing and printing of cotton fibers – 10–50% (Cellulose)
Direct: – Direct orange 34 – Direct violet – Direct black	Relatively inexpensive direct dyes, available in a full range of shades but with a high color gloss [30, 31]	Phthalocyanine, azo, nitro, benzodifuranone, and stilbene	– Anionic, water soluble – 70–95% (Cellulose)	– Cellulose fibers, cotton, viscose, paper, leather, polyamide, silk, wool, and rayon – 5–30% (Cellulose)
Basic: – Basic brown – Basic yellow 28 – Basic red 9	The cheapest basic dyes [32]	Anthraquinone, azo, hemicyanine, cyanine, oxazine, azine, triarylmethane, acridine, diazahemicyanine and xanthene	– Cationic, water soluble – 95–100 (Acrylic)	– Synthetic fibers, wool, paper, polyester inks, leather, and acrylic – 0–5% (Acrylic)
Vat: – Green 6 – Vat blue – Indigo	Vat dyes are known for better colorfastness and their application to nanofibers [84]	Indigoids and anthraquinone (including polycyclic quinones)	Colloidal, insoluble	They have an affinity for cellulose, cotton, rayon viscose, and wool
Sulfur: – Sulfur black – Sulfur blue dye – Phthalic anhydride	Sulfur dyes have no well-defined chemical structures [33, 36]	Indeterminate structures	Colloidal, insoluble	Sulfur dyes are inexpensive and are used mainly for dyeing cellulosic textile materials or mixtures of cellulose fibers and rarely silk [36, 109]

(continued)

Table 6 (continued)

Dye class and examples	General description	Chemical structure	Properties and fixation degree (%)	Applications and loss in effluent (%)
Disperse: – Disperse red – Disperse yellow – Disperse orange – Disperse blue	Dispersed dyes are poorly water-soluble compounds, they disperse in water because their commercial formulations contain surfactants necessary for the dyeing process [147]	Anthraquinone, azo, benzodifuranone, and nitro	– Very low water solubility – 90–100% (Polyester)	– Polyester, polyamide, plastic, acetate, nylon, and acrylic – 0–10% (Polyester)

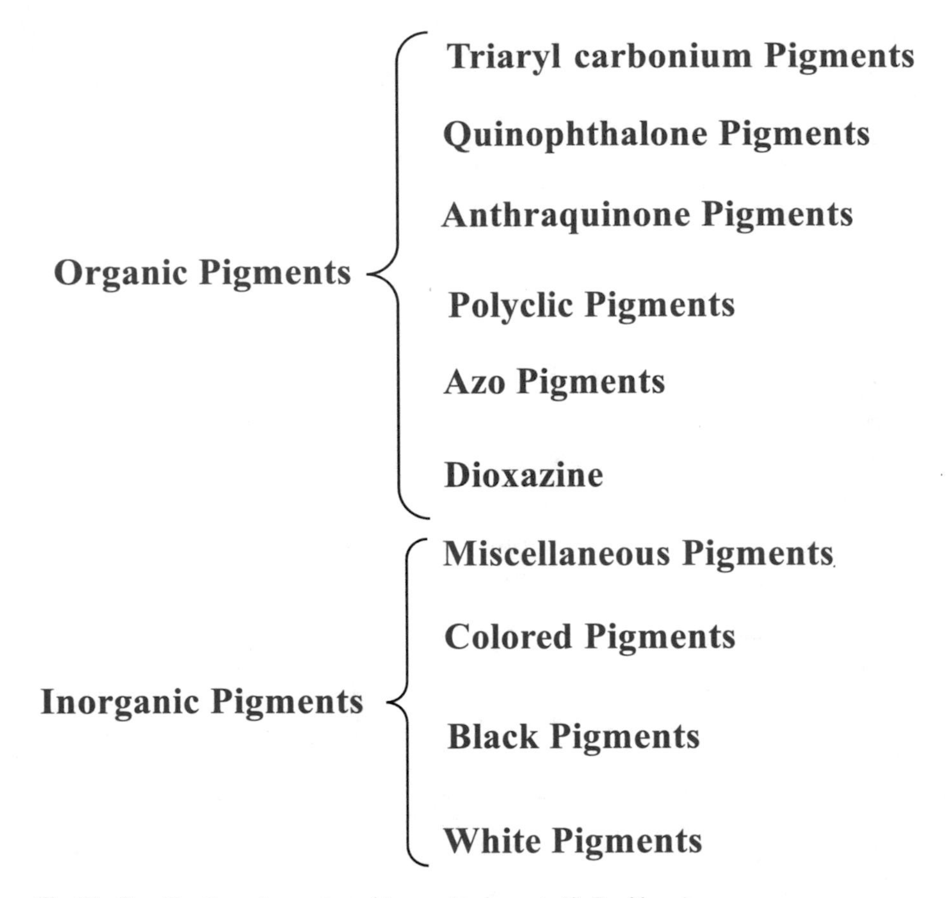

Fig. 30 Classification of organic and inorganic pigments (S. Benkhaya)

purification techniques available, membrane technology has received a lot of attention and it is one of the innovative ideas of water treatment, due to its energy saving, its moderate capital and maintenance expenditure, high efficiency, and ease of use [76]. Ultrafiltration (UF) is a pressure-driven membrane separation process that separates on the basis of molecular diameter. It is a quick and easily applicable method for fractionation and concentration steps in pure water production and water and wastewater treatment [52]. Figure 33 presents a summary of the toxics used in various processes in the textile and dye industries, different classes of textile dyes and their health effects, and sources of metals in textile effluents and dyes and base metals found in different classes of dyes).

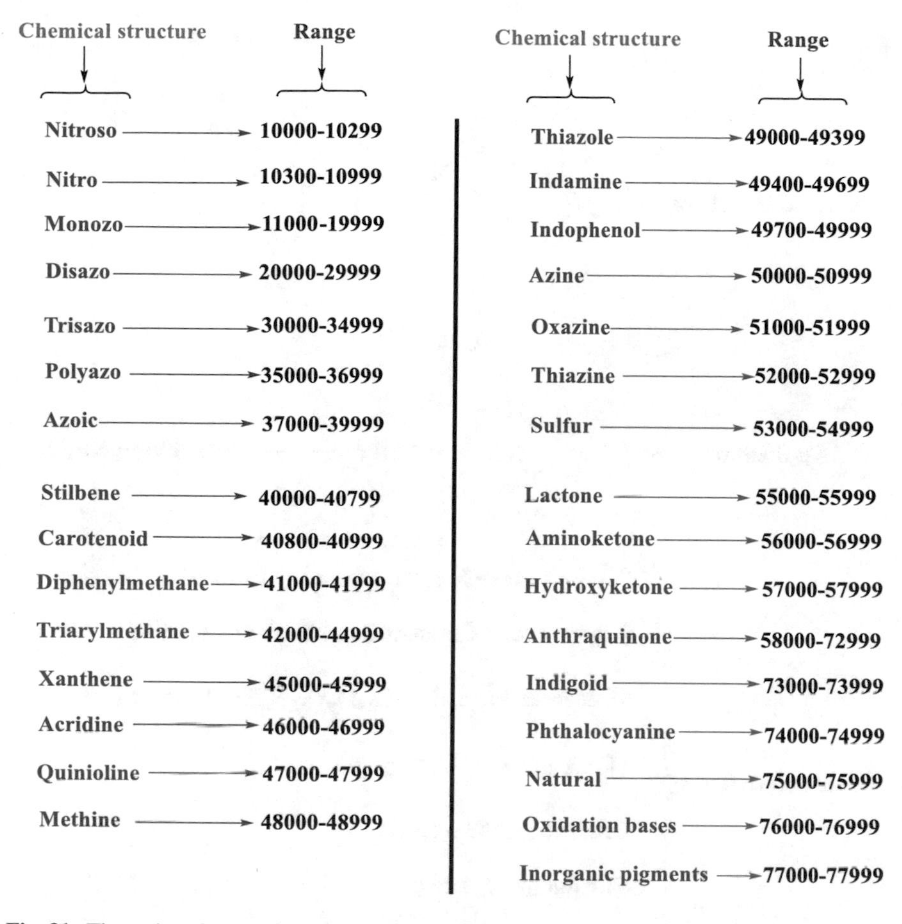

Fig. 31 The main colorant categories and C.I. constitution numbers (S. Benkhaya)

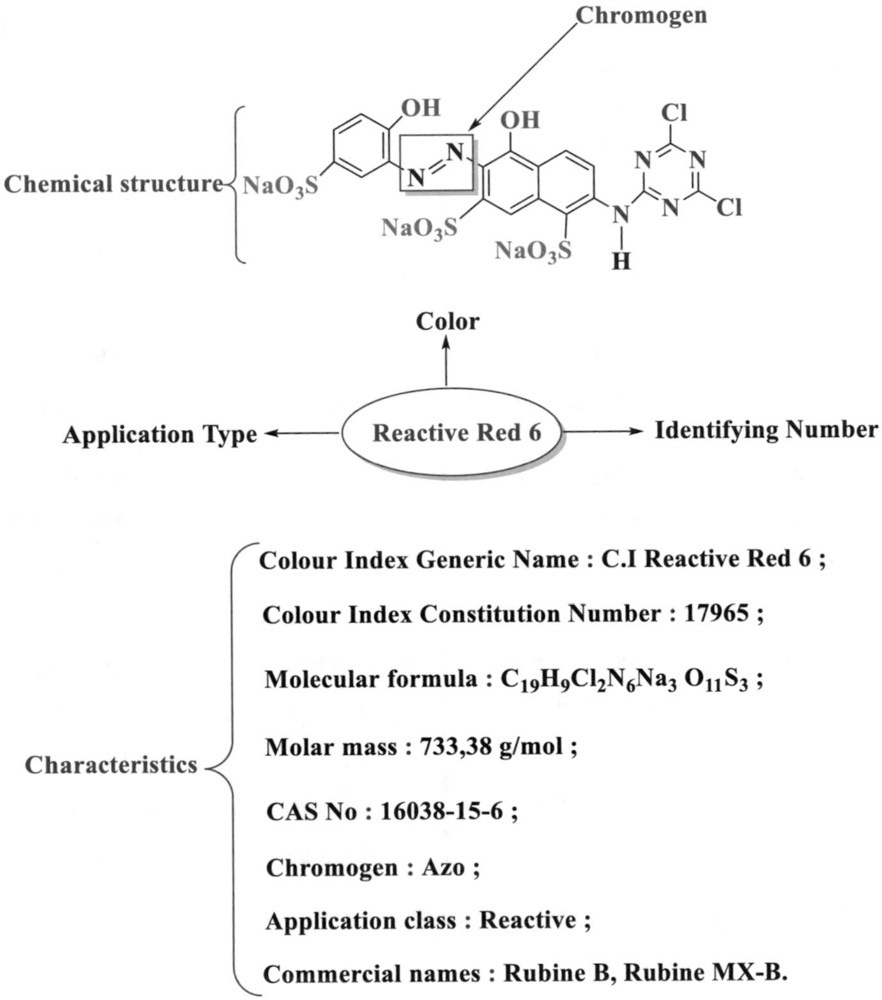

Chromogen

Chemical structure

OH OH Cl

N N N N

NaO_3S N Cl

NaO_3S N

NaO_3S H

Color

Application Type ← Reactive Red 6 → Identifying Number

Characteristics

Colour Index Generic Name : C.I Reactive Red 6 ;

Colour Index Constitution Number : 17965 ;

Molecular formula : $C_{19}H_9Cl_2N_6Na_3O_{11}S_3$;

Molar mass : 733,38 g/mol ;

CAS No : 16038-15-6 ;

Chromogen : Azo ;

Application class : Reactive ;

Commercial names : Rubine B, Rubine MX-B.

Fig. 32 Description example for the color index classification (S. Benkhaya)

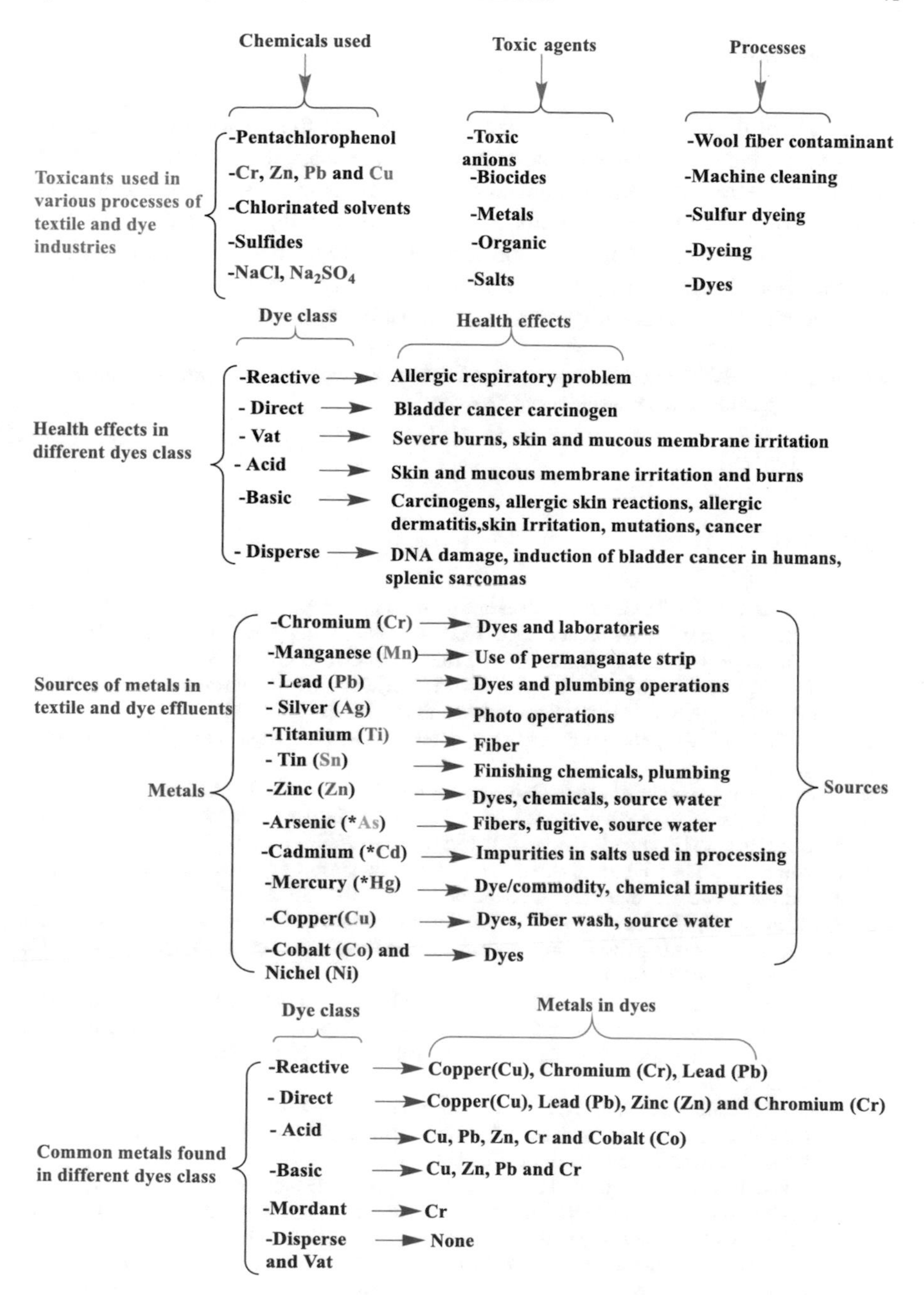

Fig. 33 Toxics from the textile and dyeing industries and their effects on human health (S. Benkhaya)

6 Conclusion

In the light of this review about textiles dyes, it is clear that many chemical classes of synthetic dyes are frequently employed on an industrial scale; anthraquinones dyes, triarylmethanes, indigoids, phenothiazine, xanthenes, and azos are among the most widespread. Some of the synthetic dyes are toxic to aquatic ecosystems. They can pose a threat to the balance of the ecosystem due to their high toxicity across different food chains. Thus, the presence of these dyes in the environment with particular regard to the aquatic environment must be monitored.

Acknowledgements The authors express their thanks to everyone who contributed from near or far to the completion of this valuable scientific work.

References

1. Abel A (2012) 16—The history of dyes and pigments: from natural dyes to high performance pigments. In: Best J (ed) Colour design. Woodhead Publishing Series in Textiles. Woodhead Publishing, pp 433–470. https://doi.org/10.1533/9780857095534.3.433
2. Al-Degs Y, Khraisheh MAM, Allen SJ, Ahmad MN (2000) Effect of carbon surface chemistry on the removal of reactive dyes from textile effluent. Water Res 34:927–935
3. Aljamali NM (2015) Review in azo compounds and its biological activity. Biochem Anal Biochem 4:1–4
4. Ameenudeen S, Unnikrishnan S, Ramalingam K (2021) Statistical optimization for the efficacious degradation of reactive azo dyes using Acinetobacter baumannii JC359. J Environ Manag 279:111512. https://doi.org/10.1016/j.jenvman.2020.111512
5. Antoniotti S, Duñach E (2002) Direct and catalytic synthesis of quinoxaline derivatives from epoxides and ene-1, 2-diamines. Tetrahedron Lett 43:3971–3973
6. Arivithamani N, Giri Dev VR (2017) Sustainable bulk scale cationization of cotton hosiery fabrics for salt-free reactive dyeing process. J Clean Prod 149:1188–1199. https://doi.org/10.1016/j.jclepro.2017.02.162
7. Arshad R, Bokhari TH, Khosa KK, Bhatti IA, Munir M, Iqbal M, Iqbal DN et al (2020) Gamma radiation induced degradation of anthraquinone Reactive Blue-19 dye using hydrogen peroxide as oxidizing agent. Radiat Phys Chem 168:108637. https://doi.org/10.1016/j.radphyschem.2019.108637
8. Asgher M, Batool S, Bhatti HN, Noreen R, Rahman SU, Javaid Asad M (2008) Laccase mediated decolorization of vat dyes by Coriolus versicolor IBL-04. Int Biodeterior Biodegrad 62:465–470. https://doi.org/10.1016/j.ibiod.2008.05.003
9. Asif Tahir M, Bhatti HN, Iqbal M (2016) Solar red and brittle blue direct dyes adsorption onto Eucalyptus angophoroides bark: equilibrium, kinetics and thermodynamic studies. J Environ Chem Eng 4:2431–2439. https://doi.org/10.1016/j.jece.2016.04.020
10. Aysha T, Zain M, Arief M, Youssef Y (2019) Synthesis and spectral properties of new fluorescent hydrazone disperse dyes and their dyeing application on polyester fabrics. Heliyon 5:e02358. https://doi.org/10.1016/j.heliyon.2019.e02358
11. Balçık U, Chormey DS, Ayyıldız MF, Bakırdere S (2020) Liquid phase microextraction based sensitive analytical strategy for the determination of 22 hazardous aromatic amine products of azo dyes in wastewater and tap water samples by GC-MS system. Microchem J 155:104712. https://doi.org/10.1016/j.microc.2020.104712

12. Bechtold T, Turcanu A (2009) Electrochemical reduction in vat dyeing: greener chemistry replaces traditional processes. J Clean Prod 17:1669–1679. https://doi.org/10.1016/j.jclepro.2009.08.004
13. Benaissa A (2012) Etude de la faisabilité d'élimination de certains colorants textiles par certains matériaux déchets d'origine naturelle. http://dspace.univ-tlemcen.dz/handle/112/1232
14. Benkhaya S, Achiou B, Ouammou M, Bennazha J, Alami Younssi S, M'rabet S, El Harfi A (2019) Preparation of low-cost composite membrane made of polysulfone/polyetherimide ultrafiltration layer and ceramic pozzolan support for dyes removal. Mater Today Commun 19:212–219. https://doi.org/10.1016/j.mtcomm.2019.02.002
15. Benkhaya S, M'rabet S, El Harfi A (2020) Classifications, properties, recent synthesis and applications of azo dyes. Heliyon 6:e03271. https://doi.org/10.1016/j.heliyon.2020.e03271
16. Benkhaya S, Mrabet S, El Harfi A (2020) A review on classifications, recent synthesis and applications of textile dyes. Inorg Chem Commun (Elsevier) 107891. https://doi.org/10.1016/j.inoche.2020.107891
17. Berradi M, Hsissou R, Khudhair M, Assouag M, Cherkaoui O, El Bachiri A, El Harfi A (2019) Textile finishing dyes and their impact on aquatic environs. Heliyon (Elsevier) 5:e02711.
18. Boudechiche N, Fares M, Ouyahia S, Yazid H, Trari M, Sadaoui Z (2019) Comparative study on removal of two basic dyes in aqueous medium by adsorption using activated carbon from Ziziphus lotus stones. Microchem J 146:1010–1018. https://doi.org/10.1016/j.microc.2019.02.010
19. Bouhelassa M (2019) Etude de la dégradation photocatalytique d'un colorant synthétique et d'un tensioactif, Université Mentouri Constantine
20. Božič M, Kokol V (2008) Ecological alternatives to the reduction and oxidation processes in dyeing with vat and sulphur dyes. Dyes Pigments 76:299–309. https://doi.org/10.1016/j.dyepig.2006.05.041
21. Burkinshaw SM, Chaccour FE, Gotsopoulos A (1997) The aftertreatment of sulphur dyes on cotton. Dyes Pigments 34:227–241. https://doi.org/10.1016/S0143-7208(96)00075-7
22. Burkinshaw SM, Chevli SN, Marfell DJ (2000) The dyeing of nylon 6,6 with sulphur dyes. Dyes Pigments 45:65–74. https://doi.org/10.1016/S0143-7208(00)00003-6
23. Burkinshaw SM, Lagonika K, Marfell DJ (2003) Sulphur dyes on nylon 6,6—Part 1: The effects of temperature and pH on dyeing. Dyes Pigments 56:251–259. https://doi.org/10.1016/S0143-7208(02)00146-8
24. Burkinshaw SM, Salihu G (2017) The role of auxiliaries in the immersion dyeing of textile fibres. Part 2: Analysis of conventional models that describe the manner by which inorganic electrolytes promote direct dye uptake on cellulosic fibres. Dyes Pigments. https://doi.org/10.1016/j.dyepig.2017.08.034
25. Burkinshaw SM, Salihu G (2017) The role of auxiliaries in the immersion dyeing of textile fibres: Part 3 Theoretical model to describe the role of inorganic electrolytes used in dyeing cellulosic fibres with direct dyes. Dyes Pigments. https://doi.org/10.1016/j.dyepig.2017.11.039
26. Burkinshaw SM, Salihu G (2019) The role of auxiliaries in the immersion dyeing of textile fibres: Part 8 Practical aspects of the role of inorganic electrolytes in dyeing cellulosic fibres with commercial reactive dyes. Dyes Pigments 161:614–627. https://doi.org/10.1016/j.dyepig.2017.09.072
27. Burkinshaw SM, Son Y-A (2010) The dyeing of supermicrofibre nylon with acid and vat dyes. Dyes Pigments 87:132–138. https://doi.org/10.1016/j.dyepig.2010.03.009
28. Cai Y, Liang Y, Navik R, Zhu W, Zhang C, Pervez MdN, Wang Q (2020) Improved reactive dye fixation on ramie fiber in liquid ammonia and optimization of fixation parameters using the Taguchi approach. Dyes Pigments 183:108734. https://doi.org/10.1016/j.dyepig.2020.108734
29. Chakraborty JN (2010) Introduction to dyeing of textiles. In: Fundamentals and practices in colouration of textiles. Elsevier, pp 1–10. https://doi.org/10.1533/9780857092823.1
30. Chakraborty JN (2010) 4—Dyeing with direct dye. In: Fundamentals and practices in colouration of textiles. Woodhead Publishing India, pp 27–42. https://doi.org/10.1533/9780857092823.27

31. Chakraborty JN (2010) 16—Dyeing with metal–complex dye. In: Fundamentals and practices in colouration of textiles. Woodhead Publishing India, pp 175–183. https://doi.org/10.1533/9780857092823.175
32. Chakraborty JN (2010) 17—Dyeing with basic dye. In: Fundamentals and practices in colouration of textiles. Woodhead Publishing India, pp 184–191. https://doi.org/10.1533/9780857092823.184
33. Chakraborty JN (2010) 5—Dyeing with sulphur dye. In: Fundamentals and practices in colouration of textiles. Woodhead Publishing India, pp 43–56. https://doi.org/10.1533/9780857092823.45
34. Chakraborty JN (2014) 5—Dyeing with sulphur dye. In: Chakraborty JN (ed) Fundamentals and practices in colouration of textiles. Woodhead Publishing India, pp 46–60. https://doi.org/10.1016/B978-93-80308-46-3.50005-4
35. Chao YC, Chung YL, Lai CC, Liao SK, Chin JC (1999) Dyeing of cotton-polyester blends with anthraquinonoid vat dyes. Dyes Pigments 40:59–71. https://doi.org/10.1016/S0143-7208(98)00033-3
36. Chebli D (2018) Traitement des eaux usées industrielles: Dégradation des colorants azoïques par un procédé intégré couplant un procédé d'oxydation avancée et un traitement biologique. PhD thesis
37. Chiarello LM, Mittersteiner M, de Jesus PC, Andreaus J, Barcellos IO (2020) Reuse of enzymatically treated reactive dyeing baths: evaluation of the number of reuse cycles. J Clean Prod 267:122033. https://doi.org/10.1016/j.jclepro.2020.122033
38. Choudhury AR (2006) Textile preparation and dyeing. Science Publishers
39. Christie R (2014) Colour chemistry. Royal Society of Chemistry
40. Clark M (2011) Fundamental principles of dyeing. In: Handbook of textile and industrial dyeing: principles, processes and types of dyes, 1st ed
41. Cote PN, Lan X, Shakhnovich AI, Domingo MJ (1999) Sulfur dyes. Google Patents. https://www.google.com/patents/US5961670
42. Crabtree RH (2009) The organometallic chemistry of the transition metals. Wiley
43. Cretescu I, Lupascu T, Buciscanu I, Balau-Mindru T, Soreanu G (2017) Low-cost sorbents for the removal of acid dyes from aqueous solutions. Process Saf Environ Prot 108:57–66. https://doi.org/10.1016/j.psep.2016.05.016
44. Dai Y, Yang B, Ding Y, Xu H, Wang B, Zhang L, Chen Z et al (2020) Real-time monitoring of multicomponent reactive dye adsorption on cotton fabrics by Raman spectroscopy. Spectrochim Acta A Mol Biomol Spectrosc 230:118051. https://doi.org/10.1016/j.saa.2020.118051
45. de Souza TNV, de Carvalho SML, Vieira MGA, da Silva MGC, Brasil D, do SB (2018) Adsorption of basic dyes onto activated carbon: experimental and theoretical investigation of chemical reactivity of basic dyes using DFT-based descriptors. Appl Surf Sci 448:662–670. https://doi.org/10.1016/j.apsusc.2018.04.087
46. Demirçalı A, Karcı F, Avinc O, Kahrıman AU, Gedik G, Bakan E (2019) The synthesis, characterization and investigation of absorption properties of disperse disazo dyes containing pyrazole and isoxazole. J Mol Struct 1181:8–13. https://doi.org/10.1016/j.molstruc.2018.12.033
47. Dong X, Gu Z, Hang C, Ke G, Jiang L, He J (2019) Study on the salt-free low-alkaline reactive cotton dyeing in high concentration of ethanol in volume. J Clean Prod 226:316–323. https://doi.org/10.1016/j.jclepro.2019.04.006
48. Duruibe JO, Ogwuegbu MOC, Egwurugwu JN (2007) Heavy metal pollution and human biotoxic effects. Int J Phys Sci 2:112–118
49. El-Aassar MR, Fakhry H, Elzain AA, Farouk H, Hafez EE (2018) Rhizofiltration system consists of chitosan and natural Arundo donax L. for removal of basic red dye. Int J Biol Macromol 120:1508–1514. https://doi.org/10.1016/j.ijbiomac.2018.09.159
50. Es-sahbany H, Berradi M, Nkhili S, Hsissou R, Allaoui M, Loutfi M, Bassir D et al (2019) Removal of heavy metals (nickel) contained in wastewater-models by the adsorption technique on natural clay. Mater Today Proc 13:866–875

51. Fang S, Feng G, Guo Y, Chen W, Qian H (2020) Synthesis and application of urethane-containing azo disperse dyes on polyamide fabrics. Dyes Pigments 176:108225. https://doi.org/10.1016/j.dyepig.2020.108225
52. Fradj AB, Boubakri A, Hafiane A, Hamouda SB (2020) Removal of azoic dyes from aqueous solutions by chitosan enhanced ultrafiltration. Results Chem (Elsevier) 2:100017
53. Franco JH, da Silva BF, de Castro AA, Ramalho TC, Pividori MI, Zanoni MVB (2018) Biotransformation of disperse dyes using nitroreductase immobilized on magnetic particles modified with tosyl group: identification of products by LC-MS-MS and theoretical studies conducted with DNA. Environ Pollut 242:863–871. https://doi.org/10.1016/j.envpol.2018.07.054
54. Förstner U, Wittmann GT (2012) Metal pollution in the aquatic environment. Springer Science & Business Media
55. Gao A, Hu L, Zhang H, Fu D, Hou A, Xie K (2018) Silicone nanomicelle dyeing using the nanoemulsion containing highly dispersed dyes for polyester fabrics. J Clean Prod 200:48–53. https://doi.org/10.1016/j.jclepro.2018.07.312
56. Garcia VSG, Rosa JM, Borrely SI (2020) Toxicity and color reduction of a textile effluent containing reactive red 239 dye by electron beam irradiation. Radiat Phys Chem 172:108765. https://doi.org/10.1016/j.radphyschem.2020.108765
57. Garg D, Majumder CB, Kumar S, Sarkar B (2019) Removal of Direct Blue-86 dye from aqueous solution using alginate encapsulated activated carbon (PnsAC-alginate) prepared from waste peanut shell. J Environ Chem Eng 7:103365. https://doi.org/10.1016/j.jece.2019.103365
58. Ghaly AE, Ananthashankar R, Alhattab M, Ramakrishnan VV (2014) Production, characterization and treatment of textile effluents: a critical review. J Chem Eng Process Technol 5:1–18
59. Ghanavatkar CW, Mishra VR, Sekar N (2020) Benzothiazole-pyridone and benzothiazole-pyrazole clubbed emissive azo dyes and dyeing application on polyester fabric: UPF, biological, photophysical and fastness properties with correlative computational assessments. Spectrochim Acta A Mol Biomol Spectrosc 230:118064. https://doi.org/10.1016/j.saa.2020.118064
60. Gita S, Hussan A, Choudhury TG (2017) Impact of textile dyes waste on aquatic environments and its treatment. Environ Ecol 35:2349–2353
61. González AS, Martínez SS (2008) Study of the sonophotocatalytic degradation of basic blue 9 industrial textile dye over slurry titanium dioxide and influencing factors. Ultrason Sonochem 15:1038–1042. https://doi.org/10.1016/j.ultsonch.2008.03.008
62. Guo G, Li X, Tian F, Liu T, Yang F, Ding K, Liu C et al (2020) Azo dye decolorization by a halotolerant consortium under microaerophilic conditions. Chemosphere 244:125510. https://doi.org/10.1016/j.chemosphere.2019.125510
63. Gupta VK et al (2009) Application of low-cost adsorbents for dye removal—a review. J Environ Manag 90:2313–2342
64. Gürses A, Açıkyıldız M, Güneş K, Gürses MS (2016) Classification of dye and pigments. In: Dyes and pigments. Springer, pp 31–45
65. Hafdi H, Joudi M, Mouldar J, Hatimi B, Nasrellah H, El Mhammedi MA, Bakasse M (2020) Design of a new low cost natural phosphate doped by nickel oxide nanoparticles for capacitive adsorption of reactive red 141 azo dye. Environ Res 184:109322. https://doi.org/10.1016/j.envres.2020.109322
66. Han R, Zhang S, Zhao W, Li X, Jian X (2009) Treating sulfur black dye wastewater with quaternized poly (phthalazinone ether sulfone ketone) nanofiltration membranes. Sep Purif Technol 67:26–30. https://doi.org/10.1016/j.seppur.2009.03.006
67. Harichandran G, Prasad S (2016) SonoFenton degradation of an azo dye, direct red. Ultrason Sonochem 29:178–185. https://doi.org/10.1016/j.ultsonch.2015.09.005
68. Hassan MM, Carr CM (2018) A critical review on recent advancements of the removal of reactive dyes from dyehouse effluent by ion-exchange adsorbents. Chemosphere 209:201–219. https://doi.org/10.1016/j.chemosphere.2018.06.043

69. Hassan MM, Saifullah K (2018) Ultrasound-assisted pre-treatment and dyeing of jute fabrics with reactive and basic dyes. Ultrason Sonochem 40:488–496. https://doi.org/10.1016/j.ultsonch.2017.07.037
70. Hihara T, Okada Y, Morita Z (2002) Photo-oxidation and -reduction of vat dyes on water-swollen cellulose and their lightfastness on dry cellulose. Dyes Pigments 53:153–177. https://doi.org/10.1016/S0143-7208(02)00017-7
71. Holme I (2006) Sir William Henry Perkin: a review of his life, work and legacy. Color Technol 122:235–251
72. Horning RH (1978) Textile dyeing wastewaters: characterization and treatment. Environmental Protection Agency, Office of Research and Development
73. Humelnicu I, Băiceanu A, Ignat M-E, Dulman V (2017) The removal of Basic Blue 41 textile dye from aqueous solution by adsorption onto natural zeolitic tuff: kinetics and thermodynamics. Process Saf Environ Prot 105:274–287. https://doi.org/10.1016/j.psep.2016.11.016
74. Hunger K et al (2003) Health and safety aspects. Ind. Dyes Chem. Prop. Appl. 625–641
75. Husain Q (2006) Potential applications of the oxidoreductive enzymes in the decolorization and detoxification of textile and other synthetic dyes from polluted water: a review. Crit Rev Biotechnol 26:201–221. https://doi.org/10.1080/07388550600969936
76. Ibrahim GS, Isloor AM, Lakshmi B (2020) Synthetic polymer-based membranes for dye and pigment removal. In: Synthetic polymeric membranes for advanced water treatment, gas separation, and energy sustainability. Elsevier, pp 39–52
77. Jabbar A, Ambreen R, S., Navaid, F.A. & Choudhary, M.I. (2019) A series of new acid dyes; study of solvatochromism, spectroscopy and their application on wool fabric. J Mol Struct 1195:161–167. https://doi.org/10.1016/j.molstruc.2019.05.019
78. Jamil A, Bokhari TH, Javed T, Mustafa R, Sajid M, Noreen S, Zuber M et al (2020) Photocatalytic degradation of disperse dye Violet-26 using TiO2 and ZnO nanomaterials and process variable optimization. J Mater Res Technol 9:1119–1128. https://doi.org/10.1016/j.jmrt.2019.11.035
79. Kant R (2011) Textile dyeing industry an environmental hazard
80. Karcı F, Bakan E (2015) New disazo pyrazole disperse dyes: synthesis, spectroscopic studies and tautomeric structures. J Mol Liq 206:309–315
81. Kariyajjanavar P, Narayana J, Nayaka YA (2013) Degradation of textile dye C.I. Vat Black 27 by electrochemical method by using carbon electrodes. J Environ Chem Eng 1:975–980. https://doi.org/10.1016/j.jece.2013.08.002
82. Khaled A, El Nemr A, El-Sikaily A, Abdelwahab O (2009) Treatment of artificial textile dye effluent containing Direct Yellow 12 by orange peel carbon. Desalination Issues 1 and 2: First international workshop between the Center for the Seawater Desalination Plant and the European Desalination Society, vol 238, pp 210–232. https://doi.org/10.1016/j.desal.2008.02.014
83. Khan S, Malik A (2014) Environmental and health effects of textile industry wastewater. In: Environmental deterioration and human health. Springer, pp 55–71
84. Khatri M, Ahmed F, Shaikh I, Phan D-N, Khan Q, Khatri Z, Lee H et al (2017) Dyeing and characterization of regenerated cellulose nanofibers with vat dyes. Carbohydr Polym 174:443–449. https://doi.org/10.1016/j.carbpol.2017.06.125
85. Khatri A, Peerzada MH, Mohsin M, White M (2015) A review on developments in dyeing cotton fabrics with reactive dyes for reducing effluent pollution. J Clean Prod 87:50–57. https://doi.org/10.1016/j.jclepro.2014.09.017
86. Khosravi A, Karimi M, Ebrahimi H, Fallah N (2020) Sequencing batch reactor/nanofiltration hybrid method for water recovery from textile wastewater contained phthalocyanine dye and anionic surfactant. J Environ Chem Eng 8:103701. https://doi.org/10.1016/j.jece.2020.103701
87. Khouni I, Louhichi G, Ghrabi A (2020) Assessing the performances of an aerobic membrane bioreactor for textile wastewater treatment: influence of dye mass loading rate and biomass concentration. Process Saf Environ Prot 135:364–382. https://doi.org/10.1016/j.psep.2020.01.011

88. Kishor R, Purchase D, Saratale GD, Saratale RG, Ferreira LFR, Bilal M, Chandra R et al (2021) Ecotoxicological and health concerns of persistent coloring pollutants of textile industry wastewater and treatment approaches for environmental safety. J Environ Chem Eng (Elsevier) 105012
89. Kuenemann MA, Szymczyk M, Chen Y, Sultana N, Hinks D, Freeman HS, Williams AJ et al (2017) Weaver's historic accessible collection of synthetic dyes: a cheminformatics analysis. Chem Sci 8:4334–4339. Royal Society of Chemistry. https://doi.org/10.1039/C7SC00567A
90. Kulandainathan MA, Muthukumaran A, Patil K, Chavan RB (2007) Potentiostatic studies on indirect electrochemical reduction of vat dyes. Dyes Pigments 73:47–54. https://doi.org/10.1016/j.dyepig.2005.10.007
91. Lacasse K (2004) Colouring. In: Textile chemicals. Springer, pp 156–372. https://doi.org/10.1007/978-3-642-18898-5_5.pdf
92. Li H, Qian H-F, Feng G (2019) Diversity-oriented synthesis of azo disperse dyes with improved fastness properties via employing Ugi four-component reaction. Dyes Pigments 165:415–420. https://doi.org/10.1016/j.dyepig.2019.02.039
93. Li Y, Tan T, Wang S, Xiao Y, Li X (2017) Highly solvatochromic fluorescence of anthraquinone dyes based on triphenylamines. Dyes Pigments 144:262–270
94. Ling W, Xinjiang S, Guoliang Z, Wenrui Z (2011) Performance of composite reverse osmosis membranes used in textile wastewater treatment and reutilization. In: 2011 international conference on computer distributed control and intelligent environmental. IEEE, pp 1611–1614
95. Liu Y, Fu J, Deng S, Zhang X, Shen F, Yang G, Peng H et al (2014) Degradation of basic and acid dyes in high-voltage pulsed discharge. J Taiwan Inst Chem Eng 45:2480–2487. https://doi.org/10.1016/j.jtice.2014.05.001
96. Liu W, Liu J, Zhang Y, Chen Y, Yang X, Duan L, Dharmarajan R et al (2019) Simultaneous determination of 20 disperse dyes in foodstuffs by ultra high performance liquid chromatography–tandem mass spectrometry. Food Chem 300:125183. https://doi.org/10.1016/j.foodchem.2019.125183
97. Luescher M (1993) Organic pigments. In: Surface coatings. Springer, pp 473–513. https://doi.org/10.1007/978-94-011-1220-8_28
98. Maliyappa MR, Keshavayya J, Mahanthappa M, Shivaraj Y, Basavarajappa KV (2020) 6-Substituted benzothiazole based dispersed azo dyes having pyrazole moiety: synthesis, characterization, electrochemical and DFT studies. J Mol Struct 1199:126959. https://doi.org/10.1016/j.molstruc.2019.126959
99. Maliyappa MR, Keshavayya J, Mallikarjuna NM, Pushpavathi I (2020) Novel substituted aniline based heterocyclic dispersed azo dyes coupling with 5-methyl-2-(6-methyl-1, 3-benzothiazol-2-yl)-2, 4-dihydro-3H-pyrazol-3-one: synthesis, structural, computational and biological studies. J Mol Struct 1205:127576. https://doi.org/10.1016/j.molstruc.2019.127576
100. Manoukian M, Tavakol H, Fashandi H (2018) Synthesis of highly uniform sulfur-doped carbon sphere using CVD method and its application for cationic dye removal in comparison with undoped product. J Environ Chem Eng 6:6904–6915. https://doi.org/10.1016/j.jece.2018.10.026
101. Mansour H, Boughzala O, Barillier D, Chekir-Ghedira L, Mosrati R (2011) Les colorants textiles sources de contamination de l'eau: CRIBLAGE de la toxicité et des méthodes de traitement. Rev Sci L'eauJournal Water Sci 24:209–238
102. Mijin DŽ, Avramov Ivić ML, Onjia AE, Grgur BN (2012) Decolorization of textile dye CI Basic Yellow 28 with electrochemically generated active chlorine. Chem Eng J 204–206:151–157. https://doi.org/10.1016/j.cej.2012.07.091
103. Miladinova PM, Vaseva RK, Lukanova VR (2016) On the synthesis and application of some mono-and dis-azo acid dyes. J Chem Technol Metall 51:249–256
104. Miladinova PM, Lukanova VR (2017) Investigations on the dyeing ability of some reactive triazine azo dyes containing tetramethylpiperidine fragment. J Chem Technol Metall 52
105. Mishra VR, Ghanavatkar CW, Sekar N (2019) UV protective heterocyclic disperse azo dyes: spectral properties, dyeing, potent antibacterial activity on dyed fabric and comparative

computational study. Spectrochim Acta A Mol Biomol Spectrosc 223:117353. https://doi.org/10.1016/j.saa.2019.117353
106. Moradnia F, Taghavi Fardood S, Ramazani A, Gupta VK (2020) Green synthesis of recyclable MgFeCrO4 spinel nanoparticles for rapid photodegradation of direct black 122 dye. J Photochem Photobiol Chem 392:112433. https://doi.org/10.1016/j.jphotochem.2020.112433
107. Nguyen TA, Fu C-C, Juang R-S (2016) Biosorption and biodegradation of a sulfur dye in high-strength dyeing wastewater by Acidithiobacillus thiooxidans. J Environ Manag 182:265–271. https://doi.org/10.1016/j.jenvman.2016.07.083
108. Nguyen TA, Fu C-C, Juang R-S (2016) Effective removal of sulfur dyes from water by biosorption and subsequent immobilized laccase degradation on crosslinked chitosan beads. Chem Eng J 304:313–324. https://doi.org/10.1016/j.cej.2016.06.102
109. Nguyen TA, Juang R-S (2013) Treatment of waters and wastewaters containing sulfur dyes: a review. Chem Eng J 219:109–117. https://doi.org/10.1016/j.cej.2012.12.102
110. Nikfar S, Jaberidoost M (2014) Dyes and colorants. In: Wexler P (ed) Encyclopedia of toxicology, 3rd ed. Academic Press, Oxford, pp 252–261. https://doi.org/10.1016/B978-0-12-386454-3.00602-3
111. Oluwaseun AC, Kola OJ, Mishra P, Singh JR, Singh AK, Cameotra SS, Micheal BO (2017) Characterization and optimization of a rhamnolipid from Pseudomonas aeruginosa C1501 with novel biosurfactant activities. Sustain Chem Pharm (Elsevier) 6:26–36
112. Ozturk E, Yetis U, Dilek FB, Demirer GN (2009) A chemical substitution study for a wet processing textile mill in Turkey. J Clean Prod 17:239–247
113. Pavithra KG, Jaikumar V (2019) Removal of colorants from wastewater: a review on sources and treatment strategies. J Ind Eng Chem (Elsevier) 75:1–19
114. Pei L, Luo Y, Saleem MA, Wang J (2021) Sustainable pilot scale reactive dyeing based on silicone oil for improving dye fixation and reducing discharges. J Clean Prod 279:123831. https://doi.org/10.1016/j.jclepro.2020.123831
115. Penthala R, Heo G, Kim H, Lee IY, Ko EH, Son Y-A (2020) Synthesis of azo and anthraquinone dyes and dyeing of nylon-6,6 in supercritical carbon dioxide. J CO2 Util 38:49–58. https://doi.org/10.1016/j.jcou.2020.01.013
116. Qin Y, Yuan M, Hu Y, Lu Y, Lin W, Ma Y, Lin X et al (2020) Preparation and interaction mechanism of Nano disperse dye using hydroxypropyl sulfonated lignin. Int J Biol Macromol 152:280–287. https://doi.org/10.1016/j.ijbiomac.2020.02.261
117. Qiu J, Tang B, Ju B, Zhang S, Jin X (2020) Clean synthesis of disperse azo dyes based on peculiar stable 2,6-dibromo-4-nitrophenyl diazonium sulfate. Dyes Pigments 173:107920. https://doi.org/10.1016/j.dyepig.2019.107920
118. Raman CD, Kanmani S (2016) Textile dye degradation using nano zero valent iron: a review. J Environ Manag 177:341–355. https://doi.org/10.1016/j.jenvman.2016.04.034
119. Rasheed T, Nabeel F, Bilal M, Iqbal HMN (2019) Biogenic synthesis and characterization of cobalt oxide nanoparticles for catalytic reduction of direct yellow-142 and methyl orange dyes. Biocatal Agric Biotechnol 19:101154. https://doi.org/10.1016/j.bcab.2019.101154
120. Ravi BN, Keshavayya J, Mallikarjuna M, Kumar V, Kandgal S (2020) Synthesis, characterization and pharmacological evaluation of 2-aminothiazole incorporated azo dyes. J Mol Struct 1204:127493. https://doi.org/10.1016/j.molstruc.2019.127493
121. Ravi BN, Keshavayya J, Mallikarjuna NM, Santhosh HM (2020) Synthesis, characterization, cyclic voltammetric and cytotoxic studies of azo dyes containing thiazole moiety. Chem Data Collect 25:100334. https://doi.org/10.1016/j.cdc.2019.100334
122. Reife A (1993) Dyes, environmental chemistry. Kirk-Othmer Encycl. Chem. Technol
123. Roessler A, Crettenand D (2004) Direct electrochemical reduction of vat dyes in a fixed bed of graphite granules. Dyes Pigments 63:29–37. https://doi.org/10.1016/j.dyepig.2004.01.005
124. Roessler A, Jin X (2003) State of the art technologies and new electrochemical methods for the reduction of vat dyes. Dyes Pigments 59:223–235. https://doi.org/10.1016/S0143-7208(03)00108-6
125. Sahu O, Singh N (2019) 13—Significance of bioadsorption process on textile industry wastewater. In: Shahid-ul-Islam, Butola BS (eds) The impact and prospects of green chemistry for

textile technology. The Textile Institute Book Series. Woodhead Publishing, pp 367–416. https://doi.org/10.1016/B978-0-08-102491-1.00013-7
126. Said B, M'rabet S, Hsissou R, Harfi AE (2020) Synthesis of new low-cost organic ultrafiltration membrane made from Polysulfone/Polyetherimide blends and its application for soluble azoic dyes removal. J Mater Res Technol. https://doi.org/10.1016/j.jmrt.2020.02.102
127. Saxena S, Raja ASM (2014) Natural dyes: sources, chemistry, application and sustainability issues. In: Roadmap to sustainable textiles and clothing. Springer, pp 37–80. https://doi.org/10.1007/978-981-287-065-0_2
128. Sessa C, Weiss R, Niessner R, Ivleva NP, Stege H (2018) Towards a surface enhanced raman scattering (SERS) spectra database for synthetic organic colourants in cultural heritage. The effect of using different metal substrates on the spectra. Microchem J 138:209–225. https://doi.org/10.1016/j.microc.2018.01.009
129. Shankarling GS, Deshmukh PP, Joglekar AR (2017) Process intensification in azo dyes. J Environ Chem Eng 5:3302–3308. https://doi.org/10.1016/j.jece.2017.05.057
130. Shen C, Pan Y, Wu D, Liu Y, Ma C, Li F, Ma H et al (2019) A crosslinking-induced precipitation process for the simultaneous removal of poly(vinyl alcohol) and reactive dye: the importance of covalent bond forming and magnesium coagulation. Chem Eng J 374:904–913. https://doi.org/10.1016/j.cej.2019.05.203
131. Shi S, Feng X, Gao L, Tang J, Guo H, Wang S (2020) Hydrolysis and carbonization of reactive dyes/cotton fiber in hydrothermal environment. Waste Manag 103:370–377. https://doi.org/10.1016/j.wasman.2019.12.052
132. Shinde S, Sekar N (2019) Synthesis, spectroscopic characteristics, dyeing performance and TD-DFT study of quinolone based red emitting acid azo dyes. Dyes Pigments 168:12–27. https://doi.org/10.1016/j.dyepig.2019.04.028
133. Shindy HA (2017) Fundamentals in the chemistry of cyanine dyes: a review. Dyes Pigments 145:505–513
134. Siddiqua UH, Ali S, Hussain T, Bhatti HN, Asghar M (2017) The dyeing process and the environment: enhanced dye fixation on cellulosic fabric using newly synthesized reactive dye. Pol J Environ Stud (Scientific Investigation Committee) 26:2215–2222. https://doi.org/10.15244/pjoes/68430
135. Siddiqua UH, Irfan M, Ali S, Sahar A, Khalid M, Mahr MS, Iqbal J (2020) Computational and experimental study of heterofunctional azo reactive dyes synthesized for cellulosic fabric. J Mol Struct 1221:128753. https://doi.org/10.1016/j.molstruc.2020.128753
136. Singh K, Arora S (2011) Removal of synthetic textile dyes from wastewaters: a critical review on present treatment technologies. Crit Rev Environ Sci Technol (Taylor & Francis) 41:807–878
137. Singha NR, Chattopadhyay PK, Dutta A, Mahapatra M, Deb M (2019) Review on additives-based structure-property alterations in dyeing of collagenic matrices. J Mol Liq 293:111470. https://doi.org/10.1016/j.molliq.2019.111470
138. Soltzberg LJ, Hagar A, Kridaratikorn S, Mattson A, Newman R (2007) MALDI-TOF mass spectrometric identification of dyes and pigments. J Am Soc Mass Spectrom (Elsevier) 18:2001–2006
139. Srinivasan S, Sadasivam SK, Gunalan S, Shanmugam G, Kothandan G (2019) Application of docking and active site analysis for enzyme linked biodegradation of textile dyes. Environ Pollut 248:599–608. https://doi.org/10.1016/j.envpol.2019.02.080
140. Stolte M, Vieth M (2001) Pathologic basis of mucosal changes in the esophagus. What the Endoscopist can (and must) see. ACTA Endosc 31:125–130
141. Sun J, Wang H, Zheng C, Wang G (2019) Synthesis of some surfactant-type acid dyes and their low-temperature dyeing properties on wool fiber. J Clean Prod 218:284–293. https://doi.org/10.1016/j.jclepro.2019.01.341
142. Tasli PT, Atay ÇK, Demirturk T, Tilki T (2020) Experimental and computational studies of newly synthesized azo dyes based materials. J Mol Struct 1201:127098. https://doi.org/10.1016/j.molstruc.2019.127098
143. Taylor JA (2000) Recent developments in reactive dyes. Color Technol 30:93–108

144. Thetford D, Chorlton AP (2004) Investigation of vat dyes as potential high performance pigments. Dyes Pigments 61:49–62. https://doi.org/10.1016/j.dyepig.2003.09.002
145. Tkaczyk A, Mitrowska K, Posyniak A (2020) Synthetic organic dyes as contaminants of the aquatic environment and their implications for ecosystems: a review. Sci Total Environ 717:137222. https://doi.org/10.1016/j.scitotenv.2020.137222
146. Turhan K, Durukan I, Ozturkcan SA, Turgut Z (2012) Decolorization of textile basic dye in aqueous solution by ozone. Dyes Pigments 92:897–901. https://doi.org/10.1016/j.dyepig.2011.07.012
147. Vacchi FI, Von der Ohe PC, de Albuquerque AF, de Vendemiatti JAS, Azevedo CCJ, Honório JG, da Silva BF et al (2016) Occurrence and risk assessment of an azo dye—the case of Disperse Red 1. Chemosphere 156:95–100. https://doi.org/10.1016/j.chemosphere.2016.04.121
148. Walters A, Santillo D, Johnston P (2005) An overview of textiles processing and related environmental concerns. Greenpeace Res. Lab. Dep. Biol. Sci. Univ. Exeter UK. http://www.greenpeace.org/seasia/th/Global/seasia/report/2008/5/textile-processing.pdf
149. Wang M, Yang J, Wang H (2001) Optimisation of the synthesis of a water-soluble sulfur black dye. Dyes Pigments 50:243–246
150. Wanyonyi WC, Onyari JM, Shiundu PM, Mulaa FJ (2019) Effective biotransformation of reactive black 5 dye using crude protease from Bacillus Cereus Strain KM201428. Energy Procedia Technol Mater Renew Energy. Environ Sustain (TMREES) 157:815–824. https://doi.org/10.1016/j.egypro.2018.11.247
151. Welham A (2000) The theory of dyeing (and the secret of life). J Soc Dye Colour 116:140–143
152. Woo SW, Kim JY, Hwang TG, Lee JM, Kim HM, Namgoong J, Yuk SB et al (2019) Effect of weakly coordinating anions on photo-stability enhancement of basic dyes in organic solvents. Dyes Pigments 160:765–771. https://doi.org/10.1016/j.dyepig.2018.07.059
153. Wu J, Li Q, Li W, Li Y, Wang G, Li A, Li H (2020) Efficient removal of acid dyes using permanent magnetic resin and its preliminary investigation for advanced treatment of dyeing effluents. J Clean Prod 251:119694. https://doi.org/10.1016/j.jclepro.2019.119694
154. Yashni G, AlGheethi A, Maya Saphira Radin Mohamed R, Nor Hidayah Arifin S, Abirama Shanmugan V, Hashim Mohd Kassim A (2020) Photocatalytic degradation of basic red 51 dye in artificial bathroom greywater using zinc oxide nanoparticles. Mater Today Proc. https://doi.org/10.1016/j.matpr.2020.01.395
155. Zhan Y, Zhao X, Wang W (2017) Synthesis of phthalimide disperse dyes and study on the interaction energy. Dyes Pigments 146:240–250. https://doi.org/10.1016/j.dyepig.2017.07.013
156. Zhang H, Wang J, Xie K, Pei L, Hou A (2020) Synthesis of novel green reactive dyes and relationship between their structures and printing properties. Dyes Pigments 174:108079. https://doi.org/10.1016/j.dyepig.2019.108079
157. Zhou X, Zhou Y, Liu J, Song S, Sun J, Zhu G, Gong H et al (2019) Study on the pollution characteristics and emission factors of PCDD/Fs from disperse dye production in China. Chemosphere 228:328–334. https://doi.org/10.1016/j.chemosphere.2019.04.136
158. Zhu Y, Wang W, Ni J, Hu B (2020) Cultivation of granules containing anaerobic decolorization and aerobic degradation cultures for the complete mineralization of azo dyes in wastewater. Chemosphere 246:125753. https://doi.org/10.1016/j.chemosphere.2019.125753
159. Zhuang M, Sanganyado E, Zhang X, Xu L, Zhu J, Liu W, Song H (2020) Azo dye degrading bacteria tolerant to extreme conditions inhabit nearshore ecosystems: optimization and degradation pathways. J Environ Manag 261:110222. https://doi.org/10.1016/j.jenvman.2020.110222
160. Zollinger H (2003) Color chemistry: syntheses, properties, and applications of organic dyes and pigments. Wiley

Degradation of Dyes Using Filamentous Fungi

Bhupinder Dhir

Abstract Dyes used in textile industry generate large amount of wastewater that needs to be treated effectively to prevent toxicity. Treatment of dye-rich wastewater by physico-chemical methods proves costly and less efficient, therefore biological methods (cost-effective and eco-friendly) have been developed. A variety of biological materials have been evaluated for their capacity to remove/treat toxic compounds such as dyes. Fungal biomass has shown high efficiency to remediate dye-rich wastewater. Fungi remove these compounds via decolorization and degradation. Reductive and oxidative enzymes assist in degradation of dyes. Many strains of filamentous fungi have shown high capacity to treat dyes. Hydrolytic enzymes secreted by filamentous fungi assist in the treatment of dyes. Lacasse enzyme produced by lignolytic fungi helps in the degradation of dyes. Enormous capacity of fungi for removal/remediation of dyes can be exploited to develop technologies for large-scale treatment of dye-containing wastewater.

Keyword Decolorization · Degradation · Dyes · Laccases · Lignolytic

1 Introduction

Dyes are chemical compounds (aromatic and heterocyclic) that impart color to materials. Dyes have been classified into different types depending on their chemical structure and chromophore groups present in them. Azo, diazo, cationic, basic, anthraquione and metal complexed dyes have been known [103]. The largest group (about 80%) of organic compounds used as colorants in textile industry has been identified as azo dyes. This is because these dyes can be synthesized easily and are chemically more stable and available in diversity of colors [25, 26].

About 7×10^5 tons of dyes are produced annually, of which approximately 2.8×10^5 tons of dyes have been released into water bodies [6, 58]. Synthetic dyes are complex structured, recalcitrant, xenobiotic, broad-spectrum chemicals, which

B. Dhir (✉)
School of Sciences, Indira Gandhi National Open University, New Delhi, India

S. S. Muthu and A. Khadir (eds.), *Dye Biodegradation, Mechanisms and Techniques*, Sustainable Textiles: Production, Processing, Manufacturing & Chemistry,
https://doi.org/10.1007/978-981-16-5932-4_2

resist degradation [19–21, 64]. The presence of one or more azo groups in azo dyes prevents their breakdown and degradation, making their accumulation persistent in the environment. The release of such high amount of dyes in the aqueous environment exerts toxic effects on living biota (flora and fauna) [8, 9, 95, 97]. In the water bodies, dyes get accumulated in the aquatic organisms, pass to other organisms through food chain and finally reach man. Effluents rich in toxic dyes need to be treated suitably to prevent toxicity exerted by these compounds [108].

The breakdown of dye releases products that prove to be a potent carcinogen. Dyes also act as causal agents of various diseases/disorders such as skin cancer, dermatitis, perforation of nasal septum, damage to mucous membrane and irritation of respiratory tract. According to reports, accidental intake of dyes in humans causes vomiting, pain, hemorrhage and diarrhea [68]. Exposure to azo dyes produces life-threatening diseases in humans such as cancer of bladder, splenic sarcomas and hepatocarcinomas. Chromosomal aberrations, lipid peroxidation and inhibition of enzyme acetylcholinesterase in mammalian cells are the other effects produced by exposure to dyes [11, 85, 114]. Toxic effects induced by various dyes have been listed in Table 1.

Adsorption, filtration, coagulation, precipitation, oxidation, reduction, photolysis and photodegradation are some of the physico-chemical methods followed for years to remove/treat dyes from wastewaters [67, 91, 113]. These methods are costly, require the addition of hazardous chemical additives and produce large amounts of sludge (problems related to secondary pollution) [42]. In biological treatment method, microbes (such as bacteria, fungi, yeast), algae and plants (bioremediation) are used [97]. This method provides an eco-friendly cost-effective alternative to other technologies [65]. Biological treatment technologies have shown capacity to degrade organic pollutants. This is because of high potential of microorganisms such as bacteria, yeasts and filamentous fungi to remove dyes [116]. The main mechanism involved in removal of dyes includes adsorption on microbial biomass, biosorption and/or enzymatic degradation (biodegradation) [10, 49, 63, 107].

Mycoremediation has emerged as a technology with high efficiency for treatment of dyes [43, 45, 50, 101, 76]. Fungi have shown high capacity to treat organic

Table 1 Toxic effect induced by various dyes

Name of dye	Toxic effects
Acid violet 7	Lipid peroxidation, chromosomal aberrations, inhibition acetyl cholinesterase activity
Methyl red	Induce mutations
Disperse blue 291	Cytotoxic, genotoxic, mutagenic, fragmentation of DNA
Congo red	Carcinogen, mutagen
Malachite green	Carcinogen, mutagen, chromosomal changes, teratogen, induce histopathological effects such as organ and tissue injury, damage and developmental abnormalities

compounds [9, 4, 18, 63]. Aggressive growth, high biomass production, extensive hyphal growth, high surface-to-cell ratio are some of the properties that help fungi to treat/degrade various dyes in high amounts. Living or dead form of fungi has shown the capacity to remove dyes and pigments. Fungi remove dyes via processes such as biosorption, biodegradation, bioaccumulation and enzymatic mineralization. The decolorization/degradation of dyes azo, anthraquinone, heterocyclic, triphenylmethane and polymeric dyes has been reported. Fungi have shown immense potential to decolourize/degrade azo, anthraquinone, heterocyclic, triphenylmethane and polymeric dyes [119]. The degradation/decolorization potential depends mainly on metabolism of fungi and expression of extracellular enzymes such as lignin peroxidase (LiP), manganese peroxidase (MnP) and laccase.

Filamentous fungi have been found to possess good capacity to degrade dyes [1, 17, 29, 57]. The strains of filamentous fungi secrete large amounts of hydrolytic enzymes and they possess an important role in degradation of dyes. The present chapter highlights the role of filamentous fungi in decoloration and degradation of various dyes and discusses the role of enzymes in their degradation. The mechanism involved in removal of dyes and factors affecting the removal are also discussed the chapter.

2 Fungal Strains with Dye Removal Capacity

About 115 fungal strains that possess good potential to degrade/decolorize dyes have been identified. Most of the strains have been found efficient to remove azo and anthraquinone dyes. These mainly include *Aspergillus niger, Aspergillus flavus, Aspergillus oryzae, Penicillium chrysogenum, Pleurotus ostreatus, Pleurotus pulmonarius, Pleurotus sajorcaju, Bjerkandera fumosa, Phanerochaete chrysosporium, Tremetes sanguinea, Tremetes versicolor, Phlebia radiata, Gonoderma lucidum, Gonoderma applanatum, Ganoderma resinaceum, Cladosporium rubrum, Laccaria fraterna* [7, 12, 24, 35, 38, 60, 87, 88, 100, 104, 122, 131]. Some species of fungi that depict high potential for removal of dyes have been listed in Table 2.

Fungal species belonging to groups—basidiomycetes and deuteromycetes—have shown high efficiency to decolorize dyes [74, 90]. The fungus *Phanerochaete chrysosporium* popularly known as “white-rot fungus” has shown an unusual high capacity to remediate/degrade dyes present in effluents released from textile industry [8, 9, 40, 50, 133]. The white-rot fungus possesses high capacity to degrade azo dyes include *Phanerochaete chrysosporium, Pycnoporus cinnabarinus, Pleurotus ostreatus, Basidiomycetous, Coriolus versicolor, Trametes versicolor, Pleurotus ostreatus* and *Coriolopsis polysona, Alternaria solani, Neurospora* sp [61, 68, 128], Mohamad et al. 2019). Fungal species mainly *Aspergillus, Penicillium, Galactomyces, Cunninghamella, Lasiodiplodia, Phanerochaete* have shown high potential to decolorize and mineralize malachite green (MG) [3, 5, 23, 56, 59, 98, 99, 126]. Studies have shown that *Funalia trogii* decolorize and degrade high concentrations of crystal violet [121]. The fungal strain, *Leptosphaerulina* sp. has shown capacity

Table 2 Removal of dyes by various fungal species

Name of dye	Fungus	References
Congo red	*Aspergillus flavus*	[17]
	Aspergillus niger	[44]
Reactive red 195, Reactive green 11	*Aspergillus niger*	[131]
Black reactive 5	*Phaenerocheate chrysosporium*	[37]
	Pleurotus eryngii	[50]
Indigo dye	*P. chrysosporium* URM6181 *Curvularia lunata* URM6179	[32]
Reactive violet 5, Light navy blue, Dark navy blue	*Hypocrea koningii*	[46]
Mordant yellow 1	*Aspergillus sp.* TS-A CGMCC	[60]
Malachite green	*T. asperellum*	[96]
	Aspergillus flavus	[14]
	P. ochrochloron	[98]
	Galactomyces geotrichum	[56]
	Penicillium simplicissimum	[30]
Thiazole yellow G	*Aspergillus niger* LAG	[13]
Basic blue	*Trichoderma harzianum*	[102]
Bromophenol blue	*Trichoderma harzianum*	[102]
Direct green	*Trichoderma harzianum*	[102]
Methylene blue	*Aspergillus sp.*	[78]
Gentian violet, crystal violet	*Aspergillus sp.*	[78]
Orange II d, Tropaeolin O	*Phaenerocheate chrysosporium*	[102]
Basic fuchsin	*Phaenerocheate chrysosporium*	[89]
Nigrosin	*Phaenerocheate chrysosporium*	[89]
Viscose orange-A	*Aspergillus fumigatus*	[94]
Direct violet-BL	*Aspergillus niger*	[94]
	Trichoderma viride	[94]

to decolorize about 90% of the dyes Novacron Red, Remazol Black and Turquoise Blue.

Immobilization of white-rot fungus on Ca-alginate beads or discs in a rotating biological contactor increases decolorization of recalcitrant dyes such as Direct Violet 51, Reactive Black 5, Ponceau Xylidine, Bismark Brown R, Orange II and Everzol Turquoise Blue G [2, 36, 62, 92]. Cultures of *Phanerochaete chrysosporium* when immobilized on cubes of polyurethane foam (PUF) increased the capacity to decolorize high amounts of polymeric dye Poly R478. Immobilized fungal cultures show high activity and resilience to environmental alterations such as changes in pH, exposure of cells or suspension cultures to toxic concentrations of compounds.

3 Mechanisms Involved in Remediation/Decolorization of Dyes

Adsorption, biosorption and enzymatic degradation have been suggested as major mechanisms involved in removal or decolorization of dyes by fungal hyphae. Living or dead biomass of fungi both possess high capacity to decolorize dyes. Decolorization of the dye occurs because of synergistic effects of mycelia and extracellular enzymes such as oxidases [103]. In the initial stages, biosorption of the dye helps in the decolorization of dyes followed by degradation in the final stages of growth.

Various dyes such as azo, anthraquinonic, heterocyclic, triphenylmethane and polymeric have shown decolorization using fungal biomass [94, 95], though the mechanism of decolorization of dyes varies according to the fungal species. The degradation of chromophore moiety of the dye molecule occurs with the help of extracelluar enzyme produced by fungi and absorption/adsorption mechanisms. The decolorization activity of fungi is regulated by various factors such as concentration of dye, amount of pellet, temperature and agitation of the media. The pH, temperature, inoculum size and NaCl concentrations in the culture conditions regulate the efficacy of the fungal strain for decolorization. Biosorption and biodegradation are the two main mechanisms that help in decolorization of azo dyes [59].

Decolorization of the dye effluent is supported by neutral culture conditions. Studies indicated that incubation time of 72 h is suitable for getting maximum decolorization. The decolorization efficiency decreases with an increase in temperature. Decolorization (more than 70%) of dyes namely Reactive Red, Navy Blue HER, Reactive Magenta B and Orange 3R has been noted in *Talaromyces funiculosum* under optimal conditions of temperature and pH [28]. Fungal species *Talaromyces versicolour* have shown complete decolorization of dyes such as Tropaeolin O, Reactive Blue 15, Congo Red, Reactive Black 5, and partial decolorization of Brilliant Red 3G-P, Brilliant Yellow 3B-A and Remazol Brilliant Blue R has been noted. Rate of decolorization of dye varies under static and shaking condition [8, 27]. The shaking conditions provide better oxygenation to the fungus and regular contact of secreted enzymes for decolorization. Moreover, agitation also helps in the growth of fungus. Carbon sources like glucose have shown to accelerate rate of decolorization. *Leptosphaerulina* sp. showed high capacity to decolorize (>90%) textile industry effluents at low pH and glucose supply.

The mechanism of removal of dyes varies for each fungal species. The fungal species such as *Penicillium pinophilum* remove dyes via biosorption at the initial stage followed by intracellular biodegradation. On the other hand, other species, *Myrothecium roridum* removal of dye takes place by extracellular biotransformation. About 99.99% decolorization of Congo red dye can be achieved by *Alternaria alternata* within 48 h whereas *Aspergillus niger* showed degradation of Procion Red MX-5B after 336 h of treatment.

3.1 Adsorption

Adsorption of dyes by fungus has been considered as the primary mechanism of its decolorization. Hydrophobic–hydrophillic interaction between the fungus and dye occurs during adsorption. About 50% of dye removal has been noted after adsorption. Treatment of biomass with organic or inorganic molecules like formaldehyde, sulfuric acid, sodium hydroxide, calcium chloride and sodium bicarbonate and exposure to high temperature enhanced adsorption capacity of fungi.

Living and dead mycelia of *Trametes versicolor* has shown potential to adsorb various dyes, viz. Acid green 27, Acid violet 7 and Indigo carmine [123].

3.2 Biosorption

Biosorption is a physicochemical method that helps in the removal of pollutants via attachment to the surface of cell membranes or other cellular components. Binding of contaminants to cell wall or membranes occurs via different mechanisms such as physical adsorption, electrostatic interaction, ion exchange, chelation and chemical precipitation [45]. Dye molecules get trapped in the inner spaces of fungal mycelium because of ion exchange hydrogen binding [127]. The process plays an important role in decolorization of dye.

The capacity of fungi to bisorb contaminants is attributed to heteropolysaccharides such as chitin, chitosan, glucan, lipid, phospholipids present on cell wall. Chitin and chitosan present in the cell wall of dead fungal biomass show high affinity for binding various dyes. Amino, hydroxyl, carboxyl, phosphate and other functional groups (sulphates, hydroxides) present on fungal biomass help in binding azo dye to cell wall by creating attractive forces [105, 106]. Filamentous fungi successfully remove toxic dyes via biosorption [16].

High-temperature exposure through autoclaving and treatment with chemicals such as 0.1 N NaOH, 0.1 M HCl and 0.1 M H_2SO_4 increase the biosorption capacity of fungus. The autoclaving brings changes in the fungal biomass by changing the surface charge. The pretreatment of biomass with acid enhances the affinity of anionic dyes to bind to fungal surface. The biosorption capacity of fungal biomass of *Lentinus sajor-caju* increased for the dye Reactive Red 120 after autoclaving (treatment at 100 °C for 10 min).

3.3 Degradation

Biodegradation is an energy-requiring process in which complex organic molecules are broken down into simpler molecules through the action of certain enzymes secreted by microorganisms. Biodegradation involves three steps—

(1) slight change in an organic molecule without change in the main structure,
(2) fragmentation of a complex organic molecule and
(3) complete mineralization, i.e. conversion of organic molecules to simpler forms such as carbon dioxide, methane, inorganic elements.

The role of reductive and oxidative enzymes in the degradation of dyes has been reported [55]. Extracellular enzymes produced by fungi play a vital role in decolorization and degradation of organic compounds such as dyes under *in vitro* conditions [33]. Laccase, azoreductases, lignin peroxidases and manganese peroxidases are the main enzymes found in fungi, which catalyze redox reactions [80, 129]. Various factors such as culture conditions, availability of nutrients, carbon source, time, pH, agitation, temperature, oxygen supply, concentration and nature of dye, additives and salts control the degradation of dyes by fungi [39, 48, 84].

Oxidative enzymes

Enzymes such as polyphenol oxidases (PPO), manganese peroxidase (MnP), lignin peroxidase (LiP), laccase (Lac), tyrosinase (Tyr), N-demethylase, dye decolorizing peroxidases and cellobiose dehydrogenase assist in degradation of azo dyes [72, 111]. The degradation of compounds occurs due to breaking of ester, amide, ether bonds and aromatic ring or the aliphatic chains of compounds by these enzymes. The enzymatic treatment of substrates results in the formation of less toxic insoluble compounds.

White-rot fungi show ligninolytic activity, which proves useful in degradation of industrial dyes and other toxic aromatic compounds. Lignin peroxidases or ligninases and manganese peroxidase (MnP) are the lignin-degrading enzymes that cause oxidative depolymerization of lignin and assist in degradation of various organo-pollutants [51]. Ligninolytic enzymes bind to the substrate thereby degrading dyes. The reduction of azo dyes by fungi depends on the enzymes such as peroxidases and phenol oxidases.

Many basidiomycetes such as *Bjerkandera sp., Agaricus bisporus, Phanerochaete flavido-alba, Ganoderma lucidium, Pleurotus pulmonarius, Trametes versicolor*, *Trametes hirsute* and *Pleurotus ostreatus* are known to possess MnPs that assist in the removal of dyes [34, 52, 54, 70, 73, 79, 130, 117]. Manganese peroxidase produced by isolates of *Clitopilus scyphoides*, *Ganoderma rasinaceum* and *Schizophyllum* showed ability to degrade recalcitrant azo dyes, sulfonephthalein dyes and kraft lignin. High amounts of dyes such as Poly-478 and Remazol Brilliant Blue R showed degradation because of MnP produced by *Lentinus edodes.*

Enzymes such as peroxidases, H_2O_2-dependent secreted by mushroom *Pleurotus ostreatus* showed high capacity for decolorizing dye, Brilliant Blue R [120]. Enzymes, horseradish peroxidase and LiP isolated from *Penicillium chrysosporium* possess the capacity to oxidize dyes such as Methylene Blue (Basic Blue 9) and Azure B. Biodegradation of malachite green (MG) by enzymes laccases and manganese peroxidase produced by fungi have been reported [14, 71, 82]. Degradation of indigo dye by laccase produced by *Trametes hirsuta* and *Sclerotium rolfsii* has been reported [31]. A new family of ligninolytic peroxidases named as versatile peroxidase (VP) has been isolated from *P. chrysosporium* [75, 93]. The enzyme peroxidases cause oxidation of anthraquinone dyes [125].

The enzymatic degradation of chemical compounds results in the formation of free radical followed by insoluble product [112]. The cleavage of carbon–carbon single bond results in degradation of the compound. The degradation of dyes and other aromatic compounds has been facilitated by free radicals generated in the process. The chain reactions get initiated when free radicals get donated or accepted electrons from other chemicals. The actions are catalyzed by enzymes peroxidase [53]. Enzyme, phenol oxidases decolorize azo dyes through a highly non-specific free radical mechanism.

In white-rot fungi, the enzymes such as Laccase (multi-copper oxidases) present on the cell surface play a major role in degradation of dyes [15, 22, 53]. The removal of dyes by enzyme laccases involves oxidation or oxidation mediated by redox mediators (e.g., ABTS) [66, 109]. Laccases oxidize the phenolic group of the azo dye resulting in formation of phenoxy radical followed by its oxidation to a carbonium ion.

Laccase enzyme produced by *Phanaerochaete pulmonarius* BPSM10 showed capacity to degrade MG. *Aspergillus niger* and *Phanaerochaete crysosporium* degraded dyes through the action of extracellular enzymes such as laccase. Degradation of Malachite green (96%) by *Aspergillus flavus, Aspergillus solani* and some white-rot fungi has been noted in liquid culture within 6 days [3, 118]. Degradation of Orange 2 acid, Orange 6 (72.8%) and many other azo dyes (45.3%) has been achieved using laccase produced by *Trametes versicolor*. The decolorization of dye such as triphenylmethane dyes has been noted in *Trametes versicolor* (Crystal violet, Bromophenol blue and Malachite green) [124]. Bleaching of these dyes by *Trametes versicolor* can be attributed to enzyme laccase, peroxidase and arsenal present within the fungal cells [86].

Reductive enzymes

Azoreductases degrade azo dyes by breaking the azo linkage bridge between the chromophoric groups to produce two arylamines. The colorless aromatic amines are produced by cleavage of the azo bond under reduced conditions [81]. The reducing equivalents such as NADPH, NADH and FADH help in breakdown of the dye.

The enzyme malachite green reductase (MG reductase) produced by fungal strains, *Penicillium pinophilum* and *Myrothecium roridum*, has shown degradation of dyes within 48 h [56]. Reduction of MG results in the formation of leucomalachite green and N-demethylated metabolites. FITR analysis suggested the formation of non-toxic metabolites such as aromatic amines, carboxylic acids, alkanes, alkenes and further conversion of aromatic amines into carboxylic acid during degradation of MG.

4 Decolorization of Dyes by White-Rot Fungi

White-rot fungi, *Phanerochaete chrysosporium* has shown high capacity for dye decolorization [41, 110]. Other white-rot fungi species viz. *Hirschioporus larincinus, Inonotus hispidus, Phlebia tremellosa* and *Coriolus versicolor* have shown an ability to decolorize dye effluent [47]. Lignin-degrading enzymes such as peroxidases viz. manganese peroxidase (MnP), lignin peroxidase (LiP) and laccases found in white-rot fungi help in decolorizaton of dyes [24, 121]. The ligninolytic enzymes (lignin modifying or degrading enzymes) are excreted extracellularly by white-rot fungi. These enzymes cause oxidation of lignin present in the fungal cell. These enzymes oxidize Mn (II) to highly reactive Mn (III) in the presence of hydrogen peroxide. Mn (III) further chelates with organic acids, which attack and oxidize lignin and other recalcitrant compounds [51,115]. The ability of fungi to decolorize dyes varies with the strain.

The decolorization of dye by white-rot fungi occurs mainly by biosorption, biodegradation, bioreactor and lignin modified enzymes. Ligninolytic enzymes mineralize dyes during degradation. Temperature of about 30 °C is the most favorable to carry out decolorization, a significant decrease has been noted at higher temperature. This is because of reduction in cell viability or inactivation of the enzymes. Acidic conditions (pH 5.5) assisted in biodegradation of methyl violet by *Aspergillus* sp. while pH higher than 6.5 inhibited decolorization [69].

The biodegradation of Congo red by *Aspergillus niger* has been reported. The cultures showed high decolorization efficiency (97%) after 6 days. About 27% of CR dye was eliminated by adsorption while 70% was treated by enzymatic biodegradation. Enzymes such as lignin peroxidase, manganese peroxidase and deaminase produced by the strain played a major role in degradation. The rate of decolorization is affected by various factors including pH. High pH values (pH > 7) changed the activity of MnP, which can be due to enzyme stability. Studies suggested that the degradation of CR dye occurs mainly by deamination and oxygenation and involves the following steps: (i) partial deamination of CR compound; (ii) asymmetric cleavage of C–N bond between the aromatic ring and the azo group by peroxidase with the loss of a sodium atom; (iii) asymmetric cleavage of C–N bond caused by peroxidase followed by deprotonation; (iv) opening of benzene ring and dehydrogenation forming intermediates; (v) action of peroxidase forming cleavage products.

Decolorization of reactive blue 25 by lignin peroxidase, laccase and tyrosinase produced by *Aspergillus ochraceus* has been reported [83]. White-rot fungus *Ganoderma tsugae* shows potential for degrading reactive black dye due to production of enzyme laccase [132]. The secretion of laccase and manganese peroxidase by white-rot fungal strain KRUS-G assists in decolorization of Remazol Brilliant Blue R

5 Conclusion and Future Perspectives

Cost-effectiveness and eco-friendly nature established fungi as a suitable material for degradation/decolorization of dyes from textile effluents. Fungi decolorize synthetic dyes via through absorption, adsorption and enzymatic degradation. Most of the studies related to degradation/decolorization of dyes have been performed under laboratory conditions. More studies are required to implement the technique in decolorization of dyes under field conditions. Identification and exploration of new fungal strains with the help of molecular techniques can prove useful in achieving remediation of dyes. The molecular tools help in identification of genes encoding enzymes that play role in degradation of complex synthetic dyes. Genetically, modified strains of fungus with high efficiency for dye degradation/decolorization of dye can be produced via technology of genetic engineering.

References

1. Al-Jawhari IFH, AL-Mansor KJ, (2017) Biological removal of Malachite Green and Congo red by some filamentous fungi. Intl J Environ Agric Biotechnol 2:723–731
2. Al-Tohamy R, Sun J, Fareed MF, Kenawy E, Ali SS (2020) Ecofriendly biodegradation of Reactive Black 5 by newly isolated *Sterigmatomyces halophilus* SSA1575, valued for textile azo dye wastewater processing and detoxification. Sci Reports10, Article number: 12370.
3. Ali H, Ahmad W, Haq T (2009) Decolourisation and degradation of malachite green by *Aspergillus flavus* and *Alternaria solani*. Afr J Biotechnol 8:1574–1576
4. Ali H (2010) Biodegradation of synthetic dyes—a review. Water Air Soil Pollut 213:251–273
5. Arunprasath T, Sudalaib S, Meenatchic R, Jeyavishnua K, Arumugama A (2019) Biodegradation of triphenylmethane dye malachite green by a newly isolated fungus strain biocatalysis agric. Biotechnol 17:672–679
6. Asad S, Amoozegar MAA, Pourbabaee A, Sarbolouki MN, Dastgheib SMM (2007) Decolorization of textile azo dyes by newly isolated halophilic and halotolerant bacteria. Bioresour Technol 98:2082–2088
7. Asse N, Ayed L, Hkiri N, Hamdi M (2018) Congo red decolorization and detoxification by *Aspergillus niger*. Bioremed Int Res Article ID 3049686
8. Asgher M, Yasmeen Q, Iqbal HMN (2013a) Enhanced decolorization of Solar Brilliant Red 80 textile dye by an indigenous white rot fungus *Schizophyllum commune* IBL-06. Saudi J Biol Sci 20:347–352
9. Asgher M, Aslam B, Iqbal HMN (2013b) Novel catalytic and effluent decolorization functionalities of sol-gel immobilized *Pleurotus ostreatus* IBL-02 manganese peroxidase produced from bio-processing of wheat straw. Chin J Catal 34:1756–1761
10. Asgher M, Shah SAH, Iqbal HMN (2016) Statistical correlation between ligninolytic enzymes secretion and Remazol Brilliant Yellow-3GL dye. Water Environ Res 88:338–345
11. Ayed L, Mahdhi A, Cheref A, Bakhrouf A (2011) Decolorization and degradation of azo dye Methyl Red by an isolated *Sphingomonas paucimobilis*: biotoxicity and metabolites characterization. Desalination 274:272–277
12. Balaraju K, Gnanadoss JJ, Muthu S, Ignacimuthu S (2008) Decolourization of azodye (congo red) by *Pleurotus ostreatus* and *Laccaria fraterna*. The ICFAI J. Life Sci 2:45–50
13. Bankole PO, Adekunle AA, Govindwar SP (2019) Demethylation and desulfonation of textile industry dye, Thiazole Yellow G by *Aspergillus niger* LAG. Biotechnol Rep 23:e00327

14. Barapatre A, Aadil KR, Jha H (2017) Biodegradation of malachite green by the ligninolytic fungus *Aspergillus flavus* clean. Soil Air Water 45:1600045
15. Bergsten-Torralba LR, Nishikawa MM, Baptista DF, Magalhaes DP, da Silva M (2009) Decolorization of different textile dyes by Penicillium simplicissimum and toxicity evaluation after fungal treatment. Braz J Microbiol 40:808–817
16. Bhatnagar A, Sillanpaa M (2010) Utilization of agroindustrial and municipal waste materials as potential adsorbents for water treatment: a review. Chem Eng J 157:277–296
17. Bhattacharya S, Das A, Mangai G, Vignesh K, Sangeetha J (2011) Mycoremediation of Congo red dye by filamentous fungi. Braz J Microbiol 42:1526–1536
18. Bilal M, Asgher M, Iqbal M, Hu H, Zhang X (2016) Chitosan beads immobilized manganese peroxidase catalytic potential for detoxification and decolorization of textile effluent. Int J Biol Macromol 89:181–189
19. Bilal M, Asgher M, Parra-Saldivar R, Hu H, Wang W, Zhang X, Iqbal HMN (2017a) Immobilized ligninolytic enzymes: an innovative and environmental responsive technology to tackle dye-based industrial pollutants—a review. Sci Total Environ 576:646–659
20. Bilal M, Iqbal HMN, Hu H, Wang W, Zhang X (2017b) Enhanced bio-catalytic performance and dye degradation potential of chitosan-encapsulated horseradish peroxidase in a packed bed reactor system. Sci Total Environ 575:1352–1360
21. Bilal M, Iqbal HMN, Hu H, Wang W, Zhang X (2017c) Development of horseradish peroxidase-based cross-linked enzyme aggregates and their environmental exploitation for bioremediation purposes. J Environ Manag 188:137–143
22. Camarero S, Ibarra D, Martínez MJ, Angel TM (2005) Lignin-derived compounds as efficient laccase mediators for decolourization of different types of recalcitrant dyes. Appl Environ Microbiol 71:1775–1784
23. Cha CJ, Doerge DR, Cerniglia CE (2001) Biotransformation of malachite green by the fungus Cunninghamella elegans. Appld. Environ. Microbiol. 67:4358–4360
24. Chagas E, Durrant L (2001) Decolourization of azo dyes by *Phanerochaete chrysosporium* and *Pleurotus sajorcaju*. Enzyme Microb Technol 29:473–477
25. Chang JS, Chou C, Lin Y, Ho J, Hu TL (2001) Kinetic characteristics of bacterial azo-dye decolorization by *Pseudomonas luteola*. Water Res 35:2841–2850
26. Chang JS, Chen BY, Lin YS (2004) Stimulation of bacterial decolorization of an azo dye by extracellular metabolites from *Escherichia coli* strain NO_3. Bioresour Technol 91:243–248
27. Chatterjee A, Singh N, Abraham J (2011) Mycoremediation of textile dyes using *Talaromyces funiculosum*. Intl J Pharm Sci Res 8:2082–2089
28. Chatterjee A, Singh N, Abraham J (2017) Mycoremediation of textile dyes using Talaromyces funiculosum JAMS1. Intl J Pharm Sci Res 8:2082–2089
29. Chatterjee S, Dey S, Sarma M, Chaudhuri P, Das S (2020) Biodegradation of Congo Red by manglicolous filamentous fungus *Aspergillus flavus* JKSC-7 isolated from Indian Sundabaran mangrove ecosystem. Appld Biochem Microbiol 56:708–717
30. Chen SH, Ting ASY (2015) Biosorption and biodegradation potential of triphenylmethane dyes by newly discovered *Penicillium simplicissimum* isolated from indoor wastewater sample. Int Biodeterior Biodegrad 103:1–7
31. Campos R, Kandelbauer A, Robra KH, Cavaco-Paulo A, Gubitz GM (2001) Indigo degradation with purified laccases from *Trametes hirsute* and *Sclerotium rolfsii*. J Biotechnol 89:131–139
32. De-Miranda RCM, Gomes EB, Pereira NJ, Marin MA, Machado KM, Gusmao NB (2013) Biotreatment of textile effluent in static bioreactor by Curvularia lunata URM 6179 and *Phanerochaete chrysosporium* URM 6181. Bioresour Technol 142:361–367
33. Dhanjal NIK, Mittu B, Chauhan A, Gupta S (2013) Biodegradation of textile dyes using fungal solates. J Environ Sci Technol 6:99–105
34. Eichlerová I, Homolka L, Lisá L, Nerud F (2005) Orange G and Remazol brilliant blue R decolorization by white rot fungi *Dichomitus squalens*, *Ischnoderma resinosum* and *Pleurotus calyptratus*. Chemosphere 60:398–404

35. El-Zaher A, Eman HF (2010) Biodegradation of reactive dyes using soil fungal isolates and *Ganoderma resinaceum*. Ann Microbiol 60:269–278
36. Enayatzamir K, Alikhani H, Yakhchali B, Tabandeh F, Rodríguez-Couto S (2010) Decolouration of azo dyes by Phanerochaete chrysosporium immobilized into alginate beads. Environ Sci Pollut Res 17:145–153
37. Enayatizamir N, Tabandeh F, Rodriguez-Couto S, Yakhchali B, Alikhani HA, Mohammadi L (2011) Biodegradation pathway and detoxification of the diazo dye Reactive Black 5 by *Phanerochaete chrysosporium*. Bioresour Technol 102:10359–10362
38. Erdal S, Taskin M (2010) Uptake of textile dye reactive Black-5 by *Penicillium chrysogenum* MT-6 isolated from cement-contaminated soil. Afr J Microbiol Res 4:618–625
39. Evangelista-Barreto NS, Albuquerque CD, Vieira RHSF, Campos-Takaki GM (2009) Cometabolic decolorization of the reactive azo dye orange II by *Geobacillus stearothermophilus* UCP 986. Text Res J 79:1266–1273
40. Faraco V, Pezzella C, Giardina P, Piscitelli A, Vanhulle S, Sannia G (2009) Decolourization of textile dyes by the white-rot fungi *Phanerochaete chrysosporium* and *Pleurotus ostreatus*. J Chem Technol Biotechnol 84:414–419
41. Ferreira VS, Magalhaes DB, Kling SH, da Silva JG, Bon EPS (2000) N-demethylation of methylene blue by lignin peroxidase from *Phanerocheate chrysosporium*. Appld Biochem Biotechnol 84:255–265
42. Forgacs E, Cserháti T, Oros G (2004) Removal of synthetic dyes from wastewater: a review. Environ Intl 30:953–971
43. Fu Y, Viraraghavan T (2001) Fungal decolorization of dye wastewaters: a review. Bioresour Technol 79:251–262
44. Fu Y, Viraraghavan T (2002) Removal of Congo red from an aqueous solution by fungus *Aspergillus niger*. Adv Environ Res 7:239–247
45. Fu Y, Viraraghavan Y (2004) Removal of Congo red from an aqueous solution by fungus *Aspergillus niger*. Adv Environ Res 7:239–247
46. Gajera HP, Bambharolia RP, Hirpara DG, Patel SV, Golakiya BA (2015) Molecular identification and characterization of novel *Hypocrea koningii* associated with azo dyes decolorization and biodegradation of textile dye effluents. Process Saf Environ Prot 98:406–416
47. Gohel JB, Parmar BP, Vyas BRM (2018) Potential of white rot fungi in the degradation of textile dyes: a review. J Cell Tissue Res 18:6545–6554
48. Grinhut T, Salame TM, Chen Y, Hadar Y (2011) Involvement of ligninolytic enzymes and Fenton-like reaction in humic acid degradation by *Trametes* sp. Appl Microbiol Biotechnol 91:1131–1140
49. Gupta VK, Rastogi A, Nayak A (2010) Adsorption studies on the removal of hexavalent chromium from aqueous solution using a low cost fertilizer industry waste material. J Colloid Interface Sci 342:135–141
50. Hadibarata T, Adnan LA, Yusoff ARM, Yuniarto A, Rubiyanto Zubir MMFA, Khudhair AB, Teh ZC, Naser MA (2013) Microbial decolorization of an azo dye Reactive Black 5 using white rot fungus *Pleurotus eryngii* F032. Water Air Soil Pollut 224:1595–1604
51. Hammel KE, Cullen D (2008) Role of fungal peroxidases in biological ligninolysis. Curr Opin Plant Biol 11:349–355
52. Hildén K, Bortfeldt R, Hofrichter M, Hatakka A, Lundell T (2008) Molecular characterization of the basidiomycete isolate *Nematoloma frowardii* b19 and its manganese peroxidase places the fungus in the corticioid genus *Phlebia*. Microbiol 154:2371–2379
53. Hofrichter M (2002) Review: lignin conversion by manganese peroxidase (MnP). Enzyme Microb Technol 30:454–466
54. Hoshino F, Kajino T, Sugiyama H, Asami O, Takahashi H (2002) Thermally stable and hydrogen peroxide tolerant manganese peroxidase (MnP) from *Lenzites betulinus*. FEBS Lett 530:249–252
55. Husain Q (2006) Potential applications of the oxidoreductive enzymes in the decolorization and detoxification of textile and other synthetic dyes from polluted water: a review. Crit Rev Biotechnol 26:201–221

56. Jadhav JP, Kalme SD, Govindwar SP (2008) Biodegradation of methyl red by *Galactomyces geotrichum* MTCC 1360. Intl Biodeterior Biodegrad 62:135–142
57. Jasińska A, Paraszkiewicz K, Sip A, Długoński J (2015) Malachite green decolorization by the filamentous fungus Myrothecium roridum–Mechanistic study and process optimization. Bioresour Technol 194:43–48
58. Jin XC, Liu GO, Xu ZH, Tao WY (2007) Decolorization of a dye industry effluent by *Aspergillus fumigatus* XC6. Appld Microbiol Biotechnol 74:239–243
59. Kanmani P, Kumar SR, Yuvraj N, Parri KA, Pethikumar V, Arul V (2011) Microbial decolorization of synthetic dyes and reactive dyes of industrial effluents by using novel fungus *Aspergillus proliferans*. Water Env Res 83:2099–2106
60. Kang Y, Xu X, Pan H, Tian J, Tanga W, Liu S (2018) Decolorization of mordant yellow 1 using *Aspergillus sp.* TS-A CGMCC 12964 by biosorption and biodegradation. Bioeng 9:222–232
61. Kapdan IK, Kargi F (2002) Biological decolorization of textile dyestuff containing wastewater by *Coriolus versicolor* in a rotating biological contactor. Enzym Microb Technol 30:195–199
62. Kapdan I, Kargi F, McMullan G, Marchant R (2000) Comparison of white-rot fungi cultures for decolorization of textile dyestuffs. Bioprocess Eng 22:347–351
63. Kaushik P, Malik A (2009) Fungal dye decolourization: recent advances and future potential. Environ Intl 35:127–141
64. Khan S, Malik A (2018) Toxicity evaluation of textile effluents and role of native soil bacterium in biodegradation of a textile dye. Environ Sci Pollut Res Intl 25:4446–4458
65. Khan R, Bhawana P, Fulekar MH (2013) Microbial decolorization and degradation of synthetic dyes: a review. Rev Environ Sci Biotechnol 12:75–97
66. Khlifi R, Belbahri L, Woodward EM, Dhouib A, Sayadi S (2010) Decolorization and detoxification of textile industry wastewater by the laccase-mediator system. J Hazard Mater 175:802–808
67. Krishnan J, Kishore AA, Suresh A, Madhumeetha B (2016) Effect of pH, inoculum dose and initial dye concentration on the removal of azo dye mixture under aerobic conditions. Int Biodeterior Biodegr 119:16–27
68. Kumar PGN, Bhat SK (2011) Fungal degradation of azo dye-red 3bn and optimization of physico-chemical parameters. Intl J Environ Sci 1:17–24
69. Kumar CG, Mongolla P, Basha A, Joseph J, Sarma VUM, Kamal A (2011) Decolorization and biotransformation of triphenylmethane Dye, Methyl Violet, by *Aspergillus sp.* isolated from Ladakh. India. J Microbiol Biotechnol 21:267–273
70. Lankinen PV, Bonnen AM, Anton LH, Wood DA, Kalkkinen N, Hatakka A, Thurston CF (2001) Characteristics and N-terminal amino acid sequence of manganese peroxidase from solid substrate cultures of *Agaricus bisporus*. Appl Microbiol Biotechnol 55:170–176
71. Levin L, Papinutti L, Forchiassin F (2004) Evaluation of Argentinean white rot fungi for their ability to produce lignin-modifying enzymes and decolorize industrial dyes. Bioresour Technol 94:169–176
72. Liers C, Bobeth C, Pecyna M, Ullrich R, Hofrichter M (2010) DyP-like peroxidases of the jelly fungus Auricularia auriculajudae oxidize nonphenolic lignin model compounds and high-redox potential dyes. Appl Microbiol Biotechnol 85:1869–1879
73. Lisov AV, Leontievsky AA, Golovleva L (2003) Hybrid Mn-peroxidase from the ligninolytic fungus *Panus tigrinus* 8/18. Isolation, substrate specificity, and catalytic cycle. Biochem 68:1027–1035
74. Machado KMG, Compart LCA, Morais RO, Rosa LH, Santos MH (2006) Biodegradation of reactive textile dyes by basidiomycetes fungi from Brazilian ecosystem. Braz J Microbiol 37:481–487
75. Manavalan T, Manavalan A, Heese K (2015) Characterization of lignocellulolytic enzymes from white-rot fungi. Curr Microbiol 70:485–498
76. Marimuthu S, Rajendran MM (2013) A review on fungal degradation of textile dye effluent. Acta Chim Pharm Indica 3:192–200
77. Mohamed BHM, Arunprasath R, Purusothaman G (2019) Biological treatment of azo dyes on effluent by *Neurospora* sp isolated and adopted from dye contaminated site. The J Textile Inst 111:1239–1245

78. Muthezhilan R, Yogananth N, Vidhya S, Jayalakshmi S (2008) Dye degrading mycoflora from industrial effluents. Res J Microbiol 3:204–208
79. Novotný C, Svobodová K, Erbanová P, Cajthaml T, Kasinath A, Lang E, Šašek V (2004) Lignolytic fungi in bioremediation: extracellular enzyme production and degradation rate. Soil Biol Biochem 36:1545–1551
80. Nyanhongo GS, Gomes J, Gubitz GM, Zvauya R, Read J, Steiner W (2002) Decolourization of textile dyes by laccases from a newly isolated strain of *Trametes modesta*. Water Res 36:1449–1456
81. Pandey A, Singh P, Iyengar L (2007) Bacterial decolorization and degradation of azo dyes. Int Biodeter Biodegr 59:73–84
82. Papinutti VL, Forchiassin F (2004) Modification of malachite green by fomes sclerodermeus and reduction toxicity to *Phanerochaete chrysosporium*. FEMS Microbiol Lett 231:205–209
83. Parshetti GK, Kalme SD, Gomare SS, Govindwar SP (2007) Biodegradation of reactive blue-25 by *Aspergillus ochraceus* NCIM-1146. Bioresour Technol 98:3638–3642
84. Pilatin S, Kunduhoglu B (2011) Decolorization of textile dyes by newly isolated *Trametes versicolor* strain. Life Sci Biotechnol 1:125–135
85. Puvaneswari N, Muthukrishnan J, Gunasekaran P (2006) Toxicity assessment and microbial degradation of azo dyes. Indian J Exp Biol 44:618–626
86. Raghukumar C, D'Souza-Ticlo D, Verma A (2008) Treatment of colored effluents with lignin-degrading enzymes: an emerging role of marine-derived fungi. Crit Rev Microbiol 34:189–206
87. Ramalingam S, Saraswathy N, Shanmugapriya S, Shakthipriyadarshani S, Sadasivam S, Sanmugaprakash M (2010) Decolorization of textile dyes by *Aspergillus tamari*, mixed fungal culture and *Pencillium purpurogenum*. J Sci Ind Res 69:151–153
88. Ramya M, Anusha B, Kalavathy S, Devilaksmi S (2007) Biodecolourisation and biodegradation of reactive blue by *Aspergillus sp*. Afr J Biotechnol 6:1441–1445
89. Rani B, Kumar V, Singh J, Bisht S, Teotia P, Sharma S, Kela R (2014) Bioremediation of dyes by fungi isolated from contaminated dye effluent sites for bio-usability. Braz J Microbiol 45:1055–1063
90. Rao RG, Ravichandran A, Kandalam G, Kumar SA, Swaraj S, Sridhar M (2019) Screening of wild basidiomycetes and evaluation of the biodegradation potential of dyes and lignin by manganese peroxidases. BioRes 14:6558–6576
91. Robinson T, McMullan G, Marchant R et al (2001) Remediation of dyes in textile effluent: a critical review on current treatment technologies with a proposed alternative. Bioresour Technol. 77:274–255
92. Rodríguez CS (2009) Dye removal by immobilised fungi. Biotechnol Adv 27:227–235
93. Ruiz-Dueñas FJ, Camarero S, Pérez-Boada M, Martinez MJ (2001) A new versatile peroxidase from *Pleurotus*. Biochem Soc Trans 29:116
94. Saranraj P, Sumathi V, Reetha D, Stella D (2010) Fungal decolourization of direct azo dyes and biodegradation of textile dye effluent. J Ecobiotechnol 2(7):12–16
95. Sen SK, Raut S, Bandyopadhyay P (2016) Fungal decolouration and degradation of azo dyes: a review. Fungal Biol Rev 30:112–133
96. Shanmugam S, Ulaganathan P, Swaminathan K, Sadhasivam S, Wu YR (2017) Enhanced biodegradation and detoxification of malachite green by *Trichoderma asperellum* laccase: degradation pathway and product analysis. Intl Biodeterior Biodegr 125:258–268
97. Sharma S, Roy S (2015) Biodegradation of Dye Reactive Black-5 by a Novel Bacterial Endophyte. Int Res J Environ Sci 4:44–53
98. Shedbalkar U, Jadhav J (2011) Detoxification of malachite green and textile industrial effluent by *Penicillium ochrochloron*. Biotechnol Bioprocess Engg 16:196–204
99. Shedbalkar U, Dhanve R, Jadhav J (2008) Biodegradation of triphenylmethane dye cotton blue by *Penicillium ochrochloron* MTCC 517. J Hazard Mater 15:472–479
100. Siddique M, Mahmmod A, Sheikh M, Gafoor A, Khaliq S, Bukhai M, Yousaf K, Rehman K, Andleeb S, Naeem MM (2012) A study on the biodegradation of some reactive textile dyes by white rot fungus (*Pleurotus ostreatus*) World Appld Sci J 18:181–185

101. Singh L (2017) Biodegradation of synthetic dyes: a mycoremediation approach for degradation/decolourization of textile dyes and effluents. J Appld Biotechnol Bioeng 3:430–435
102. Singh L, Singh VP (2010) Biodegradation of textiles dyes, Bromophenol blue and congo red by fungus-*Aspergillus flavus*. Environ. We Int J Sci Tech 5:235–242
103. Singh L, Singh VP (2012) Microbial decolourization of textile dyes by the fungus *Trichoderma harzianum*. J Pure Appld Microbiol 6:1829–1833
104. Singh L, Singh VP (2017) Decolourization of Azo (Acid red) and Anthraquinonic (Basic blue) dyes by the fungus *Aspergillus flavus*. Int J Biomed Eng Clin Sci 3:1–5
105. Solísa M, Solís A, Pérezb HI, Manjarrez N, Floresa M (2012) Microbial decolouration of azo dyes: a review. Process Biochem 47:1723–1748
106. Srinivasan A, Viraraghavan T (2010) Decolorization of dye wastewaters by biosorbents: a review. J Environ Manag 91:1915–1929
107. Srinivasan GP, Sikkanthar A, Elamaran A, Delma CR, Subramaniyan K, Somasundaram ST (2014) Biodegradation of carcinogenic textile azo dyes using bacterial isolates of mangrove sediment. J Coast Life Med 2:154–162
108. Srivastava S, Sinha R, Roy D (2004) Toxicological effects of malachite green. Aquatic Toxicol 66:319–329
109. Telke AA, Ghodake GS, Kalyani DC, Dhanve RS, Govindwar SP (2011) Biochemical characteristics of a textile dye degrading extracellular laccase from a *Bacillus sp.* ADR. Bioresour Technol 102:1752–1756
110. Tien M, Kirkm TK (1988) Lignin peroxidase of *Phanerochaete chrysosporium*. Methods Enzymol 33:569–575
111. Tilli S, Ciullini I, Scozzafava A, Briganti F (2011) Differential decolorization of textile dyes in mixtures and the joint effect of laccase and cellobiose dehydrogenase activities present in extracellular extracts from *Funalia trogii*. Enzyme Microb Tech 49:465–471
112. Torres E, Bustos-Jaimes I, Le Borgne S (2003) Potential use of oxidative enzymes for the detoxification of organic Appld. Catal B: Environ 46:1–15
113. Türgay O, Ersöz G, Atalay S et al (2011) The treatment of azo dyes found in textile industry wastewater by anaerobic biological method and chemical oxidation. Sep Purif Technol 79:26–33
114. Tsuboy MS, Angeli JPF, Mantovani MS, Knasmuller S, Umbuzeiro GA, Ribeiro LR (2007) Genotoxic, mutagenic and cytotoxic effects of the commercial dye CI Disperse Blue 291 in the human hepatic cell line HepG2. Toxicol *in vitro* 21:1650–1655
115. Van Aken B, Agathos S (2002) Implication of manganese (III), oxalate, and oxygen in the degradation of nitroaromatic compounds by manganese peroxidase (MnP). Appld Microbiol Biotechnol 58:345–351
116. Vander Zee FP, Villaverde S (2005) Combined anaerobic-aerobic treatment of azo dyes-a short review of bioreactor studies. Water Res 39:1425–1440
117. Vasina DV, Moiseenko KV, Fedorova TV, Tyazhelova TV (2017) Lignin-degrading peroxidases in white-rot fungus *Trametes hirsuta* 072. Absolute expression quantification of full multigene family. PLoS ONE 12:e0173813
118. Vasudev K (2011) Decolorization of triphenylmethane dyes by six white rot fungi isolated from nature. J Bioremed Biodegrad 2:61–66
119. Vernekar MR, Gokhale JS, Lele SS (2019) Azo dye decoloration by fungi In: Fungal bioremediation. CRC Press, eBook ISBN9781315205984
120. Vyas BRM, Molitoris HP (1995) Involvement of an extracellular H_2O_2 dependent ligninolytic activity of the white rot fungus *Pleurotus ostreatus* in the decolorization of Remazol Brilliant Blue R. Appld Environ Microbiol 61:3919–3927
121. Wesenberg D, Kyriakides I, Agathos SN (2003) White-rot fungi and their enzymes for the treatment of industrial dye effluents. Biotechnol Adv 22:161–187
122. Wilkolazka AJ, Kochnanska RJ, Malarczy KE et al (2002) Fungi and their ability to decolorize azo and anthraquinonic dyes. Enzyme Microbial Technol 30:566–572
123. Wong Y, Yu J (1999) Laccase catalyzed decolorization of synthetic dyes. Water Res 33:3512–3520

124. Yao MLC, Villanueva JDH, Tumana MLS, Caalimg JG, Bungihan ME, Dela Cruz TEE (2009) Antimicrobial activities of marine fungi isolated from seawater and marine sediments. Acta Manilana 57:19–28
125. Yang XQ, Zhao XX, Liu CY, Zheng Y, Qian SJ (2009) Decolorization of azo, triphenylmethane and anthraquinone dyes by a newly isolated *Trametes* sp. SQ01 and its laccase. Process Biochem 4:1185–1189
126. Yang Y, Wang G, Wang B, Du L, Jia X, Zhao Y (2011) Decolorization of malachite green by a newly isolated *Penicillium sp.* YW 01 and optimization of decolorization parameters. Environ Engg Sci 28:555–562
127. Yeddou-Mezenner N (2010) Kinetics and mechanism of dye biosorption onto an untreated antibiotic waste. Desalination 262:251–259
128. Yesilada O, Cing S, Asma D (2002) Decolourization of the textile dye Astrazon Red FBL by *Funalia trogii* pellets. Bioresour Technol 81:155–157
129. Zille A, Gornacka B, Rehorek A, Cavaco-Paulo A (2005) Degradation of azo dyes by *Trametes villosa* laccase over long periods of oxidative conditions. Appl Environ Microbiol 71:6711–6718
130. Zhao X, Huang X, Yao J, Zhou Y, Jia R (2015) Fungal growth and manganese peroxidase production in a deep tray solid-state bioreactor, and *in vitro* decolorization of poly R-478 by MnP. J Microbiol Biotechnol 25:803–813
131. Zope V, Kulkarni M, Chavan M (2007) Biodegradation of synthetic textile dyes reactive red 195 and reactive green 11 by *Aspergillus niger* grp: an alternative approach. J Sci Ind Res 66:411–414
132. Zubbair NA, Ajao AT, Adeyemo EO, Adeniyi DO (2018) Biotransformation and detoxification of reactive black dye by *Ganoderma tsugae.* Afr J Environ Sci Technol 12:158–171
133. Zuleta-Correa A, Merino-Restrepo A, Jiménez-Correa S, Hormaza-Anaguano A, Alonso CS (2016) Use of white rot fungi in the degradation of an azo dye from the textile industry. Dyna Rev Fac Nac Minas 83:128–135

Bioremediation of Dye Using Mesophilic Bacteria: Mechanism and Parametric Influence

Souptik Bhattacharya, Ankita Mazumder, Dwaipayan Sen, and Chiranjib Bhattacharjee

Abstract For centuries, dyes have been utilized in the tannery, textile, food, paper, cosmetic, and plastic industries. As a consequence of the fast urbanization and industrialization, the uncontrolled release of dyeing agents in the effluent is increasing. Such a release causes toxicity and pollution to the whole environment. These concerns become more critical due to the biomagnification phenomenon through various trophic levels resulting in severe toxicity in higher animals and plants including aquatic flora and fauna. Mitigation of this nuisance can be achieved by the economic application of biotechnology using safe biological agents to decolorize and degrade the dye in water bodies.

In this chapter, we reviewed the toxicity and harmful effects of various dyes along with different mechanisms and strategies of dye decolorization and degradation by biological agents while giving ampule emphasis on the mesophilic type bacteria. Further, the effect of different physicochemical parameters on dye removal efficacy was explicitly discussed. Moreover, various techniques to investigate the harmful toxic effects of the produced post degradation metabolites were also enlightened. Thus, this present chapter will deliver a quintessential perception on the feasibility of the bioremediation technique using mesophilic bacterial strains to treat dye contaminated waste streams.

Keywords Mesophilic bacteria · Biodegradation · Wastewater · Dye · Aerobic bacteria · Anaerobic bacteria, · Biomagnification · Bioreactor · Biosorption · Enzymatic degradation

---

S. Bhattacharya · A. Mazumder (✉) · C. Bhattacharjee
Department of Chemical Engineering, Jadavpur University, Kolkata 700032, India
e-mail: ankitamazumder.rs@jadavpuruniversity.in

D. Sen
Department of Chemical Engineering, Heritage Institute of Technology, Kolkata 700107, India

S. S. Muthu and A. Khadir (eds.), *Dye Biodegradation, Mechanisms and Techniques*, Sustainable Textiles: Production, Processing, Manufacturing & Chemistry,
https://doi.org/10.1007/978-981-16-5932-4_3

1 Introduction

The rapid rate of industrialization, urbanization, and scientific developments has accelerated the discharge of several unwanted pollutants in the biosphere beyond safe permissible levels [55, 68, 90, 95]. Among these toxic chemicals, the dye is a major visible contaminant whose presence is highly objectionable in the water to be utilized for either domestic or industrial purposes [15, 36, 51]. The issue of water pollution because of the discharge of untreated or partially treated dye polluted effluent into natural aquatic bodies was initially observed in the nineteenth century [44, 50, 51, 66]). Since the mid–nineteenth century, all colorants used for various dyeing and printing purposes were of natural origin. Natural dyes were mostly safe for the environment and degradable in nature as they are either plant-derived or microorganism-derived. Sources of natural dyes include roots, bark, leaves, wood, berries, lichens, shellfish, and fungi [50]. However, because of their low availability, inefficient and expensive extraction process, there was an emerging requirement in the nineteenth century for the manufacturing of a bulk quantity of cheap synthetic pigments and dyes required in several textile industrial processes. As a consequence, the synthetic dye industry evolved as a "frontier technology industry" of Victorian times. English chemist, William Henry Perkin was acknowledged as the founder of the first synthetic dye named as "Mauve" [16, 47]. Any dye molecule is comprised of mainly two groups: (1) the chromophoric group, which is accountable for the visible color and (2) the auxochrome group, responsible for the dye's solubility in aqueous phase and attraction towards the fiber materials.

Synthetic coloring agents are preferred over natural ones because of some attributes including their higher stability toward detergent, surfactant, temperature, microbial attack, and light, variation in color shades, firmness, easy and inexpensive synthesis process [70]. Nowadays, synthetic colorants are extensively utilized for leather dyeing, textile fiber dyeing, colored photography, food industry, paper printing, and as additives in various petroleum-derived products. More than ten thousand distinct pigments and dyes are being utilized in various industrial applications and their annual worldwide production is more than 0.7 MT [78]. Among the total dye production, approximately 70% are produced in India, China, Taiwan, Argentina, and Korea. In the complex industrial waste stream having different types of colorants, dye wastes are predominantly found [2]. These dyes are generally carcinogenic, teratogenic, and mutagenic in nature, which may induce chronic health hazards to both organisms and human beings [80]. Henceforth, it is obligatory to separate them from industrial wastewater before being discharged into terrestrial and aquatic environments.

To address this intricate pollutant, various conventional physico-chemical treatment methods are explored, which include flocculation, froth flotation, chemical coagulation, irradiation, ozonation, precipitation, photooxidation, adsorption, membrane filtration, ion exchange, and reverse osmosis [49, 87]. However, these techniques become limited because of operational problems, cost-expensiveness, and generation of a bulk quantity of secondary solid waste [49]. Moreover, due to the complex aromatic chemical structure, persistent, and recalcitrant nature, treatment of dye contaminated effluent by conventional physico-chemical techniques is often not so effective [56]. In this context, the biological treatment route for decolorization is a promising, economic and environment-friendly alternative [5]. Several advantages of bioremediation technology are low running cost, minimum operational difficulties, on-site application, permanent waste elimination, minimum environmental impact, and opportunity to be used in integration with other physico-chemical treatment methods [10]. Through the biodegradation process, persistent pollutants are either completely degraded or are transformed into less hazardous components. Biodegradation involves the use of the biological agent for attaining at least one of the three consequences which include (a) a small alteration in an organic molecule keeping the primary chemical structure unchanged, (b) breaking of the multipart organic molecules in such a manner that the resulted components can be reorganized for the formulation of the initial molecule, and (c) complete mineralization.

Over the last decade, numerous bacteria, yeasts, fungi, algae, and actinomycetes have been explored to bioremediate the hazardous dyes [75, 76]. Among these microorganisms, mesophilic bacteria are mostly used as the biological agent for biodegradation of dyes, because of their comparatively rapid multiplication rate under anaerobic, aerobic, facultative environmental conditions [18, 100]. The bioremediation of dyeing agents using mesophilic bacterial strains will be comprehensively reviewed in the current manuscript concerning the mechanism of biodegradation and parametric influence.

2 Types of Synthetic Dyes in Effluents

Dyes are usually categorized depending on their chemical moieties and applications. Basically, dyeing agents are comprised of two prime units (a) chromophores, which are responsible for giving the specific color to any dye, and (b) auxochromes, which complement the chromophores through intensification of color helping the water-soluble dye molecules. These chromophores are various functional groups such as carbonyl, azo, aril methane, methine, anthraquinone, nitro, etc. Moreover, auxochromes are a group of atoms that do not have the ability to provide any color, but when are present in addition to the chromophores as substituents, they either alter or intensify the color of the chromophores. Some commonly utilized auxochromes are sulfonate, amine, carboxyl, aldehyde, hydroxyl, and methyl mercaptan [28, 73, 89]. The dyes' classification depending on chemical structure and applications is highlighted in Table 1.

Table 1 Categorical classification of dyes on the basis of chemical structure, dyeing process, and applications

Type	Structural unit	Characteristics	Applications	Examples
Acid dyes	Azo, nitro, nitroso, triphenylmethane, anthraquinone, xanthene, and azine	Anionic; aqueous soluble	Dyeing of nylon, wool, modified acrylics, silk, leather, paper printing, inkjet printing, cosmetics manufacturing, food industry, etc.	Methyl orange, methyl red, congo red, orange II, orange I
Azo dyes	Azo	Insoluble dyes are synthesized in situ in the fiber through treatment with both diazoic and coupling components	Used for dyeing polyester, rayon, cellulose acetate, and cotton	Butter yellow, Disperse orange 1, aniline yellow
Basic dyes	Acridine, diazahemicyanine, cyanine, diphenylmethane, hemicyanine, thiazine, triarylmethane, oxazine, and xanthene	Cationic; water soluble	Coloring of paper, acrylic fiber, and medicine	Amine yellow, malachite green, butter yellow, methylene blue
Direct dyes	Phthalocyanine, polyazo, oxazine, stilbene,	Dyeing is performed at mild alkaline bath by adding sodium salts	Coloring of cotton, silk, rayon, wool, nylon, leather, paper. Also applied for biological staining and pH indicators	Direct black, congo red, violet 51
Disperse dyes	Azo, nitro, styryl, benzodifuranone, and anthraquinone	Non-ionic; water-insoluble	Dyeing hydrophobic fibers, polyester, nylon, cellulose acetate, synthetic polyamide, polyacrylonitriles fibers	Celliton fast pink B, disperse blue 3, violet 1
Mordant dyes	Azo and anthraquinone	A mordant is required	Applied for the appearance of black or navy shades in silks and wool	Alizarin

(continued)

Table 1 (continued)

Type	Structural unit	Characteristics	Applications	Examples
Reactive dyes	Anthraquinone, azo, phthalocyanine, triarylmethane, oxazine, formazan	A chromophore is utilized for directly reacting with fiber substrate	Coloring of cotton, wool, silk, nylon, cellulose fibers at ambient pH and temperature in home or art studio	Reactive red 120, reactive green 19, reactive violet 2, brilliant blue
Sulfur dyes	Sodium sulphide, disulfide	Treatment of the fibers in a solution comprised of sulfide compounds and organic compounds	Cost inexpensive and are generally utilized for dyeing cotton to bring dark shades	Sulfur green 12, sulfur brown 12, sulfur black 1
Vat	Indigoids, anthraquinone	Insoluble in nature. It gets reduced in alkaline environment to produce aqueous soluble metal salts. These salts will subsequently bind to the textile fiber. Further, on oxidation, they will reform to the initial insoluble dye	Used to bring indigo color	Tyrian purple, indigo blue, indigo white, bezanthrone

3 Toxicity and Harmful Effects of Dyes

Discharge of bulk quantities of synthetic dyes in the effluent can cause severe toxicity and aesthetic nuisance, which are the prime environmental concerns [78]. Improper release of dyes enhances the biochemical oxygen demand (BOD) of the aqueous system and limits the penetration of sunlight through the water surface that reduces the photosynthetic activity followed by inhibition to the proper growth of photoautotrophic organisms. Again, the color of the effluent is inconvenient to the aquatic organisms, retarding the oxygenation of the water. Moreover, the acute toxicity imparted by these colored effluents completely disturbs the ecological balance of the fauna and flora in aquatic bodies [94].

Synthetic dyes are mostly organic aromatic compounds, containing different functional groups and heavy metals. Apart from imparting toxic effects, these dyes are also mutagenic, carcinogenic, and teratogenic (because of which normal embryonic development is disrupted) to the aquatic lives [1, 65]. The azo dyes embodied with the substituent aromatic amines can cause a higher risk of chemosis, bladder cancer, contact dermatitis, skin irritation, vomiting gastritis, vertigo, hypertension, permanent blindness, exophthalmos, lacrimation, rhabdomyolysis, acute tubular necrosis supervene, respiratory distress [62]. Basic dyes are potent clastogens, which can cause mutations, tumor growth, allergy, skin allergy, dermatitis, and also cancer [79]. Cationic dyes may also induce a heart attack, shock, cyanosis, jaundice, tissue necrosis, quadriplegia in human beings [60]. Moreover, heavy metals in the dye can induce chronic toxicity, resulting in kidney failure, ulceration of the mucous membranes and skin, etc. Thus, untreated or improperly treated dye polluted effluent can introduce extreme environmental and health complications once consumed through different food chains.

4 Mechanisms of Dye Biodegradation by Biological System

The major mechanisms of dye decolorization from effluent through bioremediation route are (a) microbial biomass mediated adsorption (biosorption) and (b) inherent microbial enzyme system-mediated biodegradation [43]. In the case of the first mechanism, the adsorption of dyes can be done by either growing live microbial population or by the dead microbial cells. Adsorption of dye using biomass primarily occurs by ion exchange technique between the cell surface and the dye molecule [32, 64, 69]. When the effluent is carrying a relatively high amount of toxic pollutants or the environmental conditions are not promising or for proper cellular growth of microbes then the living microbial cells may not be much effective for dye degradation by using their inherent enzyme system. In such scenario, previously cultured microbial biomass can be applied to adsorb the dye by biosorption mechanism. On the contrary, the second mechanism involving biodegradation of recalcitrant dyes relies on the biotransformation enzymes present in various microbes, which are greatly dependent on the adaptability of microbes with the toxicity of the effluent.

Microbes such as, bacteria, fungi, and algae have a cellulosic cell wall that provides binding sites like carboxyl and hydroxyl groups essential for biosorption of dyes [93]. The dye molecules remain intact during biosorption, while during biodegradation, the primary dye structure is fragmented with the reacting enzymes, often achieving complete mineralization [71, 97]. Biosorption of dyes cannot eliminate the predicament because the dye remains adsorbed into the microbial biomass matrix. Henceforth, biosorption is specifically beneficial for such cases where dye biorecovery is a paramount concern. However, a combination of these two techniques (biomass-mediated biosorption and enzymes mediated biodegradation) is often appropriate to handle bulk quantities of dye-polluted industrial waste streams. Methods of bacterial dye biodegradation are schematically represented in Fig. 1. Extensive diversity of microorganisms which includes bacteria, algae, fungi, and yeasts are potent of biodegrading/decolorizing different dyeing agents. The isolation of new compelling microbial pure strains and understanding their dye degradation mechanism is an emerging biological research field for dye-containing effluent treatment [13].

In the case of enzymes mediated biodegradation, the oxidoreductive enzymes (reductive and oxidative) can generate reactive free radicals which can introduce complex sequences of cleavage reactions. These enzymes are most effective, where

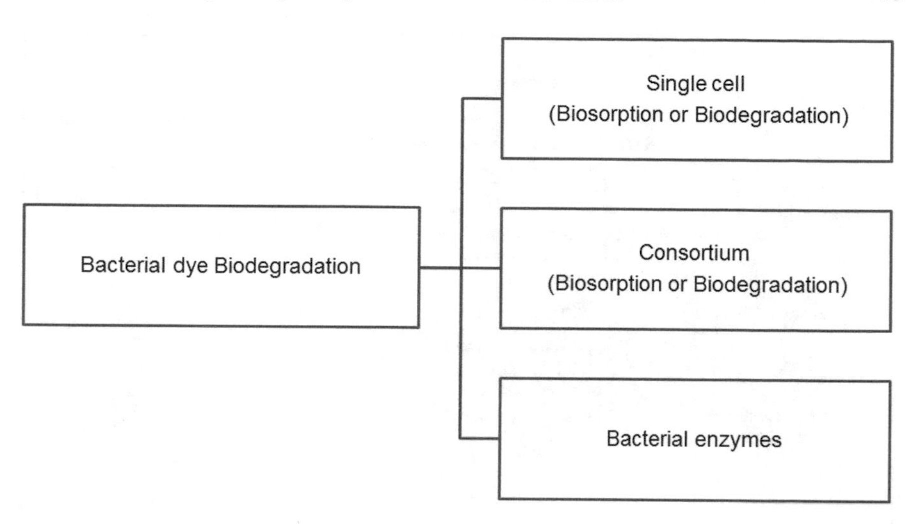

Figure. 1 Methods of bacterial dye biodegradation

the presence of the target pollutant is maximum in terms of concentration compared to other competitive pollutants [67]. Several oxidoreductive enzymes have been used by the bacteria to decolorize and biodegrade dye molecules [76]. Azoreductase is one of the prevalent reductive enzymes responsible for dye degradation. Similarly, oxidative enzymes participating in dye biodegradation are tyrosinase, peroxidases, and laccases [25, 42]. Azoreductases are responsible for carrying out the reduction reaction of the chromophoric linkage (–N = N–) present in azo dyes which helps to undertake the biodegradation of the dye and the formation of monochrome solutions as highlighted in Fig. 2 [41].

The intermediate metabolites are also degraded further aerobically or anaerobically [62]. Another enzyme named riboflavin reductase has been found to have the capacity to degrade the dyes by reducing various flavins [26, 76, 77]. The enzyme laccases (phenol oxidase) is responsible for the cleavage of the O–O bond of dioxygen to water. Peroxidases are heme-containing proteins capable of redox conversion processes, highly effective in degrading anthraquinone, a redox synthetic dye. Tyrosinase (monophenol monooxygenase) has the capability to degrade phenol group by oxidation using molecular oxygen as oxidant [33]. Monophenols are first converted to o-diphenols by hydroxylation and then o-diphenols are further oxidized to o-quinones (as shown in Fig. 3). However, the major concern here is the o-quinone, which can inhibit tyrosinase activity and regulates the reaction [76].

5 Role of Mesophilic Bacteria in Dye Biodegradation

The studies on dye biodegradation are mainly focused on bacteria because they are found to be more efficient and effective than other groups of microorganisms [45].

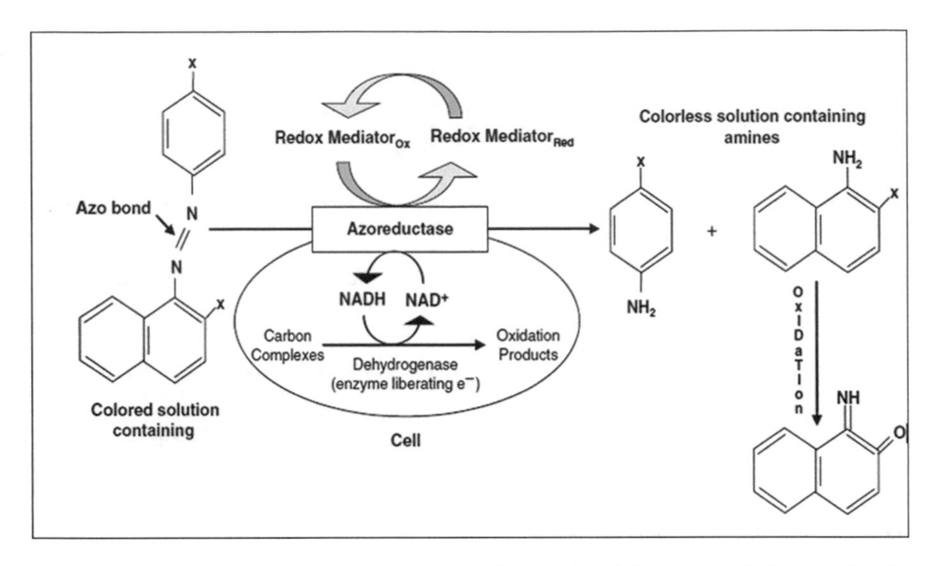

Figure. 2 Bacterial enzymatic actions for azo dye degradation. Adopted from [82] under Creative Commons Attribution License permitting unrestricted use and distribution

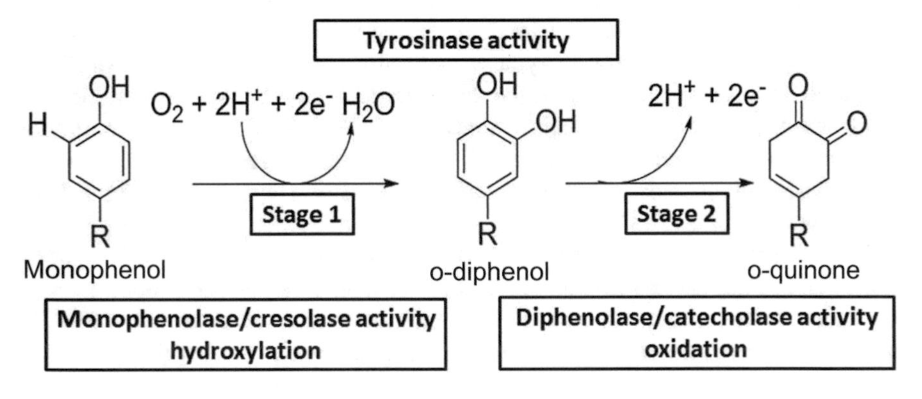

Figure. 3 Mechanism of tyrosinase activity to degrade monophenolic compounds

In comparison to the conventional chemical methods, bioremediation of dyes by bacteria is a much more environment-friendly and cost-inexpensive technique [57]. Bacteria perform the degradation of dye molecules after cleaving them into fragments with various enzymes and thus achieving complete mineralization by producing CO_2, H_2O, biomass along with other inorganics [48]. Bacterial species of *Sphingomonas xenophaga, Flexibacter filiformis, Agrobacterium tumefaciens, Alcaligenes faecalis, Ralstonia eutropha, Proteus mirabilis, Hydrogenophaga palleronii, Pseudomonas aeruginosa, Escherichia coli, Lactobacillus plantarum, Bacillus subtilis, Rhodococcus erythropolis, Bacillus licheniformis, and Serratia marcescens* were found capable to reduce azo dyes. The majorities of these species is mesophilic type bacteria and grow best in a moderate temperature ranging from 20 to 45 °C.

It was previously observed that the mesophilic bacterial strain *Pseudomonas aeruginosa* was competent enough to biodegrade the Direct Orange 39 dye at a concentration of 1,000 ppm/day [84]. *Thermomonospora sp.* and *Streptomyces sp.* were able to decolorize Poly B-411, Poly R-478, and Remazol Brilliant Blue R dyes (anthraquinonic). The *Streptomyces sp.* was also found promising in degradation of benzene derivatives by catabolic pathways. Several Streptomycetes species can also decolorize dyes such as Orange I, 3-methoxy-4-hydroxy-azobenzene-4′-sulfonic acid, and 4 (3-methoxy-4-hydroxy-phenylazo)-azobenzene-3,4′-disulfonic acid [98]. Eskandari et al. [18] showed that microbial consortia consisting of mesophilic bacterial genera such as *Pseudoarthrobacter, Gordonia, Stenotrophomonas,* and *Sphingomonas* were effective in biodegradation of Reactive Black-5 azo dye [18]. Again, [14] revealed that the yhdA gene of *Bacillus subtilis* can encode an oxidoreductase dependent on flavin mononucleotide, which can induce cleavage of the –N=N– bond present in the azo dyes with the aid of NADPH [14]. Highly promising decolorization efficiency of 96.9–99.5% for Congo red (CR) was achieved by *Vitreoscilla sp., Acinetobacter lwoffii, Pseudomonas fluorescens, Escherichia coli, Bacillus thuringiensis, Enterobacter asburiae Enterobacter. Ludwigii,* and *Enterobacter asburiae* [31].

Xenophilus azovorans, Staphylococcus aureus, Acinetobacter calcoaceticus, Bacillus sp. OY1-2, Escherichia coli, Enterococcus faecalis, Pigmentiphaga kullae K24, and *Rhodobacter sphaeroides* were extensively explored for having azoreductase enzymes to carry out the biodegradation of azo dyes [9, 88]. *Galactomyces geotrichum* MTCC 1360, *Proteus vulgaris, Micrococcus glutamicus, and Bacillus sp.* have expressed riboflavin reductase activity for the degradation of Reactive Green 19A, Brilliant Blue G, Navy Blue GL, Mordant Yellow 10, and Scarlet R dye [20, 37]. *Bacillus sp.* and *Acinetobacter calcoaceticus* expressed Lignin peroxidase activity. The phenol oxidase activity was found in *Micrococcus glutamicus* and *Pseudomonas desmolyticum.* Tyrosinase activity was observed in *P. desmolyticum* against Direct Blue-6 dye [39].

6 Strategies of Bacterial Dye Degradation

Effluents containing a cluster of structurally complex dyes are toxic for most aquatic organisms (flora and fauna) when discharged into the aquatic bodies manifesting anoxic conditions by reducing the dissolved oxygen concentrations. Several mesophilic bacteria species like *Aeromonas hydrophila, Bacillus cereus, Proteus mirabilis, Bacillus subtilis, Pseudomonas sp.,* and *Pseudomonas luteola* have showcased promising dye degradation efficacy as a single strain [38]. However, it is often tough to accomplish complete decolorization with a pure bacterial culture. Owing to the cooperative approach of the mixed bacterial cultures (consortium), an enhanced decolorization effect is often observed showing better result than single strains in decolorization and biodegradation of dyes [81]. However, it is difficult to interpret the results of mesophilic consortium-based dye degradation because the mixed bacterial cultures do not bestow the meticulous understanding of the dye metabolism mechanisms and the experimental findings are often obscure to reproduce.

It was observed that the biodegradation rate of a mesophilic bacterial consortium is generally superior to a single bacterial strain due to versatile enzymes from multiple bacterial strains that can attack the target dye molecules at distinct positions or linkage [29]. Moreover, the co-existing strains are often utilizing the metabolites produced by the decomposition of the dye molecules and achieve complete mineralization [22, 76]. However, isolation of new adaptable and formidable pure bacterial strains with multiple dye degradation capability from the wastewater and environment is very crucial for achieving greater biodegradation efficiency. It is an exciting research opportunity for future industrial effluent treatment but it takes a prolonged time and hard labor to isolate such pure cultures from textile wastewater.

Nowadays utilization of bacterial biofilms for effluent treatment is popular. Biofilms are substantially organized surface-associated microbial cells that can be comprised of both single and multiple species of bacteria. The association of the cells is mainly attributed to their self-produced extracellular polymeric substances rich in cellulose, lipids, and polysaccharides [21, 30, 59]. Biofilms are highly resistant to various environmental stress factors, including an extremely elevated concentration of toxic contaminants, temperature, pH, and salinity than their planktonic single-cell counterparts [21, 30]. Hence, dye degradation using biofilm-producing bacteria such as *Bacillus subtilis, Pseudomonas fluorescens, Acinetobacter lwoffii, Bacillus thuringiensis, Enterobacter asburiae* are a very lucrative technique for dye polluted industrial effluent treatment [31, 54].

Numerous researches have revealed that a combination of an aerobic and an anaerobic system is a particularly reasonable approach for dye biodegradation [53]. It was found that the aerobic process with agitation is appropriate for bacterial growth but maximum dye decolorization/degradation is achieved in an anaerobic system [3, 89]. Under aerobic conditions, most mesophilic bacterial species are not competent to utilize the dye as a sole source of carbon. These bacteria generally require a secondary carbon source for growth and survival. Aerobic bacteria have oxidoreductive enzymes that can disrupt the dye molecules asymmetrically or symmetrically by deamination, desulfonation, and hydroxylation process. However, the enzyme azoreductase has shown proficiency at anaerobic environment. Anaerobic degradation of different azo bonded colorants achieved with facultative anaerobic and aerobic microbes were narrated in several previous studies [19, 23, 58, 61, 99]. Most of these microbial cultures are capable of growing in aerobic atmosphere although the degradation process was accomplished only under complete anaerobic environment [38]. Adequate decolorization and removal of the recalcitrant dyes can be accomplished in the anaerobic stage and the remaining auxiliary substrates may be mineralized in a subsequent aerobic step. Thus, the integrated approach involving anaerobic and subsequent aerobic treatment is predominantly proposed for degradation of dye polluted effluents [83]. In combination treatment process color removal efficiency varies from 75 to 96% [63].

Utilization of anoxic conditions, where the concentration of dissolved oxygen is lower than 0.5 ppm has also been found promising for the effective decolorization of several colorants using both facultative anaerobic and aerobic bacteria. However, this process requires other complex nutrient sources, which amplify the operating cost [76]. Several types of bioreactor (batch and continuous mode) configurations have been used for anaerobic systems using single/mixed bacterial species for dye-containing effluent treatment [86]. These include upflow anaerobic baffled reactors, anaerobic sludge blanket, trickle-bed reactor, rotating biological contactors, and activated sludge process [6, 73].

7 Parametric Influence on the Dye Biodegradation

Several environmental and operational parameters, including dye concentration, the structure of dye, pH, temperature, supplementation of different nitrogen and carbon sources, oxygen, level of agitation, greatly influence the dye biodegradation performance. For making the treatment process highly efficient, rapid and practically feasible, prior optimization of each parameter for the bacterial remediation of dye is necessary.

7.1 Effect of Dye Concentration

The dye degradation rate progressively declines with the enhancement in the dye concentration probably because of the toxic effect of hazardous dyes to microbes or/and deficient dye to microbial cells ratio, along with the obstruction of the enzyme (azoreductase) at its active site by dye molecules [91, 92, 24]. Though, the toxicity to microorganisms is primarily related to the concentration of dye and the type of dye. Reactive and metal-complex dyes (such as Acid Black 172, Irgalan Black RBLN, Irgalan Blue 3GL, Irgalan Grey GLN) are found to exhibit exaggerated toxicity on bacterial bioremediation process [17, 52, 53].

7.2 Effect of Dye Structure

Dye possessing a simple chemical structure and less molecular weight exhibits a higher decolorization rate. Moreover, the characteristics of substituents on the aromatic ring present in the dye molecules have portrayed a significant impact on the oxidation phenomenon. For instance, with the presence of electron-giving methoxy and methyl substituents, the enzyme-mediated biodegradation of dyes is facilitated. On the contrary, the presence of electron-receiver substituents ($-SO_3H$, $-SO_2NH_2$, fluoro, chloro, nitro) at the phenyl ring's para position with reference to azo bond will restrict oxidation process and reduce the dye removal rate [34, 85]. Moreover, the susceptibility of azo bond for degradation is promoted if the substituent is present at the para position of the phenyl ring compared to the ortho and meta positions [34]. Additionally, in case of monoazo dyes, the dye degradation rate is faster than diazo and triazo dyes [35]. Metal-ion-containing dyes may impose intricacy in the biodegradation process, and eventually lowers the degradation efficacy [11].

7.3 Influence of Nitrogen and Carbon Sources

The dyes remain mostly inadequate in nitrogen and carbon sources, because of which dye biodegradation without additional supplement of these nitrogen and carbon sources is challenging. Single bacterium as well as consortium generally demands either or both carbohydrates and multipart organic sources (yeast extract, peptone) for effective degradation [45, 72]. Since among different carbon sources, glucose is readily available and highly effective for microbial metabolism, its inclusion enhances the efficiency of biodegradation [4]. Peptone, yeast extract, and urea are good nitrogen sources from organic origin that can be supplemented to restore the NADH, that plays the role of an electron donator to undertake the reduction of dyes using microbial agents, and thereby higher dye degradation can be achieved [12].

7.4 Effect of pH

The pH level controls the transport of dye molecules across the membrane of microorganisms' cells. This is contemplated as the rate-controlling step of the degradation process [7]. Microbial dye degradation rate is greater at optimized pH and follows a declining trend at extreme alkaline or acidic pH 6–10 pH is optimum for bacterial decolorization of dye [27, 46]. However, as most of the textile industrial processes are undertaken at alkaline (high pH) conditions, the sustainability or tolerance of mesophilic bacterial strains to high pH condition is recommended markedly.

7.5 Effect of Temperature

Temperature is another vital parameter for any processes related to microbial vitality [8], including the remediation of dye polluted wastewater by mesophiles. Studies related to microbial decolorization reported that the decolorization rate enhances up to a certain optimum temperature range (25–40 °C). However, enhancing the temperature beyond the optimum value will drastically reduce the dye biodegradation rate. This is probably attributed to the denaturation of azoreductase enzyme or the damage of cellular integrity at extreme temperatures. In case of dye degradation at extremely high temperature, utilization of thermophilic bacterial strains shows better performances compared to the mesophilic ones [74, 75].

7.6 Impact of Oxygen and Shaking

The degradation of dyes was more proficient under strict anaerobic conditions, though it can also be performed in semi-anaerobic environment [96]. When dye degradation process is operated at anaerobic conditions, activities of reductive enzyme are higher, which facilitates to break the complex dye structures. Dissolved oxygen acts as an inhibiting agent to the reduction process of dye. This inhibition effect of oxygen can be indirectly validated by comparing the efficacy of decolorization process performed under shaking and static environment [40]. Inefficient decolorization and degradation were evidenced at shaking/agitated environment, as improved oxygenation was provided through shaking. This indirectly manifests that the oxygen imposes inhibition effect to microbial-induced degradation mechanism.

8 Toxicity of Dye Degradation Products

In some particular cases, the degraded products of dyes are found hazardous, mutagenic, and carcinogenic type. The anaerobic bacterial population present in the lower gastrointestinal tract of mammals can reduce the ingested dye molecules. In the intestinal tract, the reduction of azo dyes by anaerobic bacteria can generate acyloxy amines as a dye degradation product, which is carcinogenic and often leads to bladder cancer. The acyloxy amines are converted to carbonium and nitrenium ions that can attach with RNA and DNA of somatic cells provoking the mutations to form malignant tumors [47]. Several other moieties like benzidine, 1-amino-2-naphthol, o-tolidine, and 1, 4-phenylenediamine are also harmful [43]. Similar compounds may be generated during dye-contaminated effluent treatment and may cause toxic health hazards to both plants and animals. Hence, it is advisable to inspect various toxicity levels (Fig. 4) of the degraded dye products after bioremediation and before effluent discharge.

9 Future Prospective

Accumulation of dye in wastewater creates environmental pollution and health-related problems to plants and animal kingdom present in the biosphere. Biodegradation of dyes present in effluents using diverse group of mesophilic bacterial strains has evolved as a promising strategy. As environmental policies are becoming stringent by the regulating authorities, a compelling requirement to develop ecofriendly,

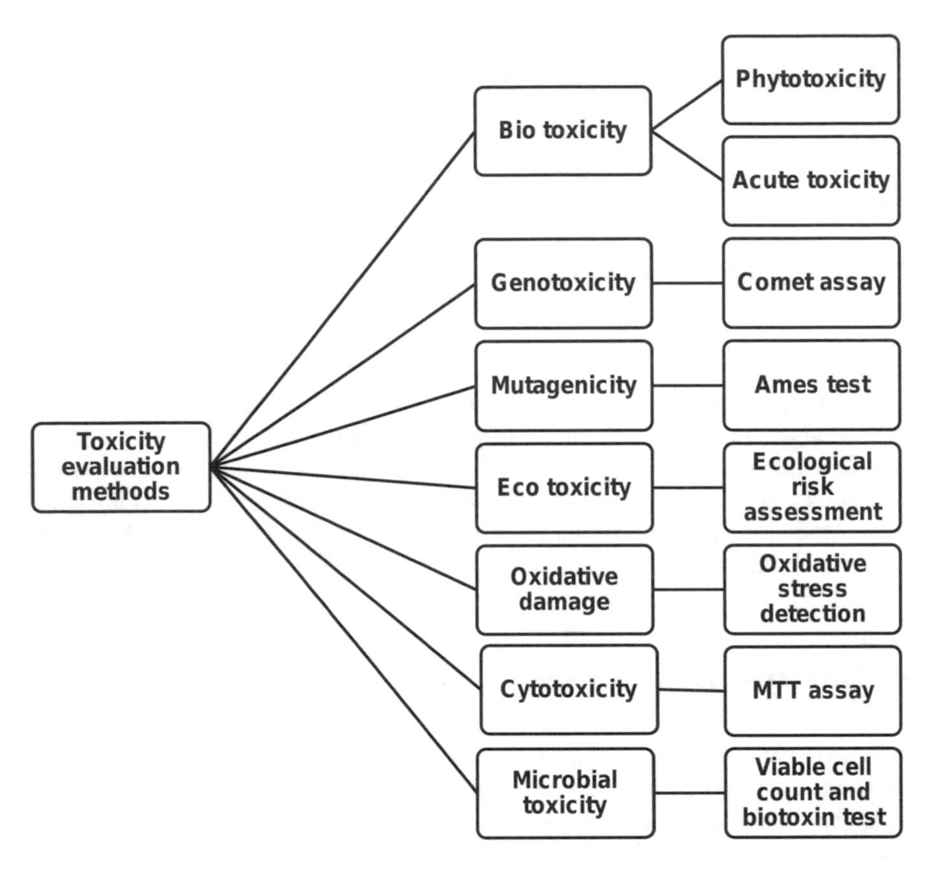

Figure. 4 Various types of toxicity assessment methods for the degraded dye products

cost-effective, and technically efficient treatment methods is paramount. Bioremediation using single and mixed bacterial consortium are environmentally benign and cost-effective strategy for the decolorization of effluents discharged by industrial facilities.

Further improvement of the degradation potential of mesophilic bacteria may be achieved through divulging them gradually to elevated concentration of dyeing agents, where they will adapt and evolve. Also, genetic modification of mesophilic bacteria is another interesting tool for improving the dye degradation potential. Hence, research on the regulation of genes and proteins present in various bacteria and the critical analysis of their effects may be further explored for selecting the microbial agent with greater biodegradation proficiencies. Presently, many esteemed laboratories worldwide including sophisticated laboratories in developing countries like India are actively involved in the progressive research for better dye biodegradation using various mesophilic bacterial strains.

Acknowledgement The authors would like to convey their acknowledgement to the Council of Scientific & Industrial Research (CSIR), New Delhi, India for awarding the senior research fellowship (file no. 09/096(0879)/2017-EMR-I).

References

1. Alves de Lima RO et al (2007) Mutagenic and carcinogenic potential of a textile azo dye processing plant effluent that impacts a drinking water source. Mutat Res Toxicol Environ Mutagen 626(1–2):53–60. https://doi.org/10.1016/j.mrgentox.2006.08.002
2. Anjaneyulu Y, Sreedhara Chary N, Samuel Suman Raj D (2005) Decolourization of industrial effluents—available methods and emerging technologies—a review. Rev Environ Sci Bio/Technol 4(4):245–273. https://doi.org/10.1007/s11157-005-1246-z
3. Barathi S, Aruljothi KN, Karthik C, Padikasan IA (2020) Optimization for enhanced ecofriendly decolorization and detoxification of Reactive Blue160 textile dye by Bacillus subtilis. Biotechnol Reports 28e00522. https://doi.org/10.1016/j.btre.2020.e00522
4. Bardi L, and Marzona M (2010) Factors affecting the complete mineralization of azo dyes. Springer Berlin Heidelberg, 195–210. https://doi.org/10.1007/698_2009_50
5. Behera M, Nayak J, Banerjee S, Chakrabortty S, Tripathy SK (2021) A review on the treatment of textile industry waste effluents towards the development of efficient mitigation strategy: an integrated system design approach. J Environ Chem Eng 9(4). https://doi.org/10.1016/j.jece.2021.105277
6. Bharti V et al (2019) Biodegradation of methylene blue dye in a batch and continuous mode using biochar as packing media. Environ Res 171356–364. https://doi.org/10.1016/j.envres.2019.01.051
7. Bhattacharya A, Goyal N, Gupta A (2017) Degradation of azo dye methyl red by alkaliphilic, halotolerant Nesterenkonia lacusekhoensis EMLA3: application in alkaline and salt-rich dyeing effluent treatment. Extremophiles 21(3):479–490. https://doi.org/10.1007/s00792-017-0918-2
8. Bhattacharya S, Gupta D, Sen D, Bhattacharjee C (2021) Development of Micellized Antimicrobial Thiosulfinate: a contemporary way of drug stability enhancement. pp 83–89. https://doi.org/10.1007/978-981-15-7409-2_8
9. Blümel S, Stolz A (2003) Cloning and characterization of the gene coding for the aerobic azoreductase from Pigmentiphaga kullae K24. Appl Microbiol Biotechnol 62(2–3):186–190. https://doi.org/10.1007/s00253-003-1316-5
10. Boopathy R (2000) Factors limiting bioremediation technologies. Bioresour Technol 74(1):63–67. https://doi.org/10.1016/S0960-8524(99)00144-3
11. Brown MA, De Vito SC (1993) Predicting azo dye toxicity. Crit Rev Environ Sci Technol 23(3):249–324. https://doi.org/10.1080/10643389309388453
12. Chang JS, Kuo TS, Chao YP, Ho JY, Lin PJ (2000) Azo dye decolorization with a mutant Escherichia coli strain. Biotechnol Lett 22(9):807–812. https://doi.org/10.1023/1005624707777
13. Chen B-Y et al (2010) Assessment upon azo dye decolorization and bioelectricity generation by Proteus hauseri. Bioresour Technol 101(12):4737–4741. https://doi.org/10.1016/j.biortech.2010.01.133
14. Deller S et al (2006) Characterization of a thermostable NADPH:FMN oxidoreductase from the mesophilic Bacterium Bacillus subtilis†. Biochemistry 45(23):7083–7091. https://doi.org/10.1021/bi052478r
15. Donkadokula NY, Kola AK, Naz I, Saroj D (2020) A review on advanced physico-chemical and biological textile dye wastewater treatment techniques. Rev Environ Sci Bio/Technology 19(3):543–560. https://doi.org/10.1007/s11157-020-09543-z

16. Druding SC (1982) Dye History from 2600 BC to the 20th century. www.unitedcolor.com. Accessed 27 January 2021
17. Du L-N et al. (2012) Biosorption of the metal-complex dye Acid Black 172 by live and heat-treated biomass of Pseudomonas sp. strain DY1: kinetics and sorption mechanisms. J Hazard Mater 205–20647–54. https://doi.org/10.1016/j.jhazmat.2011.12.001
18. Eskandari F, Shahnavaz B, Mashreghi M (2019) Optimization of complete RB-5 azo dye decolorization using novel cold-adapted and mesophilic bacterial consortia. J Environ Manage 24191–98. https://doi.org/10.1016/j.jenvman.2019.03.125
19. Fatima M, Farooq R, Lindström RW, Saeed M (2017) A review on biocatalytic decomposition of azo dyes and electrons recovery. J Mol Liq 246275–281. https://doi.org/10.1016/j.molliq.2017.09.063
20. Field JA, Brady J (2003) Riboflavin as a redox mediator accelerating the reduction of the azo dye Mordant Yellow 10 by anaerobic granular sludge. Water Sci Technol 48(6):187–193. https://doi.org/10.2166/wst.2003.0393
21. Flemming H-C et al (2016) Biofilms: an emergent form of bacterial life. Nat Rev Microbiol 14(9):563–575. https://doi.org/10.1038/nrmicro.2016.94
22. Forgacs E, Cserháti T, Oros G (2004) Removal of synthetic dyes from wastewaters: a review. Environ Int 30(7):953–971. https://doi.org/10.1016/j.envint.2004.02.001
23. Franca RDG et al (2020) Oerskovia paurometabola can efficiently decolorize azo dye Acid Red 14 and remove its recalcitrant metabolite. Ecotoxicol Environ Saf 191110007. https://doi.org/10.1016/j.ecoenv.2019.110007
24. Garg N, Garg A, Mukherji S (2020) Eco-friendly decolorization and degradation of reactive yellow 145 textile dye by Pseudomonas aeruginosa and Thiosphaera pantotropha. J Environ Manage 263110383. https://doi.org/10.1016/j.jenvman.2020.110383
25. Gkaniatsou E et al (2018) Enzyme encapsulation in mesoporous metal-organic frameworks for selective biodegradation of harmful dye molecules. Angew Chemie Int Ed 57(49):16141–16146. https://doi.org/10.1002/anie.201811327
26. Gomaa OM, Fapetu S, Kyazze G, Keshavarz T (2017) The role of riboflavin in decolourisation of Congo red and bioelectricity production using Shewanella oneidensis-MR1 under MFC and non-MFC conditions. World J Microbiol Biotechnol 33(3):56. https://doi.org/10.1007/s11274-017-2223-8
27. Guo J et al (2007) Biocalalyst effects of immobilized anthraquinone on the anaerobic reduction of azo dyes by the salt-tolerant bacteria. Water Res 41(2):426–432. https://doi.org/10.1016/j.watres.2006.10.022
28. Gupta VK, Suhas, (2009) Application of low-cost adsorbents for dye removal—a review. J Environ Manage 90(8):2313–2342. https://doi.org/10.1016/j.jenvman.2008.11.017
29. Hameed BB, Ismail ZZ (2018) Decolorization, biodegradation and detoxification of reactive red azo dye using non-adapted immobilized mixed cells. Biochem Eng J 13771–77. https://doi.org/10.1016/j.bej.2018.05.018
30. Haque F, Sajid M, Cameotra SS, Battacharyya MS (2017) Anti-biofilm activity of a sophorolipid-amphotericin B niosomal formulation against Candida albicans. Biofouling 33(9):768–779. https://doi.org/10.1080/08927014.2017.1363191
31. Haque MM, Haque MA, Mosharaf MK, Marcus PK (2020) Novel bacterial biofilm consortia that degrade and detoxify the carcinogenic diazo dye Congo red. Arch Microbiol. https://doi.org/10.1007/s00203-020-02044-1
32. Hassan MM, Carr CM (2018) A critical review on recent advancements of the removal of reactive dyes from dyehouse effluent by ion-exchange adsorbents. Chemosphere 209201–209219. https://doi.org/10.1016/j.chemosphere.2018.06.043
33. Hofrichter M et al (2020) Fungal peroxygenases: a phylogenetically old superfamily of heme enzymes with promiscuity for oxygen transfer reactions. pp 369–403. https://doi.org/10.1007/978-3-030-29541-7_14.
34. Hsueh C-C, Chen B-Y, Yen C-Y (2009) Understanding effects of chemical structure on azo dye decolorization characteristics by Aeromonas hydrophila. J Hazard Mater 167(1–3):995–1001. https://doi.org/10.1016/j.jhazmat.2009.01.077

35. Hu T-L (2001) Kinetics of azoreductase and assessment of toxicity of metabolic products from azo dyes by Pseudomonas luteola. Water Sci Technol 43(2):261–269. https://doi.org/10.2166/wst.2001.0098
36. Ismail M et al (2019) Pollution, toxicity and carcinogenicity of organic dyes and their catalytic bio-remediation. Curr Pharm Des 25(34):3645–3663. https://doi.org/10.2174/1381612825666191021142026
37. Jadhav SU, Jadhav MU, Kagalkar AN, Govindwar SP (2008) Decolorization of Brilliant blue G dye mediated by degradation of the microbial consortium of Galactomyces geotrichum and Bacillus sp. J Chinese Inst Chem Eng 39(6):563–570. https://doi.org/10.1016/j.jcice.2008.06.003
38. Jamee R, Siddique R (2019) Biodegradation of synthetic dyes of textile effluent by microorganisms: an environmentally and economically sustainable approach. Eur J Microbiol Immunol 9(4):114–118. https://doi.org/10.1556/1886.2019.00018
39. Kalme SD, Parshetti GK, Jadhav SU, Govindwar SP (2007) Biodegradation of benzidine based dye Direct Blue-6 by Pseudomonas desmolyticum NCIM 2112. Bioresour Technol 98(7):1405–1410. https://doi.org/10.1016/j.biortech.2006.05.023
40. Kalyani DC, Patil PS, Jadhav JP, and Govindwar SP (2008) Biodegradation of reactive textile dye Red BLI by an isolated bacterium Pseudomonas sp. SUK1. Bioresour Technol 99(11):4635–4641. https://doi.org/10.1016/j.biortech.2007.06.058
41. Kandelbauer A, Guebitz GM (2005) Bioremediation for the decolorization of textile dyes—a review. In: Environ chem. Springer, Berlin/Heidelberg, pp 269–288. https://doi.org/10.1007/3-540-26531-7_26
42. Kandelbauer A, Guebitz GM (no date) Bioremediation for the decolorization of textile dyes—a review. In: Environ chem. Springer, Berlin/Heidelberg, pp 269–288. https://doi.org/10.1007/3-540-26531-7_26
43. Khan R, Bhawana P, Fulekar MH (2013) Microbial decolorization and degradation of synthetic dyes: a review. Rev Environ Sci Bio/Technology 12(1):75–97. https://doi.org/10.1007/s11157-012-9287-6
44. Khan S, Malik A (2018) Toxicity evaluation of textile effluents and role of native soil bacterium in biodegradation of a textile dye. Environ Sci Pollut Res 25(5):4446–4458. https://doi.org/10.1007/s11356-017-0783-7
45. Khehra MS, Saini HS, Sharma DK, Chadha BS, Chimni SS (2005) Comparative studies on potential of consortium and constituent pure bacterial isolates to decolorize azo dyes. Water Res 39(20):5135–5141. https://doi.org/10.1016/j.watres.2005.09.033
46. Kılıç NK, Nielsen JL, Yüce M, Dönmez G (2007) Characterization of a simple bacterial consortium for effective treatment of wastewaters with reactive dyes and Cr(VI). Chemosphere 67(4):826–831. https://doi.org/10.1016/j.chemosphere.2006.08.041
47. Kodam KM, Kolekar YM (2015) Bacterial degradation of textile dyes, pp 243–266. https://doi.org/10.1007/978-3-319-10942-8_11
48. Kolekar YM et al (2012) Decolorization and biodegradation of azo dye, reactive blue 59 by aerobic granules. Bioresour Technol 104818–822. https://doi.org/10.1016/j.biortech.2011.11.046
49. Kumar CG, Mongolla P (2015) Microbial Degradation of basic dyes in wastewaters, pp 85–110. https://doi.org/10.1007/978-3-319-10942-8_4
50. Kumar Gupta V (2020) Fundamentals of natural dyes and its application on textile substrates. In: Chem technol nat synth dye pigment. IntechOpen. https://doi.org/10.5772/intechopen.89964
51. Lellis B, Fávaro-Polonio CZ, Pamphile JA, Polonio JC (2019) Effects of textile dyes on health and the environment and bioremediation potential of living organisms. Biotechnol Res Innov 3(2):275–290. https://doi.org/10.1016/j.biori.2019.09.001
52. Li T, Guthrie JT (2010) Colour removal from aqueous solutions of metal-complex azo dyes using bacterial cells of Shewanella strain J18 143. Bioresour Technol 101(12):4291–4295. https://doi.org/10.1016/j.biortech.2010.01.024

53. Libra JA, Borchert M, Vigelahn L, Storm T (2004) Two stage biological treatment of a diazo reactive textile dye and the fate of the dye metabolites. Chemosphere 56(2):167–180. https://doi.org/10.1016/j.chemosphere.2004.02.012
54. Limoli DH, Jones CJ, Wozniak DJ (2015) Bacterial extracellular polysaccharides in biofilm formation and function. In: Microb biofilms. ASM Press, Washington, DC, USA, pp 223–247. https://doi.org/10.1128/9781555817466.ch11
55. Mazumder A, Das S, Sen D, Bhattacharjee C (2020a) Kinetic analysis and parametric optimization for bioaugmentation of oil from oily wastewater with hydrocarbonoclastic Rhodococcus pyridinivorans F5 strain. Environ Technol Innov 17100630. https://doi.org/10.1016/j.eti.2020.100630
56. Mazumder A, Bhattacharya S, Bhattacharjee C (2020b) Role of nano-photocatalysis in heavy metal detoxification, pp 1–33. https://doi.org/10.1007/978-3-030-12619-3_1
57. McMullan G et al (2001) Microbial decolourisation and degradation of textile dyes. Appl Microbiol Biotechnol 56(1–2):81–87. https://doi.org/10.1007/s002530000587
58. Moosvi S, Keharia H, Madamwar D (2005) Decolourization of textile dye reactive violet 5 by a newly isolated bacterial consortium RVM 11.1. World J Microbiol Biotechnol 21(5):667–672. https://doi.org/10.1007/s11274-004-3612-3
59. Mosharaf MK et al (2018) Metal-adapted bacteria isolated from wastewaters produce biofilms by expressing proteinaceous curli fimbriae and cellulose nanofibers. Front Microbiol 9. https://doi.org/10.3389/fmicb.2018.01334
60. Nohynek GJ, Fautz R, Benech-Kieffer F, Toutain H (2004) Toxicity and human health risk of hair dyes. Food Chem Toxicol 42(4):517–543. https://doi.org/10.1016/j.fct.2003.11.003
61. Oliveira JMS et al (2020) Intermittent aeration strategy for azo dye biodegradation: a suitable alternative to conventional biological treatments? J Hazard Mater 385121558. https://doi.org/10.1016/j.jhazmat.2019.121558
62. Pandey A, Singh P, Iyengar L (2007) Bacterial decolorization and degradation of azo dyes. Int Biodeterior Biodegradation 59(2):73–84. https://doi.org/10.1016/j.ibiod.2006.08.006
63. Pearce C (2003) The removal of colour from textile wastewater using whole bacterial cells: a review. Dye Pigment 58(3):179–196. https://doi.org/10.1016/S0143-7208(03)00064-0
64. Peng H, Guo J (2020) Removal of chromium from wastewater by membrane filtration, chemical precipitation, ion exchange, adsorption electrocoagulation, electrochemical reduction, electrodialysis, electrodeionization, photocatalysis and nanotechnology: a review. Environ Chem Lett 18(6):2055–2068. https://doi.org/10.1007/s10311-020-01058-x
65. Penninks A, Baert K, Levorato S, Binaglia M (2017) Dyes in aquaculture and reference points for action. EFSA J 15(7). https://doi.org/10.2903/j.efsa.2017.4920
66. Petroviciu I et al (2019) Dyes and biological sources in nineteenth to twentieth century ethnographic textiles from Transylvania. Romania Herit Sci 7(1):15. https://doi.org/10.1186/s40494-019-0255-0
67. Putatunda S, Bhattacharya S, Sen D, Bhattacharjee C (2019) A review on the application of different treatment processes for emulsified oily wastewater. Int J Environ Sci Technol 16(5):2525–2536. https://doi.org/10.1007/s13762-018-2055-6
68. Qin H, Su Q, Khu S-T, Tang N (2014) Water quality changes during rapid urbanization in the Shenzhen river catchment: an integrated view of socio-economic and infrastructure development. Sustainability 6(10):7433–7451. https://doi.org/10.3390/su6107433
69. Robinson T, McMullan G, Marchant R, Nigam P (2001) Remediation of dyes in textile effluent: a critical review on current treatment technologies with a proposed alternative. Bioresour Technol 77(3):247–255. https://doi.org/10.1016/S0960-8524(00)00080-8
70. Rodríguez-Couto S (2009) Dye removal by immobilised fungi. Biotechnol Adv 27(3):227–235. https://doi.org/10.1016/j.biotechadv.2008.12.001
71. Routoula E, Patwardhan SV (2020) Degradation of anthraquinone dyes from effluents: a review focusing on enzymatic dye degradation with industrial potential. Environ Sci Technol 54(2):647–664. https://doi.org/10.1021/acs.est.9b03737
72. Sani RK, Banerjee UC (1999) Decolorization of triphenylmethane dyes and textile and dyestuff effluent by Kurthia sp. Enzyme Microb Technol 24(7):433–437. https://doi.org/10.1016/S0141-0229(98)00159-8

73. dos Santos AB, Cervantes FJ, van Lier JB (2007) Review paper on current technologies for decolourisation of textile wastewaters: perspectives for anaerobic biotechnology. Bioresour Technol 98(12):2369–2385. https://doi.org/10.1016/j.biortech.2006.11.013
74. Saratale RG, Saratale GD, Chang JS, and Govindwar SP (2009a) Ecofriendly degradation of sulfonated diazo dye C.I. Reactive green 19A using micrococcus glutamicus NCIM-2168. Bioresour Technol 100(17):3897–3905. https://doi.org/10.1016/j.biortech.2009.03.051
75. Saratale RG, Saratale GD, Kalyani DC, Chang JS, Govindwar SP (2009b) Enhanced decolorization and biodegradation of textile azo dye Scarlet R by using developed microbial consortium-GR. Bioresour Technol 100(9):2493–2500. https://doi.org/10.1016/j.biortech.2008.12.013
76. Saratale RG, Saratale GD, Chang JS, Govindwar SP (2011) Bacterial decolorization and degradation of azo dyes: a review. J Taiwan Inst Chem Eng 42(1):138–157. https://doi.org/10.1016/j.jtice.2010.06.006
77. Saratale RG, Rajesh Banu J, Shin H-S, Bharagava RN, Saratale GD (2020) Textile industry wastewaters as major sources of environmental contamination: bioremediation approaches for its degradation and detoxification. Bioremed Ind Waste Environ Saf. Springer, Singapore, pp 135–167. https://doi.org/10.1007/978-981-13-1891-7_7
78. Sarkar S, Sarkar S, Das SS, Bhattacharjee C (2019) Colour removal from industrial wastewater using acid-modified tea-leaves, a domestic waste. Desalin WATER Treat 161188–202. https://doi.org/10.5004/dwt.2019.24236
79. Sarma GK, Sen Gupta S, Bhattacharyya KG (2019) Removal of hazardous basic dyes from aqueous solution by adsorption onto kaolinite and acid-treated kaolinite: kinetics, isotherm and mechanistic study. SN Appl Sci 1(3):211. https://doi.org/10.1007/s42452-019-0216-y
80. Satapathy MK, Das P (2014) Optimization of crystal violet dye removal using novel soil-silver nanocomposite as nanoadsorbent using response surface methodology. J Environ Chem Eng 2(1):708–714. https://doi.org/10.1016/j.jece.2013.11.012
81. Shanmugam BK, Easwaran SN, Mohanakrishnan AS, Kalyanaraman C, Mahadevan S (2019) Biodegradation of tannery dye effluent using Fenton's reagent and bacterial consortium: a biocalorimetric investigation. J Environ Manage 242106–113. https://doi.org/10.1016/j.jenvman.2019.04.075
82. Sharma H, Shirkot P (2019) Bioremediation of azo dyes using biogenic iron nanoparticles. J Microbiol Exp 7(1). https://doi.org/10.15406/jmen.2019.07.00232
83. Shoukat R, Khan SJ, Jamal Y (2019) Hybrid anaerobic-aerobic biological treatment for real textile wastewater. J Water Process Eng 29100804. https://doi.org/10.1016/j.jwpe.2019.100804
84. Singh L, Singh VP (2015a) Textile dyes degradation: a microbial approach for biodegradation of pollutants, pp 187–204. https://doi.org/10.1007/978-3-319-10942-8_9
85. Singh L, Singh VP (2015b) Textile dyes degradation: a microbial approach for biodegradation of pollutants characterization of dye degrading fungi and metabolites produced by some endophytic fungi view project textile dyes degradation: a microbial approach for biodegradation of pollutants. Springer 9783319109411:187–204. https://doi.org/10.1007/978-3-319-10942-8_9
86. Sonwani RK, Swain G, Giri BS, Singh RS, Rai BN (2020) Biodegradation of Congo red dye in a moving bed biofilm reactor: performance evaluation and kinetic modeling. Bioresour Technol 302122811. https://doi.org/10.1016/j.biortech.2020.122811
87. Srinivasan A, Viraraghavan T (2010) Decolorization of dye wastewaters by biosorbents: a review. J Environ Manage 91(10):1915–1929. https://doi.org/10.1016/j.jenvman.2010.05.003
88. Suzuki T, Timofei S, Kurunczi L, Dietze U, Schüürmann G (2001) Correlation of aerobic biodegradability of sulfonated azo dyes with the chemical structure. Chemosphere 45(1):1–9. https://doi.org/10.1016/S0045-6535(01)00074-1
89. Sweety (2018) Bioremediation of textile dyes: appraisal of conventional and biological approaches. In: Phytobiont ecosyst restit. Springer, Singapore, pp 459–487. https://doi.org/10.1007/978-981-13-1187-1_23.

90. Tao Y, Li F, Crittenden JC, Lu Z, Sun X (2016) Environmental impacts of China's urbanization from 2000 to 2010 and management implications. Environ Manage 57(2):498–507. https://doi.org/10.1007/s00267-015-0614-x
91. Tony BD, Goyal D, Khanna S (2009a) Decolorization of Direct Red 28 by mixed bacterial culture in an up-flow immobilized bioreactor. J Ind Microbiol Biotechnol 36(7):955–960. https://doi.org/10.1007/s10295-009-0574-3
92. Tony BD, Goyal D, Khanna S (2009b) Decolorization of textile azo dyes by aerobic bacterial consortium. Int Biodeterior Biodegr 63(4):462–469. https://doi.org/10.1016/j.ibiod.2009.01.003
93. Tsai W-T, Chen H-R (2010) Removal of malachite green from aqueous solution using low-cost chlorella-based biomass. J Hazard Mater 175(1–3):844–849. https://doi.org/10.1016/j.jhazmat.2009.10.087
94. Venkata Mohan S, Chandrasekhar Rao N, Karthikeyan J (2002) Adsorptive removal of direct azo dye from aqueous phase onto coal based sorbents: a kinetic and mechanistic study. J Hazard Mater 90(2):189–204. https://doi.org/10.1016/S0304-3894(01)00348-X
95. Wan G, Wang C (2014) Unprecedented urbanisation in Asia and its impacts on the environment. Aust Econ Rev 47(3):378–385. https://doi.org/10.1111/1467-8462.12076
96. Wang Z, Yin Q, Gu M, He K, Wu G (2018) Enhanced azo dye Reactive Red 2 degradation in anaerobic reactors by dosing conductive material of ferroferric oxide. J Hazard Mater 357226–234. https://doi.org/10.1016/j.jhazmat.2018.06.005
97. Wanyonyi WC, Onyari JM, Shiundu PM, Mulaa FJ (2017) Biodegradation and detoxification of malachite green dye using novel enzymes from bacillus cereus strain KM201428: kinetic and metabolite analysis. Energy Procedia 11938–51. https://doi.org/10.1016/j.egypro.2017.07.044
98. Zhou W, Zimmermann W (1993) Decolorization of industrial effluents containing reactive dyes by actinomycetes. FEMS Microbiol Lett 107(2–3):157–161. https://doi.org/10.1111/j.1574-6968.1993.tb06023.x
99. Zhu Y, Wang W, Ni J, Hu B (2020) Cultivation of granules containing anaerobic decolorization and aerobic degradation cultures for the complete mineralization of azo dyes in wastewater. Chemosphere 246125753. https://doi.org/10.1016/j.chemosphere.2019.125753
100. Zhuang M et al (2020) High throughput sediment DNA sequencing reveals azo dye degrading bacteria inhabit nearshore sediments. Microorganisms 8(2):233. https://doi.org/10.3390/microorganisms8020233

Mechanism and Techniques of Dye Removal by Microflora

Bishal Singh and Evangeline Christina

Abstract Dye is one of the integral parts of human civilization. It has been used in day-to-day life from prehistoric periods. There are more than ten thousand different types of dyes present in the market which are used in industries related to food, textile, paint, cosmetics, paper and pharmaceuticals. Most of the recent dyes are synthetic in nature and have xenobiotic, toxic, mutagenic and cancer-causing properties. There are various classes of dyes based on their chemical structure or based on their mode of action, some common classes of dyes which are used in industries are azo dyes, vat dyes, acidic dyes, basic dyes, reactive dyes, disperse dyes and others Theses dyes after being used in the various process are discharged to various water resources by various industries without proper treatment or by partial treatment, which leads to water pollution and affects the aquatic ecosystem and human health. Several conventional physicochemical methods based on principles of coagulation, membrane filtration, oxidation, reverse osmosis and others have been used but these methods are associated with several drawbacks related to cost, complexity, end product and efficiency. These limitations can be overcome by using biological methods in which various microflora having suitable properties are used. Methods using microorganisms are commercially viable, have a low initial investment, simple and ecologically suitable. Degradation of dyes by microflora can be achieved by either biosorption or enzymatic action. There are several oxidizing and reducing enzymes produced by microflora are used in dye decolourization with effective result. This chapter focuses on various biological methods for dye decolourization, advantages of using the biological method over conventional methods and the future in the field of dye removing by microflora.

Keywords Dyes · Bioremediation · Microflora · Biosorption · Decolourization

---

B. Singh
Department of Microbiology and Cell Biology, Indian Institute of Science, Bangalore, Karnataka, India

E. Christina (✉)
Department of Molecular Biology and Genetic Engineering, School of Bioengineering and Biosciences, Lovely Professional University, Phagwara, Punjab, India
e-mail: evangeline.23827@lpu.co.in

S. S. Muthu and A. Khadir (eds.), *Dye Biodegradation, Mechanisms and Techniques*, Sustainable Textiles: Production, Processing, Manufacturing & Chemistry,
https://doi.org/10.1007/978-981-16-5932-4_4

1 Introduction

Dyes are aromatic compounds which are used in our day to day life and have a deep impact on human civilization, it has been used in a various way and has a serious impact in our daily life. More than a million tons of dyes are synthesized and consumed in various industries every year. Dyes can be utilized to produce various goods from industries associated with paper, textile, leather food, cosmetics, pharmaceutical and paints [1]. Harmful effects of these dyes on the environment and human health have been found in several reports and studies, which showed toxic, allergic, mutagenic and carcinogenic nature of dyes to animal, plants and humans beings [2]. Due to partial degradation of azo dye (most commonly used dye in industries), aromatic amines are generated which are harmful to both animals and plants [3]. The hazardous nature of these dye in the environment is enhanced due to their association with other pollutants released by industries in water resources without proper treatment. In case of some dyes, they can combine with metals like copper, nickel, iron, boron, cobalt and others to form a complex compound which helps to provide them stability and increases the fixation process. These metal-dye complex are not only harmful in nature but also have the ability to accumulate and increase the organic load. The exotoxic nature and bioaccumulating properties of dye in aquatic life ultimately reach to humans through the food chain and lead to several diseases like skin allergy, dermatitis, skin cancer, asthma, respiratory tract irritation, kidney damage, liver damage, genetic mutation, photodynamic damage and several different types of cancers [4].

There are several techniques and approaches to degrade or decolourize dyes. some of the old, traditional, conventional and well understood and some are new, complex and need a good establishment. Most traditional methods are physiochemical methods which use the principle of flocculation, precipitation, oxidation, reduction, electrolysis, membrane filtration, reverse osmosis and others. These methods are almost understood, but have various drawbacks related to cost-effectiveness, source utilization, the large amount of sludge production, toxic compound production and harmful effect on environmental. These drawbacks of physicochemical approaches forced to develop new methods of dye degradation with the help of microbes. It is because these microbes can survive in any condition, have highly adaptive nature, consume fewer resources and energy and do not harm ecology. These microbes which are the microflora are isolated, identified and studied exhibit highly successful dye removal. It has been also reported that this microorganism produces several enzymes which have the potentials for dye degradation. It is considered that microbes are an alternative and highly effective solution for biodegradation of dyes globally. When these microflorae are coupled with other methods like membrane technology, their ability to degrade dyes will enhance more. The exploration in the field of genetics, metabolomics and proteomic not only enhances the existing approaches but also widens the field for biodegradation of dyes by microflora [5].

This chapter aims at reviewing the current trends in dye removal by microflora. Firstly, the major dyes used in industries and their harmful effects are discussed. The involvement of microflora in dye removal has opened a new area in biodegradation process of dye, with ecofriendly and highly effective results.

2 Major Dyes Used in Industries

Dyes are organic colourants, which are used in our day-to-day life from prehistoric periods. It has been used for various purposes from food additive to textile industries. These can be natural or synthetic, the first evidence of using dye was dated 4000 years ago in Egyptian mummy which was indigo [6]. More than ten thousand of different dyes are being used in industries related to textile, paper, food, leather, paint, pharmaceuticals and cosmetics. Due to advancements in techniques and competition in market industries are mostly using synthetic dyes like azo dyes, direct dyes, reactive dyes to develop a product which is more attractive to customers. In the last decade, various environmental issues have been reported due to dye which has affected several textile dye industries.

Dyes isolated from natural sources require mordants during their application as they have low binding affinities. This reason for natural dye makes them not suitable for industrial applications [7]. Whereas synthetic dyes are derived from natural dyes and are modified version, which are most commonly used in industries. The first synthetic dye was produced in 1871 as picric acid from natural dye indigo. Classification of dyes can be done based upon the chemical structure, chromophore present and their actions. Based upon the chemical structure it can be classified as dyes with Azo group, dyes with Nitro group, dyes with Nitroso group, dyes with Indigo group, dyes with Phthalein group, dyes with Triphenylmethane group and dyes with Anthraquinone group as given in Table 1, and based upon the mode of action as Azoic dyes, Vat dyes, Direct dyes, Dispersed dyes, Sulphur dyes, Reactive dyes, Acidic dyes and Basic dyes.

Table 1 Summarizes the class of dyes based on their chemical structure along with examples, molecular weight and C.I. number

S. No.	Class	Examples	Molecular wt (g/mol)	C.I. number
1	Azo	Acid red B	446.40	14,720
2	Nitro	Mordant green 4	277.23	10,005
3	Indigo	Indigo	262.27	10,215
4	Anthraquinone	Acid blue 25	416.40	62,055
5	Triphenylmethane	Crystal violet	407.97	42,555
6	Phthalein	O-cresolphthalein	346.40	68,995
7	Nitroso	Picric acid	229.10	10,305

Table 2 Summarizes the classification of azo dyes based upon the number and type of linkage

S. No.	Classes	Example	Molecular wt (g/mol)	C.I. number
1	Monoazo	Methyl orange	327.33	13,025
2	Diazo	Red ponceau S	760.6	103,116
3	Triazo	Direct blue 71	1029.87	34,140
4	Poliazo	Direct red 80	1373.05	35,780
5	Azo lakes	Lithol rubine Bk	424.44	15,850
6	Napthol	Napthol yellow S	358.19	10,316
7	Benzimidazolone	Benzimidazolone yellow H3G	405.33	11,781

2.1 *Azoic Dyes*

Dyes which contain azo linkage in their chemical structure are called as azo dyes. The number of azo linkages can be different in numbers so, the classification of azo dye is also done based on the number of azo bonds present in the molecule as monoazo dyes, diazo dyes, triazo dyes and polyazo dyes. These dyes are considered as the largest class among dyes which consist of seventy percent among all dyes used in industries related to food, textile, leather, paper, cosmetics, paint and pharmaceuticals [8]. Azo dyes can be obtained naturally and synthetically in diverse forms. According to a report published by [9], there are more than two thousand different types of azo dyes used in industries and more than a million-ton azo dyes are produced yearly. The simple coupling reaction of azo dyes, diverse structural variations, suitable adaptability property based on different application makes them one of the suitable dyes for industrial application. More than fifty thousand tons of these dyes are discarded from various industries in environment yearly. Azo dyes on partial degradation are transformed to aromatic compound like azoamine. These azoamines are harmful and toxic in nature. Some common examples of azo dyes are Allura red AC, Alcian yellow, acid orange, acid red, etc (Table 2).

2.2 *Vat Dyes*

These are the dyes named based on the method of application in textile industries. This dye is very effective in colouring fibres made from cellulose, mostly cotton. They show the property of unparalleled fastness in context to various agents like detergent, bleach and light, which is due to the insoluble nature of the vat dye in water. When these dyes are applied for other fibres which may be synthetic, shows no effective result. The binding affinity of vat dye with fibres is due to their selective diffusion behavior. This selective nature of this dye arises nonuniform shades when used in synthetic fibres.

The best example of vat dyes is indigo, which is a natural dye, isolated from the indigo plant Indigofera. There are various derivatives of indigo dye based on the modification by halogenation process which is further classified in various classes as indigoid, thiondigoid, anthraquinones and others. The preferable characteristics of vat dyes are light and wet fastness which is suitable for textile industries.

2.3 Sulphur Dyes

Sulphur dyes are usually utilized in paper and textile industries. These dyes were first reported in 1966 and were found very effective for cotton and cellulose fibres. It is reported that the annual production of sulphur dye is 120,000 tons which is the highest percentage of production among groups of dyes. Black sulphur is the most commonly used dye for cellulosic fibres [10]. Classification of sulphur dye is based on the method used for sulfurization and the initial material used [11]. The initial compound during the synthesis of sulphur dyes includes mostly aromatic compounds like benzene, azobenzenes with a functional group like hydroxy, nitro, nitroso and others. Whereas the method of sulfurization includes reactions like oxidation, reduction, substitution and ring formation. Sulphur dyes are commonly used for fibres made with cellulose, synthetic fibres blend with cotton, some paper industries and leather industries [12]. These dyes are cost-effective, excellent in light and wet fastness ensures the wide application and heavy use of these dyes.

2.4 Acidic Dyes

These dyes are predominantly used and have higher efficiency in acidic environmental condition at the pH range of 3–7, these are generally used along with formic or acetic acids. The acidic strength of environment applied depends upon the individual property of dye as these dyes are diverse in nature. Some of the acidic dyes have metal complexes and are mostly used to colour nylon, wool and silk [13]. The binding property of acidic dyes depends on the sulfonated group, which makes them polar in nature and water-soluble. The interaction between dyes and wool is due to ionic force and Vander wall force, these forces of interaction provide the fastness property to these dyes. Indian ink, Nigrosoin and Congo red are the most common examples of acid dyes.

2.5 Basic Dyes

Basic dyes are mostly used to impart colour to paper, nylon, acrylic polymers and some modified polyesters. As these dyes have poor mobility they are often used with retarders. These are polar dyes and can easily dissolve in water. It produces cations which interact with the fibres and polymers by electrostatic force. These dyes can be positively charged or can have delocalized charge as present in triarylmethane, xanthenes and acridine dyes. Methylene blue, Thionine and Crystal violet are the most common examples of this class.

2.6 Disperse Dyes

These dyes are synthetic in nature which targets hydrophobic compounds [11]. These are mostly used as a commercial mixture in large amount along with the large quantity of water, as they are insoluble or partially soluble in nature, most of the dyes remain in the water bath and generate a large volume of wastewater. Due to non-ionic nature of these dyes, they form an aqueous dispersion and are commonly used for the substrates like polyester, nylon, acrylic fibre, cellulose acetate [14].

2.7 Reactive Dyes

Dyes with the functional group and polar nature are known as Reactive dyes. They mostly contain functional group like dichloro-s-triazine which can make the covalent bond with their substrates when applied in textile industries [15]. These dyes show excellent light and wash fastness on cotton which makes them desirable for industrial uses [16]. These dyes are commonly used to stain wool, cotton and nylon. One of the major problems associated with these dyes is poor dye fixation.

3 Harmful Impact of Dyes in the Environment

Dyes are used in various manufacturing industries and the waste generated from it are discarded in water resources. These dyes are causative agents for environmental degradation and human diseases. It not only affects ecological niches but also harms water resources and animal health [4]. The dyes are recalcitrant in nature and they bioaccumulate, sediment and then gets mixed up in the public water supply chain [17]. In case of partial degradation of azo dyes, aromatic azoamines are generated. These amines are toxic and cancer-causing in nature [17]. Xenobiotic nature of some dyes impacts the ecosystem and hinders the structure and functionality of both aquatic flora, fauna and human health when it is exposed for a longer duration. In most cases, dyes are used with metal complex for their higher binding affinity. Metal associated which have a higher half-life and contains metals like copper, nickel, cobalt, chromium and lead [18]. when these metal associated dyes are relished in an aquatic habitat, it is bioaccumulated in fish tissues and ultimately reach to the human organ through food chain causing various pathophysiological condition [4]. It is found that chromium present in the dye causes oxidative stress to plants and harm the growth by affecting photosynthesis and carbon dioxide assimilation (Fig. 1; Table 3).

Synthetic dyes such as 2-Naphtylamine and benzidine, used in textile industries are responsible for higher incidence and prevalence rate of disease like bladder cancer [19]. The harmful effect of these textile dye range from dermatitis to problem related to the CNS [4]. Some dyes can affect at the cellular level where it can hinder the function of enzymes. In the case of textile workers, dermatitis, rhinitis, asthma, conjunctivitis and other allergic reactions are seen due to exposure with dye [11]. The genotoxicity of these dyes can lead to chromosomal aberration and have the potential for mutagenesis [11]. Dyes like Azure-B, commonly used in industries

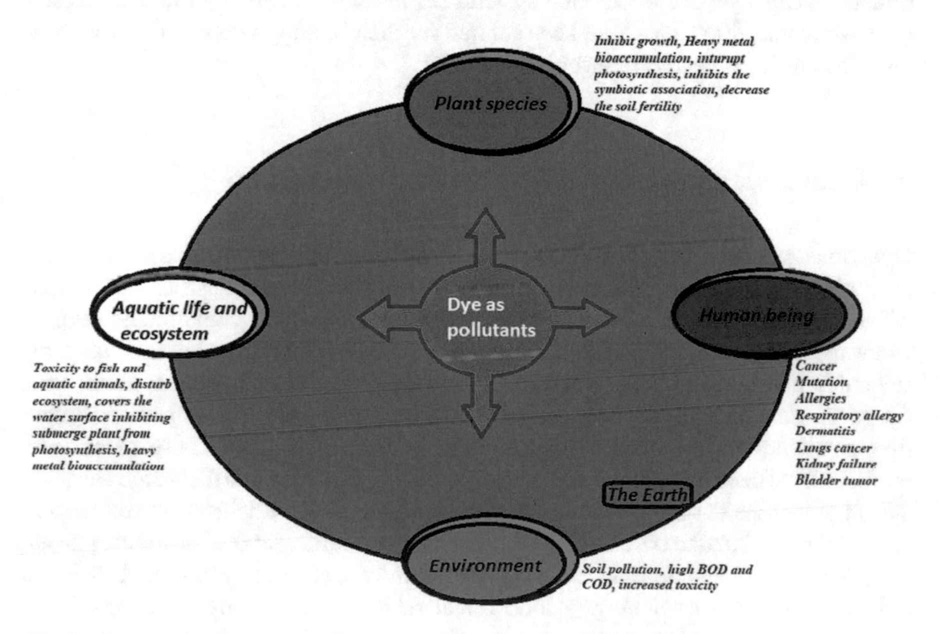

Fig. 1 Shows the harmful effect of dyes to various habitat and living organism

Table 3 Summarizes the metals associated with dyes and examples related to it

S. No.	Metal in complex	Example	C.I. number
1	Iron (Fe^{2+})	Phthalocyanine	23,925
2	Nickel (Ni)	BDN	691,182
3	Boron (B^{3+})	Borondibenzopyromethene	131,818
4	Cobalt (Co^{3+})	Acid black 180	13,710
5	Copper (Cu^{2+})	Reactive blue13	181,575
6	Chromium (Cr^{3+})	Acid black 172	23,976

can intercalate DNA structure, interact with the cell membrane and other cellular organelles. This dye also affects the function of cellular enzymes like monoamine oxidase A and glutathione reductase which plays important role in human behaviour and cellular homeostasis respectively. Dispersed Red 1 dye shows the mutagenic property in hepatoma cells and in human lymphocytes during several studies. Orange 1 dye can induce DNA damage by supporting basepair substitution and a frameshift mutation.

Most azo and the nitro group-containing dyes have shown neoplastic as well as carcinogenic properties some of the examples are Sudan I cause neoplastic liver nodule, Red 9 causes sarcoma in the liver, mammary gland, bladder and hematopoietic system, crystal violet can cause mitotic poisoning which leads to hepatocarcinoma, uterus, ovary and bladder carcinoma [20]. Various dyes have significant effects on the digestive tract, renal system, respiratory system and multiple organs [19]. More than 4000 dyes have been studied for their toxicity out of which 100 dyes have the ability to develop cancer in human [21].

4 Conventional Methods for Dye Decolourization

Dyes released from various industries as an affluent have various harmful effects on both ecosystem and human health. In most of the cases, industries discharged effluent without proper treatment or partial treatment which contaminate several water resources. There are several conventional physiochemical processes used by several industries to treat their discharged some of them are adsorption, membrane filtration, ozonization, electrochemical oxidation, photo-electrocatalysis, coagulation, flocculation, advanced oxidation and photocatalysis [22]. These mentioned methods have been proved effective in dye removing process from discharge wastewater in industries but they have several drawbacks associated with these techniques which can be related to cost, complex infrastructure, inefficient results, toxic byproducts, amount of sludge produced and production of a secondary pollutant. Several industries use two or more physiochemical methods commonly to reduce these drawbacks and get effective results. In spite of combining various physicochemical methods, the results obtained at the end are not promising due to recalcitrant and resistive nature of dye to the degradation process [23]. In the present scenario, the need and based on futuristic approach wastewater treatment process are not only focusing towards the quality of ecosystem, human health but also they are concerned towards utilizing and reusing the treated wastewater in day to day life to fulfil the growing demand of water. These physiochemical processes of dye degradation resulted in the production of aromatic amino compounds from azo dyes. These azo amines have the potential to induce mutation can alter human DNA and can lead to various disease like cancer. Some approaches to conventional physicochemical methods used in dye decolourization are explained briefly (Fig. 2).

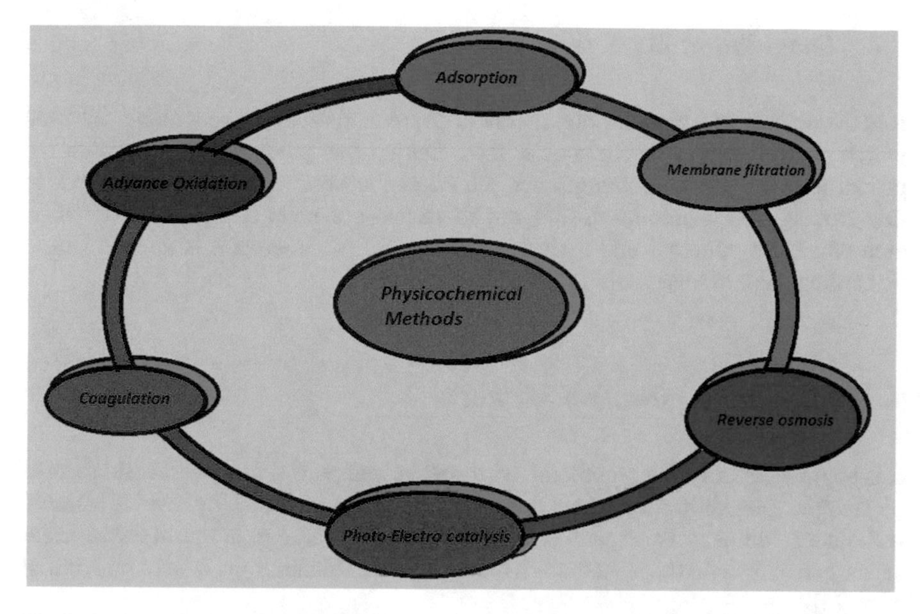

Fig. 2 Shows different physicochemical methods used in dye removing from waste water

4.1 Adsorption Methods

It is the process which is based on the principle of adsorption where activated charcoal or carbon are used along with cobs, sawdust or other absorbing materials to remove dyes. In this process, the efficiency of the result depends on the quality of absorbents used and based on the quality of activated carbon. After the process of dye removal, the absorbent is a pollutant that needed to be disposed of properly [24].

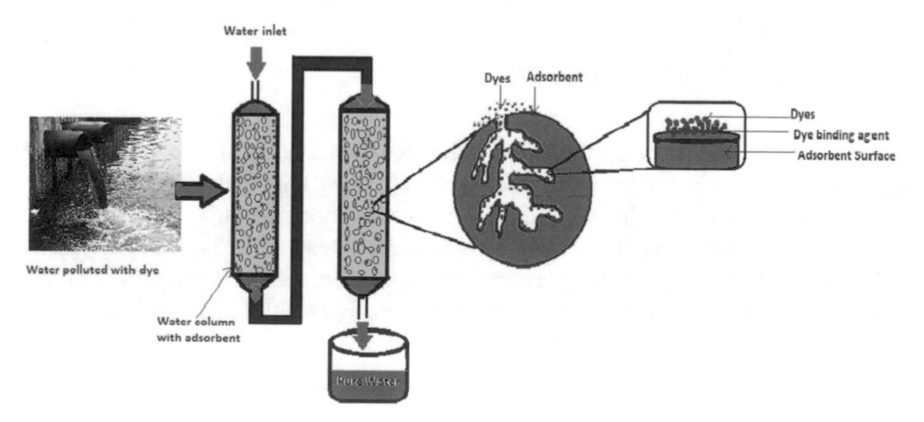

Fig. 3 Shows the technology and principle of adsorption process for purification of dye contaminated water

4.2 Coagulation Methods

It is the process in which various chemical compounds such as coagulants are used which generally form precipitate or flocs. During this process, the flocs with the proper size, weight and strength are settled at the bottom which is removed. In this process, a large amount of sludge with various chemical compounds are generated which are pollutant and toxic in nature so further treatment is needed before discarding those sludge [23].

4.3 Membrane Filtration Methods

It is the process in which membrane with specialized pore size is used which works on the principle of filtration where dyes are filtered by membrane by reverse osmosis techniques. In this process nano or microfiltration technology are used for efficient separation of dye. As these methods include advanced technology, it has a high initial cost [25].

4.4 Oxidation Methods

It is the process in which dyes are oxidized to degrade or transformed into various state, with the help of various methods and compounds like ozone, hydrogen peroxide, Fenton's reagent, photolysis with ultraviolet light and sonolysis. The major drawback of these methods is limited lifetime, ineffective for insoluble dyes and produce a large amount of sludge [23].

4.5 Electrochemical Methods

There are several electrochemical approaches for dyes degradation but most of the common approach is electrochemical oxidation of dyes, where electrochemical cell is used along with electrodes to oxidize the dye at the anode. As this process depends on the various parameters like pH, minerals content, temperature, the voltage applied, the concentration of dyes, anode material and operating condition, it is difficult and complex to standardize these processes for an effective result [22].

5 Involvement of Microflora in Dye Removing

5.1 Bacteria

Bacteria are tiny microorganisms with higher capabilities in fields of bioremediation. Various studies have been done to identify bacteria with the capabilities which can decolourise dyes present in the environment [26]. These types of bacteria are present and can be isolated from various niches like soil, water, animal excreta and contaminated food. These microorganisms having high efficiency in the process of decolourization and degradation of dyes are easy to culture and maintain [27]. Bacteria can decompose the dyes either by aerobic or anaerobic mechanism or both to give an effective result [28]. As these microbes are useful for a wide range of dyes, eco-friendly in nature, cost-effective and produce less sludge are considering as one of the effective methods for dye bioremediation.

5.1.1 Pure Culture of Bacteria

Bacteria used in decolourization process of dyes mostly used their enzyme which breaks several bonds in dye and helps in the degradation process. Harmful aromatic azoamine which are generated during the degradation of dyes in case of the physico-chemical process is completely removed in aerobic or anaerobic condition by bacteria [29]. Most of the common type of dye used in industries are azo dyes and these are found resistive in nature during bacterial degradation process in aerobic condition as presence of oxygen prevents breakage of the azo bond [30] but there are several bacteria like *Bacillus subtillis, Bacillus cereus, Aeromonas hydrophillia, Acinetobacter sp., S. hominis, S. aureus* which can degrade azo dyes effectively [31–33]. In some cases, bacteria have developed oxygen insensitive enzymes like azorductase which can degrade azo dyes even in an oxygen-rich environment [34]. Extremophiles can grow and degrade dyes in extreme environmental' conditions with high pH, temperature, salinity and presence of xenobiotic compound. Some bacteria like *Staphylococcus Exigubacterium, Aeromonas hydrophila* can degrade dyes in high salt concentration, *Geobacillus stearothermophilus* UCP 986 and *Bacillus badius* can degrade dyes in high temperature [28]. There are several bacterial species identified and studied which have dye degradation ability which is briefly given in the supplementary Table 4.

5.1.2 Consortium of Bacteria

Utilization of single bacteria species might not provide suitable result but the use of bacteria consortium gives effective result in biodegradation and mineralization process. In this process, multiple bacteria play a synergistic role in dye degradation [36]. The consortium is made of prepared either by a combination of two or

Table 4 Summarizes different types of bacteria associated with dye degradation along with their efficiency, suitable environmental condition and mode of action

Common bacteria	Process and enzymes used	pH, Temp (°C)	Common dyes used	Decolourization efficiency (%)	References
Rhodopseudomonas palustris	Anaerobic condition	– , 25	Reactive red	100	[25]
Acinetobacter baumanni	Oxidation, reduction process	7, 37	Congo red	99.10	[26]
Alcaligenes sp.	Azoreductase enzymes	7, 25	Reactive red	90	[27]
Alishewanella sp.	Azoreductase and DCIP reductase enzymes	7, 37	Reactive blue	95	[34]
Bacillus cereus	Aerobic condition	7, 35	Cibacron red	67–81	[29]
B. halodurans	Aerobic biodegradation	9, 37	Acid black	90	[3]
B. pumilus	Anaerobic degradation	8, 30	Navy blue	95	[2]
P. aeruginosae	DCIP reductase enzymes	7, 40	Remazol red	97	[3]
B. subtilis	Unknown method	8.5, 35	Reactive red	97	[29]
Acinetobacter sp.	Microaerophilic condition	7, 30	Dispersed orange	90.20	[30]
B. laterosporous	Oxidation, reduction	7, 30	Golden yellow HER	87	[4]
Comamonas sp.	Oxidation process	6.5, 40	Direct red 5B	100	[3]
E. faecalis	Oxidation and reduction process	7, 40	Direct red	100	[32]
F. stearothermophilus	Aeration condition	5.5, 50	Orange II	98	[32]
M. glutamicus	Oxidation and reduction process	6.8, 37	Reactive green	100	[35]
S. hominis	Unknown reason	7, 35	Acid orange	94	[32]
M. luteus	Unknown reason	8, 37	Direct orange	96	[36]

(continued)

Table 4 (continued)

Common bacteria	Process and enzymes used	pH, Temp (°C)	Common dyes used	Decolourization efficiency (%)	References
P. entomophilla	Azoreductase enzyme	5–9, 37	Reactive black	93	[4]
S. aureus	Azoreductase enzymes	– , 37	Sudan III	97	[31]
Bacillus sp.	Aerobic degradation	7, 40	Congo red	85	[2]
Shewanella sp.	Aerobic condition	– , –	Acid orange	98	[36]

more species of bacteria, fungi or both [35]. The harmful compounds generated as the intermediate product during the degradation process are transformed into nontoxic compounds as complementary bacteria used in this process. Some bacterial consortium like *Enterobacter cloacae* and *Enterococcus casseliflavus* combinedly degrade Orange II in 15 min [38], *Bacillus odysseyi, Morganella morgani* and *Proteus sp.*, combinedly can degrade reactive blue in less than 3 h [39], *Bacillus flexus, Bacillus cereus, Bacillus cytotoxicus* can degrade Direct Blue 151 and Red within 5 days. In the same way, other microorganisms which can be used to make consortium are *Stenotrophomonas acidaminiphila, Pseudomonas fluorescence, Bacillus cereus, Pseudomonas putida, Micrococcus luteus, Paenibacillus polymyxa, Providencia sp., Pseudomonas aeuroginosa, Arthrobacter, B. cerreus, Pseodomonas sp., B. megaterium, Rizobium, M. glutamicus* [40–42]. These bacteria in the consortium can degrade and decolourize various dyes like Acid Red 88, Reactive Violet 5R and others.

Table 5 Summarizes different types of bacterial consortium involves in dye degradation along with their dyes substrate, efficiency, suitable environmental condition and mode of action

Common consortium	Mechanism and enzymes used	pH, Temp (°C)	Common dye used	Efficiency (%)	References
A. Caviae, Protues mirabilish and *R. globerulus*	Unknown	7, 37	Azo dyes, acid orange	90	[29]
Enterococcus casseliflavus, E. cloacae	Aerobic condition	7, 37	Orange II	100	[38]
Pseudomonas, Rizobium and*Arthrobacter*	Aerobic condition	7, 37	Acid orange	100	[41]
Proteobacteria sp.	Azoreductase	– , –	Reactive blue	55.50	[35]
P. vulgarish, M. glutamicus	Degradation	3, 37	Reactive green	100	[36]
B. Megaterium, B. vallismortis, B. cereus, B. pumilus, B. subtilis	Aerobic degradation	– , 37	Congo red, acid blue	70–96	[40]
P. vulgarish, M. glutamicus	Reduction	7, 37	Scarlet R	100	[36]
Penicillium sp. Exiguobacterium sp.	Anaerobic degradation	7, 37	Reactive dark blue	97	[42]
Pseudomonas sp., Aspergilus ochraceus	Tyrosinase, azoreductase	– , 30	Reactive navy blue	80	[41]
B. odysseyi, M. morganii, proteus sp.	Lignin peroxidase, laccase	– , 30	Reactive blue	100	[39]

5.2 Fungi

The process of bioremediation with the help of fungus is known as mycoremediation. There are several fungi which have the capability to degrade various dyes either by biosorption or by utilization of several enzymes. Broadly fungi used for degradation or decolourization of dyes are classified as filamentous fungus and yeast. Utilization of fungi in the field of bioremediation has been initiated and various study has been done to identify different species and strains of fungi which are important in this approach.

5.2.1 Filamentous Fungi

The process of biodegradation of dyes with the help of fungi has been studied and found that various soluble, insoluble, phenolic and nonphenolic dye can be decolourized effectively by them [43]. It has also been demonstrated that degradation of aromatic dyes by fungus is the secondary metabolic response when there is a lack of nutrient source, fungi used these compounds as an alternative resource of energy. These fungi can convert various organic compounds with the help of multiple cellular

Table 6 Summarizes different types of fungi associated with dye degradation along with their dye substrate, efficiency, suitable environmental condition and the mechanism used by them

Common fungus	Process and enzymes used	pH, Temp (°C)	Common dye used	Efficiency (%)	References
Trametes sp.	Laccase and Mnperoxidase enzymes	4.5, 30	Solar brilliant red	100	[46]
Aspergillus niger	Degradation process	3, –	Congo red	99	[47]
G. oxysporum	Aerobic, degradation process	– , 24	Yellow GAD	100	[43]
C. elegans	Adsorption process	5.6, 28	Reactive orange, red and black	93	[51]
Ganoderma sp.	Laccase enzyme	5.5, 28	Malachite green	75–96	[50]
Armillaria sp.	Laccase enzyme	4, 40	Reactive black	80	[44]
P. chyryosporium	Lignin peroxidase enzyme, degradation process	4.5, 30	Direct red	100	[45]
Q. eryngii	Laccase and lignin peroxidase enzymes	3, 40	Reactive black	94	[44]
P. Sajor-caju	Adsorption, degradation process	– , –	Reactive blue	100	[8]
Coriolus versicolor	Degradation process	– , 30	Acid orange	85	[50]
T. trogi	Laccase and Mnperoxidase enzymes	– , –	Orange	100	[48]

enzymes by the process of conversion reaction which includes hydroxylation of dyes. There are various filamentous fungi having the capability of degradation of various dyes, some of them are like *P. chryososporium*, *Curvularia lunata* can degrade Indigo dyes up to 95%, *Hypocrea koingii* can degrade five dyes including reactive violet, Red–black, Dark Navy and others. There are other fungi which produce several oxidizing and reducing enzymes which degrade dyes in a non-specific manner. Most common examples of enzyme-producing fungi are *Trametes sp., Armillaria sp., P. chyryosporium*, white-rot fungus and ligninolytic fungus of basidiomycetes class [44–46].

It is well understood and accepted that fungi are very efficient in the bioremediation process of various dyes, where they use various approaches like biosorption or enzymatic degradation of dyes [47]. The efficiency of the result is affected by various factors some can be abiotic like pH, temperature, carbon source, time, nutrient accessibility, salt concentration, oxygen concentration and nitrogen sources [48]. it has been also found that utilization of filamentous fungi for textile effluent containing synthetic dye in large amount cause problem sometimes.

5.2.2 Yeast

Yeast is also a type of fungi which have a higher impact on the process of biodegradation and fermentation. Yeasts have adopted various methods for decolourization of various dyes some of the approaches are adsorption based and some are associated with enzymatic degradation [49]. It has been found that utilization of yeast in the biodegradation process has advantageous effect over utilization of bacteria during the process as the yeast can grow rapidly as bacteria and can also grow in adverse environmental conditions [50]. Some dye-degrading fungi are *Candida zeylanoides* can degrade several azo dyes, Ascomycetes yeast species like *Candida tropicalis* (Violet 3), *Debaryomyces polymorphous* (Reactive Black 5) and *Issatchenkia occidentalis. Candida oleophila* can degrade reactive black [8]. Most suitable condition for yeast for degradation of dyes is an acidic environment, as it has been observed that various yeast like *Candida albicans* degrades Direct violet dye mostly at 2.5 pH, *Candida tropicalis* degrade violet 3 at 4 pH with best results.

5.3 *Algae*

Algae are one of the most influencing living micro-flora which have a higher impact in process of bioremediation which is highly used to treat various textile effluent containing various pollutants among which dyes are one of the major parts. It has been proved from various study that algae can offer a solution for the global environmental problem. In most of the cases algae used three different approaches for decolourization of dyes where they utilize chromophore to develop biomass, it has been seen

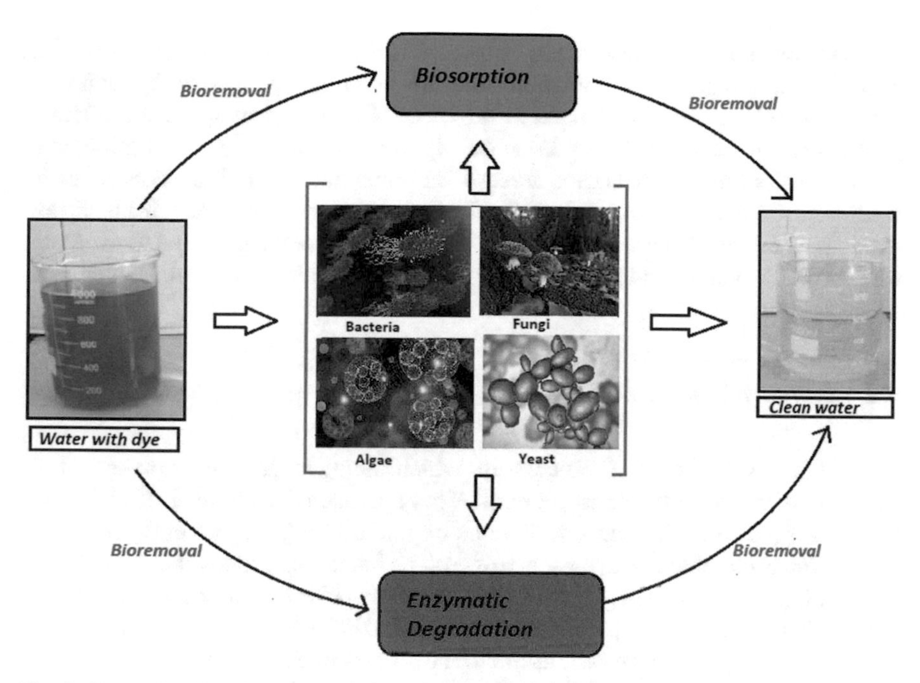

Fig. 4 Shows the role of microflora in dye decolourization

that algal biomass and growth is not inhibited due to presence of pollutant-like dyes in an industrial effluent [52].

Algae decolourizes dyes by the process of adsorption or enzymatic reaction. Dye degrading algae mostly belongs to Blue-green algae, diatoms and green algae groups. Mostly *Oscillatoria* and C*hlorella* have higher potential for colour decolourization and production of CO_2 and H_2O [53]. In some case of algae they produce enzymes as azoreductase which breaks the azo bond present in azo dyes, some species of algae like *Oscillatoria curviceps* produces enzymes like azoreductase, polyphenol oxidase and laccase which can degrade acid Black dye, *Chara* and *Scenedesmus obliquus* from green algae can also enzymatically degrade Congo Red and Crystal violet, *Chara vulgaris* and *S. quadricauda* can degrade a large range of textile dye when they are immobilized on alginate [54]. In the case of bacterial and fungal bioremediation process of dyes, various supplements are needed to be supplied like carbon source, oxygen but in the case of algal bioremediation, no extra supplement is needed.

6 Methods of Dye Removal by Microflora

The biological process of dye degradation and decolourization is considered as one of the best approaches in removing dyes from textile effluent, industrial waste and contaminated water resources. This approach utilizes various micro-flora like bacteria, fungi, algae or combination of all for an effective outcome by degrading almost every type of dyes. In most of the cases of a biological approach, microorganisms remove dyes by two approaches which can be either by biosorption or by biodegradation. In both scenarios, it has been found that utilization of microbes for wastewater treatment are ecologically suitable and requires a minimum initial investment. The biological approach has been also found to have the potential to overcome various disadvantages and drawbacks possessed by the physicochemical approach.

6.1 Biosorption by Microflora

Biosorption is the natural process in which living organisms uptake certain compounds and accumulate as biomass. This approach for removal of toxic dyes from water resources is successfully done by microflora [55]. The phenomenon of biosorption is possible due to presence of various functional groups such as –COOH, –OH, $-PO_4$, $-NH_2$ and others in the lipids and heteropolysaccharides, in the cell wall of microflora which makes them polar in nature and helps in interaction with charged group of dyes [2]. There are various microbial species like *Corynebacterium glutamicum* can absorb Reactive Red 4 dye, *Bacillus weihenstephannsis* can remove congo red by this approach'. Similarly, fungi are also used as dye decolourizer by this approach. In addition to these algae are considered as one of the best biosorbents of dyes with highly effective result. The choice of algae for dye degradation is due to their availability in diverse habitat, cell wall with higher surface area and with higher interactive affinity with dyes, which can attract dyes electrostatically and forms different types of complex [56]. In case of dead algae, there is no nutrient demand and can be used, stored for longer duration and can be regenerated with the help of organic solvents when needed so, they are considered more effective tools in comparison to living algae [57]. In case of living algae large amount of nutrient sources are required to sustain the development and physiological process, to do so heavy investment are required to maintain them in bioreactors of wastewater treatment plants but once they are dehydrated or dried their physiological process is stopped at that environmental condition where they either form spores or adapt different mechanism to sustain that environment. In this condition, these algae can be regenerated when they are required in bioreactors by providing suitable environmental condition. These inactive forms of algae are easy to manage as they do not require the extra resources of nutrient and have higher storage life due to their adaptability to environment, so these forms are considered as highly effective tools in process of dye degradation where they can be used whenever they are needed.

The phenomenon of biosorption not only depends on the charged groups present in the cell wall of microbes but also on the pH of the media or water containing dyes, temperature of the environment, ionic strength of the solution, the time of contact in between dye and the microorganism, the material used as absorbent, type and structure of dyes present in contaminated water, the concentration of dyes, inhibitors present in the water and the type of microorganism used for degradation process [51]. This process of dye removal is selective, cost-effective, efficient and work in low concentration with very appreciable results. These all benefits of biosorption method for dyes degradation proved it to be a better approach over presently existing physiochemical methods. Only drawbacks associated with this method is an early saturation and no control over the process. In the process of biosorption, the microflora used for dye degradation have low volume capacity, i.e. they can only adsorb very less amount of dye due to higher surface by volume ratio leading to less saturation time of the bed, which also result in frequent replacement of bed causing increase cost and effort in dye removing technology [58]. Different microflora uses different methods for Biosorpion, as these all methods depend upon the type of microorganism used, their physiological condition and other abiotic factors. These all biotic and abiotic factors cannot be fully controlled in an in-vitro condition like bioreactors and thus can not be fully controlled [59].

6.2 *Enzymatic Decolourization by Microflora*

One of the most effective approaches in dyes decolourization and degradation is the utilization of various enzymes for wastewater treatment. There are already coexisting several physiochemical approaches which work on principles of coagulation, adsorption, chemical treatment and ionic extraction for removing dyes but due to their various drawbacks and problems liked with them related to cost, toxic compound as a byproduct. In the case of the physicochemical process, a large amount of sludge is generated which also pollutes the environment. Enzymatic methods became an alternative approach for removing several dyes. There are several enzymes produced by various microbes which break bonds of dyes and showed various chemical reaction for dye degradation [37]. The enzymes used for dyes decolourization are broadly divided as reducing and oxidizing based on their action (Fig. 5).

Fig. 5 Shows the enzymatic degradation of dyes from waste water

6.2.1 Reducing Enzyme

Enzymes are proteins made from different amino acids and have catalytic activity for various physiological processes. These enzymes which are effective in dye decolourization are mostly produced by microbes like bacteria, algae, fungi and yeast [28]. These reductase enzymes are highly effective in reduced conditions. one of the best examples of these enzymes is azoreductase which breaks azo bond present in azo dye and breaks the compound into aromatic amines which further degrade into nontoxic compounds. The function of these reductase enzymes depends on various reducing equivalents like Nicotinamide adenine dinucleotide phosphate (NADPH), Nicotinamide adenine dinucleotide hydride (NADH) and Flavin adenine dinucleotide (FADH). It has been also found that in various situation dyes are bound with several other functional groups leading in increase molecular weight which inhibits the transportation of those compound through cell membrane and in that case it has been found that those complex forms of dye are degraded by reductase enzyme which suggests that decolourization of dyes with enzymes do not depend upon the effectiveness of cellular intake of dyes [60]. There are various enzymes identified based upon reducing equivalence and are called as NADH-DCIP reductases, FMN-dependent reductase, NADH-dependent reductases, NADPH-dependent reductase and FMN-independent reductase. Whereas it has been also found that NADHDCIP reductase is a marker enzyme of bacteria and fungi having the potential to degrade various xenobiotic compound [61].

6.2.2 Oxidizing Enzymes

In the same way as reducing enzymes degrade dyes by the process of reduction, oxidizing enzymes decolourize the dye by an oxidation process. There are several oxidizing enzymes like cellobiose dehydrogenase, dye decolourizing peroxidase, laccase, tyrosinase, lignin peroxidase, N-demethylase, manganese peroxidase and polyphenol oxidase secreted by microbes which are effective in dyes degradation. These oxidizing enzymes are found in different species of bacteria, filamentous fungi, yeast and algae, where they breakdown various dyes during their metabolic process [32]. These microbes convert the dye into less toxic product and then remove from wastewater in form of radical and insoluble product [62]. one of the most common examples of oxidizing enzymes is peroxidase, which is iron-containing enzyme and generally found in the various microorganism. In the same way lignin peroxidase and horseradish peroxidase found in *Penicillium chrysosporium* can oxidize Methylene blue and Azure B dye. In recent studies, versatile peroxidase has been isolated and placed with lignin peroxidase and manganese peroxidase in ligninolytic peroxidase family [63]. Versatile peroxidase can oxidise Mn^{2+} to Mn^{3+}, whereas laccases are copper oxidizing enzyme and can degrade copper-containing dyes, these are nonspecific in nature and can oxidize various dyes even in absence of electron acceptors like oxygen. Enzymes like polyphenol oxidase have a tetrameric structure with four copper atoms and two binding sites in each molecule of tetrameric structure, which helps in removing various dyes like from textile effluents. Multiple bacteria secreting these type of enzymes are *G. geotrichum*, *B. laterosporus*, consortium GG-BL which uses different types of the enzyme for the degradation of multiple dyes.

7 Advantages Due to the Microflora Over Conventional Physicochemical Methods for Dye Degradation

There are several conventional physicochemical methods used by several industries for decolourization and degradation of various dyes which are released as an effluent which is latter discharged in water resources after proper treatment. Traditional methods for treatment of waste dyes can be based on membrane filtration, adsorption, coagulation, flocculation, oxidation and electrochemical. As these methods have various drawbacks related to them, which is overcome by biological methods where various microorganism are used for dyes degradation. It has been seen that the utilization of these microflorae is highly cost-effective as the operating cost, infrastructure cost and maintenance of resources are less [64]. In the case of physicochemical methods used for removing dye are economically not viable, as they need well-established infrastructure, advanced equipment and have high maintenance cost [36]. In the case of physicochemical methods, the large volume of sludge is generated as an end product which is difficult to manage and dispose but in case of the biological method this challenge is not faced.

When these methods are compared based on their impact on the environment it is found that during the processing of dyes with physicochemical methods, various toxic intermediate products and final products are generated due to partial degradation of lower efficiency. They also produce a large volume of sludge which act as secondary pollutant contaminating various natural resources and aquatic life [65]. Whereas in case of microbial approach they are considered eco-friendly as living microorganism or enzyme released by microorganism are used in these methods for decolourization of dye without production of any toxic substance [64]. biological methods utilize very less amount of water and energy while decomposition and degradation of dyes in comparison of physicochemical methods. It has been also revealed from several studies that physicochemical methods are less efficient in dye removing in comparison to biological methods as it has been seen that these physicochemical cannot completely degrade dye, metal-dye complex and recalcitrant dyes and produce intermediate products which are further needed to be processed [65], Whereas complete degradation and mineralization of dyes are done in case of biological methods under certain environmental condition [66]. These studies suggest that microbial methods of dye degradation utilize fewer resources and energy with an effective result without harming the environment or without producing pollutant in comparison to traditional physicochemical methods.

8 Future of Dye Degradation Method by the Microflora

Microflora is considered as one of the effective solutions for bioremediation. These organisms are eco-friendly, highly adaptive and can be found in almost every habitat throughout the globe. The efficiency and capability of these organisms are highly appreciable, due to their positive impact in dyes degradation process it is needed to develop and standardize various methods and technologies for effective outcomes. As these biological methods involve various microorganism, which cannot be controlled during the process of degradation but their growth and metabolic activity can be enhanced by providing exact nutritional requirement and environmental condition by the process of biostimulation. In some cases, the hybrid technology can be also developed in which both microbial and physicochemical technology are hybridized for complete degradation of dyes or complex of dyes [37]. it has been studied that the combination of techniques can overcome each other's limitation and can provide an effective result. These microbes can be also combined with membrane technology which will increase the surface area and exposure time (contact time). As the higher surface area and higher contact time, higher the rate of degradation.

In recent time the molecular approach towards the exploration of the microbial world by the techniques associated with metagenomics, transcriptomics, proteomics and metabolomics. These approaches to study and explore microorganism not only provides sufficient data to look and understand microorganism deeply at the molecular level but it will also help to develop various strategy to manipulate other microbes with the help of genetic engineering and metabolic engineering [67]. Multiple crucial genes, proteins and enzymes can be identified and used to build strategy or method through which complete degradation of an industrial effluent containing dye can be achieved [5]. Nanotechnology coupled with the microbial method can be another strategy for complete degradation of pollutants like dyes. These all approaches for the improvement of microbial associated biodegradation of dyes is to achieve cost-effective, highly efficient clean technology in the upcoming future.

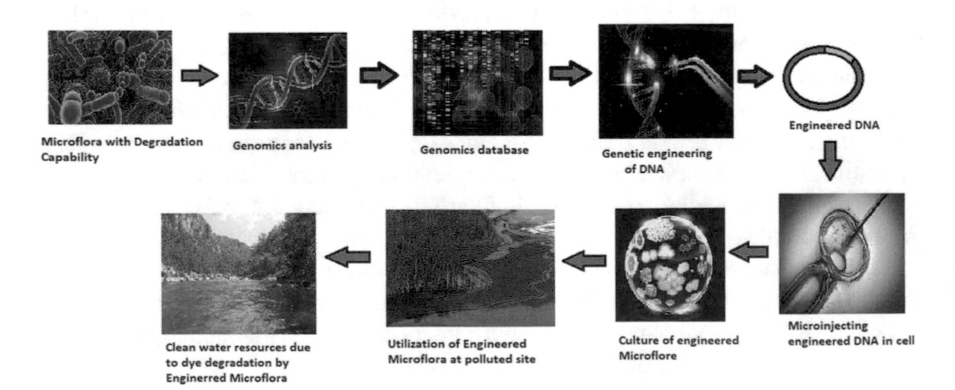

Fig. 6 Shows the role of genome engineering in dyes degradation technology to enhance the efficiency of microbes

References

1. Hussain RA, Badshah A, Raza B, Saba S (2016) Functional metal sulfides and selenides for the removal of hazardous dyes from Water. J Photochem Photobiol B 159:33–41
2. Das A, Mishra S, Verma VK (2015) Enhanced biodecolorization of textile dye remazol navy blue using an isolated bacterial strain Bacillus pumilus HKG212 under improved culture conditions. J Biochem Technol 6(3):962–969
3. Jadhav UU, Dawkar VV, Ghodake GS, Govindwar SP (2008) Biodegradation of Direct Red 5B, a textile dyes by newly isolated Comamonas sp. UVS. J Hazard Mater 158:507–516. https://doi.org/10.1016/j.jhazmat.2008.01.099
4. Khan S, Malik A (2018) Toxicity evaluation of textile effluents and role of native soil bacterium in biodegradation of a textile dye. 25(5):4446–4458
5. Zhang L, Pan J, Liu L, Song K, Wang Q (2019) Combined physical and chemical activation of sludge-based adsorbent enhances Cr (VI) removal from wastewater. J Clean Prod 238:117904
6. Berton G, Gordon S (1983) Immunology 49:705
7. Agarwal P (2009) Application of natural dyes on textiles
8. Lucas MS, Dias AA, Sampaio A, Amaral C, Peres JA (2007) Water Res 41:1103–1109
9. Fatima M, Farooq R, Lindström RW, Saeed M (2017) A review on biocatalytic decomposition of azo dyes and electrons recovery. J Mol Liq 246:275–281. ISSN: 0167-7322. https://doi.org/10.1016/j.molliq.2017.09.063
10. Wang M, Yang J, Wang H (2001) Dyes Pigments 50:243–246
11. Benkhay S, M'rabet S, El Harfi A (2020) A review on classifications, recent synthesis and applications of textile dyes. Inorg Chem Commun 115:107891. ISSN:1387-7003
12. Shore J (1995) Cellulosics dyeing, society of dyers and colourists
13. Nunn DM (1979) The dyeing of synthetic-polymer and acetate fibres. Dyers Co. Publications Trust
14. Clark M (2011) Handbook of textile and industrial dyeing: principles, processes and types of dyes.
15. Gaffer HE (2013) Carbohyd Polym 97:138–142
16. Gao Y, Cranston R (2008) Text Res J 78:60–72
17. Giri BS, Raza N, Roy K, Kim KH, Rai BN et al (2018) Recent advancements in bioremediation of dye: current status and challenges. Bioresour Technol 253:355–367
18. Christie RM (2001) Colour chemistry. Royal Society of Chemistry, United Kingdom
19. Christie RM (2007) Environmental aspects of textile dyeing. Elsevier
20. Pohanish RP (2017) Sittig's handbook of toxic hazardous chemicals and carcinogens. Elsevier, Amsterdam; William Andrew, Cambridge
21. Lacasse K, Baumann W (2012) Textile chemicals: environmental data and facts. Springer, Dortmund
22. Gupta VK, Khamparia S, Tyagi I, Jaspal D, Malviya A (2015) Decolorization of mixture of dyes: a critical review. Glob J Environ Sci Manag 1(1):71–94
23. Ayanda OS, Nelana SM, Naidoo EB (2018) Ultrasonic degradation of aqueous phenolsulfonphthalein (PSP) in the presence of nano-Fe/H_2O_2. Ultrason Sonochem 47:29–35
24. Zhao B, Shang Y, Xiao W, Dou C, Han R (2014) Adsorption of Congo red from solution using cationic surfactant modified wheat straw in column model. J Environ Chem Eng 2:40–45
25. Ahmad AL, Harris WA, Ooi BS (2012) Removal of dye from wastewater of textile industry using membrane technology. J Teknologi 36:31–44
26. Celik L, Ozturk A, Abdullah MI (2012) Biodegradation of Reactive Red 195 azo dye by the bacterium Rhodopseudomonas palustris 51ATA. Afr J Microbiol Res 6:120–126. https://doi.org/10.5897/AJMR11.1059
27. Ning X, Yang C, Wang Y, Yang Z, Wang J, Li R (2014) Decolorization and biodegradation of the azo dye Congo Red by an isolated Acinetobacter baumannii YNWH 226. Biotechnol Bioprocess Eng 19:687–695. https://doi.org/10.1007/s12257-013-0729-y

28. Misal SA, Lingojwar DP, Shinde RM, Gawai KR (2011) Purification and characterization of azoreductase from alkaliphilic strain Bacillus badius. Process Biochem 46:1264–1269. https://doi.org/10.1016/j.procbio.2011.02.013
29. Joshi T, Iyengar L, Singh K, Garg S (2008) Isolation, identification and application of novel bacterial consortium TJ-1 for the decolorization of structurally different azo dyes. Bioresour Technol 99:7115–7121. https://doi.org/10.1016/j.biortech.2007.12.074
30. Ola IO, Akintokun AK, Akpan I, Omomowo IO, Areo VO (2010) Aerobic decolorization of two reactive azo dyes under varying carbon and nitrogen source by Bacillus cereus. Afr J Biotechnol 9:672–677. https://doi.org/10.5897/AJB09.1374
31. Zhiqiang C, Wenjie Z, Jiangtao M, Jinyan C (2015) Biodegradation of azo dye Disperse Orange S-RL by a newly isolated strain Acinetobacter sp. SRL8. Water Environ Res 87:516–523. https://doi.org/10.2175/106143014X13975035526068
32. Pan H, Feng J, Cerniglia CE, Chen H (2011) Effects of Orange II and Sudan III azo dyes and their metabolites on Staphylococcus aureus. J Ind Microbiol Biotechnol 38:1729–1738. https://doi.org/10.1007/s10295-011-0962-3
33. Singh RP, Singh PK, Singh RL (2014) Bacterial decolorization of textile azo dye Acid Orange by Staphylococcus hominis RMLRT03. Toxicol Int 21:160–166. https://doi.org/10.4103/0971-6580.139797
34. Lim SL, Chu WL, Phang SM (2010) Use of Chlorella vulgaris for bioremediation of textile wastewater. Bioresour Technol 101(19):7314–7322. https://doi.org/10.1016/j.biortech.2010.04.092
35. Kolekar YM, Nemade HN, Markad VL (2012) Decolorization and biodegradation of azo dye Reactive Blue 59 by aerobic granules. Bioresour Technol 104:818–822. https://doi.org/10.1016/j.biortech.2011.11.046
36. Saratale RG, Saratale GD, Chang JS, Govindwar SP (2010) Decolorization and biodegradation of reactive dyes and dye wastewater by a developed bacterial consortium. Biodegradation 21:999–1015. https://doi.org/10.1007/s10532-010-9360-1
37. Singh RL, Singh PK, Singh RP (2015) Enzymatic decolorization and degradation of azo dyes—a review. Int Biodeterior Biodegrad 104:21–31. https://doi.org/10.1016/j.ibiod.2015.04.027
38. Chan GF, Rashid NAA, Koay LL, Chang SY, Tan WL (2011) Identification and optimization of novel NAR-1 bacterial consortium for the biodegradation of Orange II. Insight Biotechnol 1:7–16. https://doi.org/10.5567/IBIOT-IK.2011.7.16
39. Patil PS, Shedbalkar UU, Kalyani DC, Jadhav JP (2008) Biodegradation of Reactive Blue 59 by isolated bacterial consortium PMB11. J Ind Microbiol Biotechnol 35:1181–1190. https://doi.org/10.1007/s10295-008-0398-6
40. Tony BD, Goyal D, Khanna S (2009) Decolorization of textile azo dyes by aerobic bacterial consortium. Int Biodeterior Biodegrad 63:462–469. https://doi.org/10.1016/j.ibiod.2009.01.003
41. Ruiz-Arias A, Juarez-Ramirez C, De los Cobos-Vasconcelos D, Ruiz-Ordaz N, Salmeron-Alcocer A, Ahuatzi-Chacon D, Galindez-Mayer J (2010) Aerobic biodegradation of a sulfonated phenylazonaphthol dye by a bacterial community immobilized in a multistage packed-bed BAC reactor. Appl Biochem Biotechnol 162:1689–1707. https://doi.org/10.1007/s12010-010-8950-z
42. Qu Y, Shi S, Ma F, Yan B (2010) Decolorization of Reactive Dark Blue K-R by the synergism of fungus and bacterium using response surface methodology. Bioresour Technol 101:8016–8023. https://doi.org/10.1016/j.biortech.2010.05.025
43. Porri A, Baroncelli R, Guglielminetti L, Sarrocco S (2011) Fusarium oxysporum degradation and detoxification of a new textile glyco-conjugate azo dye (GAD). Fungal Biol 115:30–37. https://doi.org/10.1016/j.funbio.2010.10.001
44. Hadibarata T, Yusoff ARM, Aris A, Salmiati HT, Kristanti RA (2012) Decolorization of azo, triphenylmethane and anthraquinone dyes by laccase of a newly isolated Armillaria sp. F022. Water Air Soil Pollut 223:1045–1054. https://doi.org/10.1007/s11270-011-0922-6
45. Sen K, Pakshirajan K, Santra SB (2012) Modelling the biomass growth and enzyme secretion by the white rot fungus Phanerochaete chrysosporium: a stochastic based approach. Appl Biochem Biotechnol 167:705–713. https://doi.org/10.1007/s12010-012-9720-x

46. Asgher M, Yasmeen Q, Iqbal HMN (2013) Enhanced decolorization of solar Brilliant Red 80 textile dye by an indigenous white rot fungus Schizophyllum commune IBL-06. Saudi J Biol Sci 20:347–352. https://doi.org/10.1016/j.sjbs.2013.03.004
47. Karthikeyan K, Nanthakumar K, Shanthi K, Lakshmanaperumalsamy P (2010) Response surface methodology for optimization of culture conditions for dye decolorization by a fungus Aspergillus niger HM11 isolated from dye affected soil. Iran J Microbiol 2:213–222
48. Grinhut T, Salame TM, Chen Y, Hadar Y (2011) Involvement of ligninolytic enzymes and Fenton-like reaction in humic acid degradation by Trametes sp. Appl Microbiol Biotechnol 91:1131–1140. https://doi.org/10.1007/s00253-011-3300-9
49. Yu Z, Wen X (2005) Screening and identification of yeasts for decoloring synthetic dyes in industrial wastewater. Int Biodeterior Biodegrad 56:109–114. https://doi.org/10.1016/j.ibiod.2005.05.006
50. Hai FI, Yamamoto K, Nakajima F, Fukushi K (2012) Application of a GAC-coated hollow fiber module to couple enzymatic degradation of dye on membrane to whole cell biodegradation within a membrane bioreactor. J Membr Sci 389:67–75. https://doi.org/10.1016/j.memsci.2011.10.016
51. Ambrosio ST, Vilar JC, da Silva CAA, Okada K, Nascimento AE, Longo RL (2012) A biosorption isotherm model for the removal of reactive azo dyes by inactivated mycelia of Cunninghamella elegans UCP542. Molecules 17:452–462. https://doi.org/10.3390/molecules17010452
52. Dubey SK, Dubey J, Mehra S, Tiwari P, Bishwas AJ (2011) Potential use of cyanobacterial species in bioremediation of industrial effluents. Afr J Biotechnol 10:1125–1132
53. Acuner E, Dilek FB (2004) Treatment of Tectilon Yellow 2G by Chlorella vulgaris. Process Biochem 39:623–631. https://doi.org/10.1016/S0032-9592(03)00138-9
54. Chu WL, Yike-Chu S, Siew-Moi P (2009) Use of immobilised Chlorella vulgaris for the removal of color from textile dyes. J Appl Phycol 21:641–648. https://doi.org/10.1007/s10811-008-9396-3
55. Bhatnagar A, Sillanpaa M (2010) Utilization of agroindustrial and municipal waste materials as potential adsorbents for water treatment: a review. Chem Eng J 157:277–296. https://doi.org/10.1016/j.cej.2010.01.00
56. Donmez G, Asku Z (2002) Removal of chromium(VI) from saline wastewater by Dunaliella species. Process Biochem 38:751–762. https://doi.org/10.1016/S0032-9592(02)00204-2
57. Fu Y, Viraraghavan T (2001) Fungal decolorization of dye wastewaters: a review. Bioresour Technol 79:251–262. https://doi.org/10.1016/S0960-8524(01)00028-1
58. Hammaini A, Ballester A, Blazquez MI, Gonzalez F, Munoz J (2002) Effect of the presence of lead on the biosorption of copper, cadmium and zinc by activated sludge. Hydrometallurgy 67:109–116. https://doi.org/10.1016/S0304-386X(02)00157-3
59. Aksu Z, Donmez G (2003) A comparative study on the biosorption characteristics of some yeast for Remazol Blue reactive dye. Chemosphere 50:1075–1083. https://doi.org/10.1016/S0045-6535(02)00623-9
60. Pearce CI, Lloyd JR, Guthrie JT (2003) The removal of color from textile wastewater using whole bacterial cells: a review. Dyes Pigments 58:179–196. https://doi.org/10.1016/S0143-7208(03)00064-0
61. Bhosale S, Saratale G, Govindwar S (2006) Mixed function oxidase in Cunninghamella blakesleeana (NCIM-687). J Basic Microb 46:444–448. https://doi.org/10.1002/jobm.200510117
62. Torres E, Bustos-Jaimes I, Le Borgne S (2003) Potential use of oxidative enzymes for the detoxificat -ion of organicpollutants. Appl Catal B Environ 46:1–15. https://doi.org/10.1016/S0926-3373(03)00228-5
63. Martinez AT (2002) Molecular biology and structure function of lignin-degrading heme peroxidases. Enzyme Microb Technol 30:425–444. https://doi.org/10.1016/S0141-0229(01)00521-X
64. Dong H, Guo T, Zhang W, Ying H, Wang P, Wang Y, Chen Y (2019) Biochemical characterization of a novel azoreductase from Streptomyces sp.: application in eco-friendly decolorization of azo dye wastewater. Int J Biol Macromol 140:1037–1046

65. Guo G, Li X, Tian F, Liu T, Yang F, Ding K, Wang C (2020) Azo dye decolorization by a halotolerant consortium under microaerophilic conditions. Chemosphere 244:125510
66. Rathod J, Dhebar S, Archana G (2017) Efficient approach to enhance whole cell azo dye decolorization by heterologous overexpression of Enterococcus sp. L2 azoreductase (azoA) and Mycobacterium vaccae formate dehydrogenase (FDH) in different bacterial systems. Int Biodeterior Biodegrad 124:91–100
67. An Q, Cheng J, Wang Y, Zhu M (2020) Performance and energy recovery of single and two stage biogas production from paper sludge: Clostridium thermocellumaugmentation and microbial community analysis. Renew Energy 148:214–222

Dye Degradation by Fungi

Vinay Kumar, Garima Singh, and S. K. Dwivedi

Abstract Dye pollution is rising drastically due to massive use in different types of industrial activities. Synthetic dye pollution is the major issue of the environment at present due to their recalcitrant, toxic, carcinogenic and mutagenic behavior. Dye pollution affects aquatic life by impairing the sunlight permeability and can damage the aquatic ecosystem terribly. Therefore, treatment of dye-containing wastewater is required. Bioremediation seems as safe and ecofriendly approach for wastewater treatment. Fungi have massive dye decolorization potential and can be used for treatment. Studies have reported many fungal species for decolorization of dye so, for better understanding of their application in wastewater treatment process, the involved mechanism in dye decolorization should be known. This review focused on application process of fungi in dye decolorization, role of enzyme, protein, genes and surface functional group in degradation and biosorption process. The toxicity of fungal degraded dye end products is also reviewed in this chapter which is an important aspect in fungal application for dye contaminated wastewater treatment.

Keywords Dye · Fungi · Enzyme · Fungal dye degradation · Degradation mechanism · Toxicity

1 Introduction

Dyes have a great impact on aquatic life which are generated from different types of industries such textile, paper and pulp, color and printing points, leather, paints, food and cosmetic. On the basis of chemical structure of chromophore, there are almost 25 types of dye classes and over one thousand dyes are utilized in textile production for a variety of color fabrics [2, 120]. Disposal of these effluents from industries containing dye into the natural water bodies causes artistic damages and can influence the life form by declining the light permeability that affects the photosynthetic activity and availability of oxygen in the water bodies. Some of the dyes are also persistent and

V. Kumar (✉) · G. Singh · S. K. Dwivedi (✉)
Department of Environmental Science, School for Environmental Sciences, Babasaheb Bhimrao Ambedkar University, Lucknow 226025, India

S. S. Muthu and A. Khadir (eds.), *Dye Biodegradation, Mechanisms and Techniques*, Sustainable Textiles: Production, Processing, Manufacturing & Chemistry,
https://doi.org/10.1007/978-981-16-5932-4_5

highly toxic to terrestrial as well as aquatic fauna of the environment [61]. Toxic, mutagenic and carcinogenic features of the many dyes are widely reported in the studies. It has been also found in the studies that exposure of some of the dye such azo dyes, malachite green, congo red, etc., increase the chances of chromosomal fractures, fertility loss, and can affect the respiratory enzymes in living organisms [2, 33, 113].

Due to high dissolution of dyes into the water, the physicochemical methods are less effective for their removal from the wastewater [123] and many of the downsides such as generation of huge quantity of toxic sludge, use of huge amount of chemical and energy and requirement of skilled manpower are also coupled with these techniques ([69, 61, 64–73, 101]). Microbes (fungi and bacteria) are used in microbial bioremediation processes as an eco-friendly and sustainable way for management of dye polluted wastewater. Many of the bacterial and fungal species have been reported with the potential to degrade complex dye molecules in simpler non/less toxic compounds or break down into carbon dioxide, water and others. They are also serving as decomposer in ecosystem functioning as an essential biotic component of the ecosystem that's why their utilization in treatment of dye polluted wastewater is more eco-friendly than known physicochemical techniques.

Fungi being an active agent of ecosystem as saprophytes produces many of the enzymes for instance laccase, lignin peroxidase, manganese peroxidase, etc., which can potentially catalyze various types of dye molecules (congo red, malachite green, methylene blue, etc.). Jasinska et al. [61] investigated the malachite degradation mechanism and potential of fungus *Myrothecium roridum* which produces laccase for dye degradation. In another report, *Aspergillus terreus* GS28 degraded Congo red and was found to extracellularly secrete laccase and manganese peroxidase [109]. Many of the review articles are available that deal with microbial bioremediation of dye contaminated water and wastewater. There are several types of mechanisms that have been explained in the literature for decolorization/degradation of synthetic dyes by fungi. Therefore, in this review fungi application in the meadow of bioremediation and fungal dye degradation/decolorization mechanisms are discussed. The toxicity of fungal degraded various dye end products is also narrated.

2 Dye Pollution Sources and Impact

This era is dealing with the use of fashionable clothes, paper, cosmetic, etc., that increases the heavy load on industrial activities for manufacturing of these products. The production of varieties of colors in high quantity via natural processes is unable to fill the demands of the present which necessitate the alternative methods to produce varieties of color in huge amount that can fill the need of the present. In the chemical synthesis of dye, varieties of color can be multiplied easily in huge amounts in very short duration of time. The chemically synthesized dyes have high brightness and binding capacity with substrate and are utilized in multiple industries for coloring purposes. Most of the dyes are exploited in the textile, carpet, paper and printing industries for production of varieties of colored fibers and papers.

Due to huge application in industrial processes, dyes are produced in high amount with the generated effluents and disposed-off into the fresh water bodies like rivers, lakes and ponds without proper treatment. Synthetic dyes are exploited in diverse industries like textile, carpet, printing, paper and pulp, cosmetic, paint, laundry and leather are generated on large scale and due to their toxicity, they are related to environmental degradation. Among the dyes, Azo dye is the class of dye which is extensively being used. There are over 3,000 dyes belong to the class of azo dye [123] and it is expected that roughly 280,000 tons of synthetic dye are released from textile industries annually around the world [83, 144]. The chemical structures of some of these synthetic dyes are presented in Fig. 1. Based on their origin, dyes are categorized in two types: *Natural dyes* and *Synthetic dyes.* Natural dyes have no significant

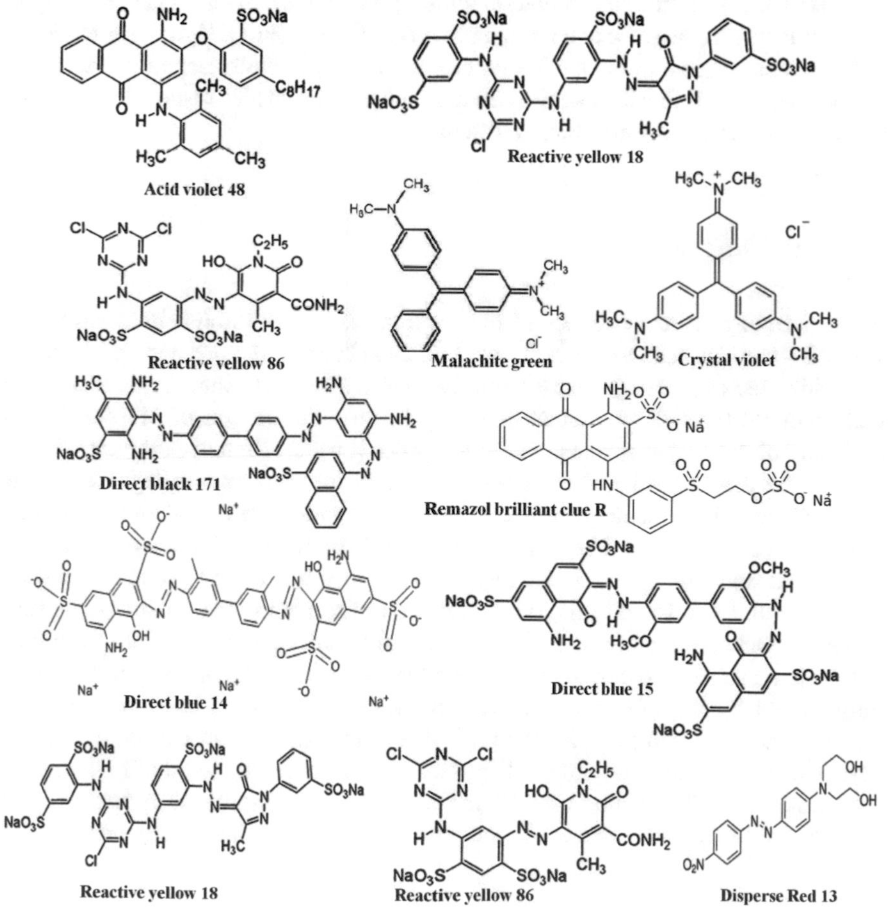

Fig. 1 Chemical structure of some synthetic dyes

impact on the environment, while synthetic dye possesses great concern. Synthetic dyes contains one or more than one chromophore such as acridine, anthraquinone, Azo (-N=N-), diazonium, oxazin, nitro, thiazin, phthalocyanine and triarylmethane which generate various color by absorbing the light in visible region (400–700 nm) ([31], [115]).

Accumulation of dyes into water bodies reduces the permeability of light and affects the aquatic ecosystem drastically by reducing the photosynthetic process. In addition, accumulated dye can also raise the chemical oxygen demand (COD) along with biological oxygen demand (BOD) of fresh water body and generate a noxious environment that turns into a degraded ecosystem [66, 105]. Synthetic dyes are extremely toxic and can influence the growth and metabolic process of the living creatures such as Reactive brilliant red can cause disturbance in the function of human serum albumin [80], disperse orange 1 can enhance the frequencies of micronuclei in lymphocytes and HepG2 cells which cause DNA damage [103], orasol navy blue 2RB can increase frame shift mutation without metabolic activation [124]. Many of the dyes have carcinogenic and mutagenic properties (Table 1). Dyes have tendency to be recalcitrant in aerobic environment which leads to their accumulation in soil and sediment at the physicochemical treatment location and also responsible for their transport to municipal water supply system.

3 Why Fungi?

Fungi normally occur in almost all the environmental conditions either normal or stressed situations such as drought, alkaline and acidic in the presence of contaminants (like heavy metal, pesticides and dye, etc.). The wastewater is accomplished with multiple types of pollutants that may hamper the growth of bioremediators but many of the fungal species exhibit tolerance toward different pollutants. Being saprophytes they use and produce numerous extracellular enzymes to degrade organic substrate and utilize the degraded substrate as energy and nutrients source for their growth and development. The extracellular release of cluster of enzymes associated with the degradation of dye is the basic criteria to select them to treat wastewater polluted with dyes. In addition, the application of growing form of fungi has self-replenishment ability which promises no need for addition of consortium from time to time and can be used continuously for a long time. From technical point of view, fungi are easily separable from the treated wastewater if used in bioreactor, easy to handle and use, can grow in low graded substrate and do not cause any type of environmental damages after their disposal. There is no requirement of high energy, chemicals and skilled manpower with the use of fungi for treatment perspective that make it more sustainable and environmental friendly than other techniques.

Table 1 Effect of dyes on living beings

Dye	Toxicity	References
Acid violet 7	Lipid peroxidation, Aberration in chromosome	Ben Mansour et al. [84]
Disperse red-1	Functional change in human lymphocytes	Chequer et al. [27]
Reactive black-5	Decline activity of urease cause ammonification in arginine rate of terrestrial ecosystem	Topac et al. [121]
Disperse blue-291	Mutagenic, cytotoxic and genotypic effects	Tsuboy et al. [122]
Malachite green	Carcinogenesis and Mutagenesis	Jasińska et al. [61]
Congo red	Carcinogen and Mutagen	Asses et al. [9]
Mordent red-73	Lead to release of the toxic chromium (IV) salt into the environment	Elmorsi et al. [39]
Disperse red-1	DNA damage caused which increase the number of micronuclei in lymphocyte of human being	Chequer et al. [27]
Disperse orange-1	Enhance the frequencies of micronuclei in lymphocytes sand HepG2 cells which cause DNA damage	Ferraz et al. [43]
Astrazon blue FGRL	Alteration of few enzymatic activity and increase in glutathione reductase	Gongord et al. [48]
Sudan I	Liver and urinary bladder carcinogen in mammals	Stiborová et al. [114]
Direct black 38	Urinary bladder cancer, Liver carcinogen	Robens et al. [103]
Disperse red 13	DNA damage in human hepatoma cells	Oliveira et al. [91]
Direct blue 6	Teratogenes is in rats during pregnancy	IARC [57]
Direct red 28	Carcinogen	Ding et al. [37]
Basic red 9	Bacterial DNA damage and hypertrophy of thyroid in Mice	IARC [57]
Reactive orange 16	Mutagenic effect	Novotný et al. [89]
Rodamine 6G	Mutagenesis	Nestmann et al. [88]
Acid blue 80	Incensement of apoptosis in epithelial cells line RTL-W1	Bae and Freeman [10]
Benzopurpurine 4B	Endocrine disrupting agents	Bazin et al. [16]
Methyl orange	Carcinogenic and mutagenic effect	Purnomo et al. [98]
Methylene blue	Increase chemical oxygen demand which lead to death of aquatic organism	Rizqi and Purnomo [102]
Orasol navy blue 2RB	Without metabolic activation increased frame shift mutation	Venturini and Tamaro [124]
Disperse orange 37	Mutagenic response	Lima et al. [34]

4 Fungi and Dye Degradation

Disposal of inadequately treated dyes containing wastewater generated from industrial activities causes high pollution load on natural water body that affect the ecosystem functioning and results into degraded ecosystem by various ways. Dyes have been reported to cause various types of abnormalities in human beings as well as environment. Due to dye's toxicity, carcinogenic and mutagenic characteristics their removal from wastewater is necessary. Many attempts have been made by the researchers for dye decolorization/removal from polluted water. Biological method is more appropriate and significant than other known methods due to their applicability and environmental friendliness. In this regards many of the fungal species have been explored to remove different types of pollutants including synthetic dyes. Fungi exhibit several types of enzymes that have the degradation potential of synthetic dyes and other organic contaminants. In addition, fungal cell is made up of lipid, protein and carbohydrates that provide attractive characteristics to its cell surface that is biosorption features. Adsorptive characteristics of fungi give a unique feature to it and increase its potential for removal and degradation of pollutants. The fungi that have been reported to decolorize different types of dyes belong to the class ascomycetes, basidiomycetes and duteromycetes especially white rot-ligninolytic fungi which have high potential to produce various enzymes including laccase, manganese peroxidase and lignin peroxidase. These enzymes play crucial function as a biocatalyst in the dye degradation process.

There are numerous reports available in public domain on degradation of synthetic dyes by diverse types of fungal species (Table 2) ([8, 93, 58, 106, 81, 104]). *Aspergillus*, *Trichderma*, *Phanerochaete* and *Pleurotus* species are widely studied in the degradation of various class of synthetic dyes including direct, disperse, azo and anthraqinone ([106, 109], [129, 44, 17, 82, 130]). The mechanisms engaged in the decolorization of dye molecules by fungi are degradation and adsorption (biosorption) [6, 9, 19, 109]. In the biological management of dyes containing water, fungi can be applied in various ways which are as follows:

4.1 *Use of Growing Culture*

This is a very common method for use of fungi in the degradation/decolorization of dye and mostly performed in batch study. In this method, fungi growth and dye decolorization has simultaneously happened where degradation is performed by releasing extracellular enzymes and some amount of dye is also adsorbed on the surface of fungi that enhance the decolorization performance of fungi. In an investigation, *Aspergillus terreus* GS28 was utilized for direct blue-1 decolorization in liquid culture. Almost 99.2% dye decolorization was recorded under the optimized condition at 7th day of incubation while decolorization happened via sorption and degradation. Sorption was driven by surface functional group and degradation was

Table 2 Dye degradation potential of some fungi and their degraded dye end products

Dye	Enzyme	Degraded product	Instrumental analysis	References
Mordant yellow 1	MnP, LiP and Laccase	Methyl 2-fluoro-5- nitrobenzoate, benzene and 2-methylpropionic acid	GC–MS	Liu et al. [82]
Congo red	Laccase and MnP	Naphthalene amine, biphenyl amine, biphenyl	FTIR and GC–MS	Si et al. [108]
Azo dye	Laccase	Biphenyl and naphthalene diazonium	HPLC	Krishnamoorthy et al. [67]
Procion red MX-5B	Azo-reductase	Primary and secondary amines	UV–Vis, FTIR	Almeida and Corso [6]
Congo red	MnP	Napthylamine and Benzidine	GC–MS	Wang et al. [126]
Congo red	LiP and MnP	Napthalene sulfonate and Cycloheptadienylium	LC–MS	Asses et al. [9]
Congo red	MnP	Aromatic amines	HPLC and FTIR	Chakraborty et al. [25]
Acid red 3R	MnP and LiP	4-aminonapthalene-1-sulfonic acid, 8-amino-7-hydroxynapthelene-1 and 3-disulfonic acid	GC–MS	He et al. [50]
Congo red	Laccase	Napthalene	Mass spectrometry and FTIR	Iark et al. [58]
Scarlet RR	LiP, Laccase and LiP	N-(1l3-chlorinin-2-yl)-2-{methyl[(4-oxo-3,4-dihydroquinolin-2- yl)methyl] amino}acetamide, N-(1l3-chlorinin-2-yl)-2-{[(4-oxo-3,4-dihydronaphthalen-2-yl)methyl]amino}acetamide, N-ethyl-1l3-chlorinin-2-amine and 5-({[2-(1l3-chlorinin-2-ylamino) ethyl]amino}methyl)cyclohexa-2,4-dien-1-one	FTIR, HPLC and GC–MS	Bankole et al. [13]
Malachite green	Laccase and MnP	N-demethylated primary and secondary arylamines	UV–Vis, FTIR and LC–MS	Barapatre et al. [15]

(continued)

Table 2 (continued)

Dye	Enzyme	Degraded product	Instrumental analysis	References
Brilliant green	Laccase	N-(diethylamino) phenyl) (phenyl) methylene) cyclohexadienylidine) N-methylthanaminium	UV–Vis, FTIR and LC–MS	Kumar et al. [70]
Rubine GFL	Laccase and Azo –reductase	4-[(2-methyl-4-nitrophenyl) diazenyl] phenol, 1-(2-methyl-4-nitrophenyl)-2-phenyl diazene, (2-methyl-4-nitrophenyl) diazene,	FTIR, GC–MS	Lade et al. [77]
Direct blue-1	Laccase and MnP	[4,5-Diazotricyclo [4.3.0.0 (3,7) non-4-in-2-one] and [1,2-Benzene Dicarboxylic Acid, 3-Nitro)]	GC–MS	Singh and Dwivedi [109]
Acid red 97	Laccase	3-(2-hydroxy-1-naphthylazo) benzenesulfonic acid, [(Phenol,2,6-bis(1,1-Diemethylethyl)-] and Naphthalene 1,2-dione	LC–MS	Pandi et al. [93]
Reactive blue-25	Laccase and LiP	phthalimide, di-iso-butyl phthalate	FTIR and GC–MS	Parashetti et al. [94]
Reactive red- 120		2-aminobenzenesulfonic acid and 3-methanesulfinylbut-3-en-2-one	HPLC and FTIR	Su and Lin (2013)
Disperse red 3B	LiP and MnP	Diisobotyl phthalate, 4-Hydroxy-2-butan-one, ammonia	UV–Vis, FTIR and GC–MS	Tang et al. [116]
Reactive green	Laccase, LiP and DCIP reductase	Benzoic acid, 2(-1-oxopropyl)	UV–Vis, HPLC and FTIR	Sinha and Osborne (2016)
Disperse blue 2BLN	MnP	Dibutyl phthalate and 2-(N-methyl-p-phenylenediamine) ethanol	FTIR and GC–MS	Pan et al. [92]

performed by manganese peroxidase and lignin peroxidase [109]. Similarly, brilliant green decolorization performance of three fungal species *Pleurotus forida*, *Pleurotus eryngii* and *Pleurotus sajor-caju* was reported by Naraian et al. (2018). The decolorization efficiency for brilliant green was 99, 91, 87% by *P. forida*, *P. eryngii* and *P. sajorcaju*, respectively. There are several other fungal species also have been reported for the degradation/decolorization of different type of dyes [19, 22, 25, 76, 116].

4.2 Use of Immobilized Fungi

In the immobilized form of fungi, mostly fungus inoculums are immobilized on the surface of any solid medium [74, 68]. This process is basically applied when fungi are used in bioreactor for decolorization of dyes. Immobilization of fungus on solid surface also enhanced the applicability of fungi and can be used in continuous treatment of dye-containing wastewater for a long time without making more effort. Sometimes it is also known as solid–liquid-phase decolorization. Andleeb et al. [7] examined the Drimarene blue K_2RL dye biodegradation potential of *Aspergillus flavus* SA2 in immobilized form in lab-scale fluidized bed reactor (FBR). The fungus was immobilized on sand with size of 0.2 mm and investigated for biodegradation. In the FBR, *Aspergillus flavus* SA2 was able to remove 71.3% of color from the simulated textile water by biodegradation and bio-decolorization mechanisms including 85.57% and 84.70% removal of BOD and COD, respectively. Recently, in an investigation, Alam et al. [5] immobilized the fungus *Trametes hirsuta* D7 in light expanded clay aggregate and utilized for decolorization of acid blue 129, reactive blue 4 and remazol brilliant blue R (RBBR). The immobilized *Trametes hirsuta* D7 showed high degradation performance for the acid blue 129, RBBR and reactive blue 4 with decolorization performance of 96%, 95% and 90%, respectively. These studies showed that fungi can give more effective results in the degradation/decolorization of dye when applied in immobilized form. Some other studies were also investigated on synthetic dye decolorization/degradation performance of various fungi as well fungal-originated enzymes such as MnP, laccase and LiP ([3, 35, 118, 125]).

4.3 Use of Fungal Extract Containing Enzymes

The ligninolytic white-rot fungi are reported to generate diverse types of enzyme having the potential for degradation/decolorization of dyes. These enzymes are LiP, MnP and laccase which have been extensively reported in the studies. In an investigation, *Phanerochaete chrysosporium* CDBB 686 extract containing MnP, LiP, laccase obtained from fermentation of corncob was used for the decolorization of congo red, poly R-478 and methyl red. Fungal extract successfully decolorized Congo red, Poly R-478 and methyl red with decolorization efficiency of 41.8%, 56.8% and 69.7%, respectively [111]. Akpinar and Urek [4] investigated the laccase production capability of *Pleurotus eryngii* by using solid-state bioprocess utilizing Peach waste of the fruit juice industry. The obtained laccase containing extract of *P. eryngii* was investigated for methyl orange degradation and reported 43% decolorization efficiency. Many other similar reports are also available on the dye decolorization performance of fungal extract ([30, 62, 78 87]).

4.4 Use of Isolated Enzymes from Fungi

The enzymes MnP, LiP and laccase are isolated in the different studies from liquid culture of different types of fungi and were employed in decolorization/degradation of dyes and various other organic pollutants. Bouacem et al. [19] isolated two peroxidases (LiP BA45 and MnP BA30) from the fungus *Bjerkandera adusta* strain CX-9. Both enzymes LiP BA45 and MnP BA30 were monomer with molecular mass of 30.12 and 45.22 kDa and highly active at pH 3.0 and 70 °C, pH 4.0 and 50 °C, respectively. Both enzymes were analyzed for decolorization of synthetic dyes remazol brilliant violet 5R, acid blue 158, cibacet brilliant blue BG, reactive dye remazol, brilliant blue reactive, polymeric dye R, methyl green and indigo carmine significantly decolorizes, while MnP BA30 enzymes was highly effective than LiP BA45 in synthetic dye decolorization performance. Similarly, Chairin et al. [24] purified laccase from liquid culture of *Trametes polyzona* WR 710-1, a white rot fungus. Purified laccase was exploited for decolorization of synthetic dyes namely bromophenol blue, acridine orange, remazol brilliant blue R, relative black 5, methyl orange and congo red. Purified laccase showed high efficiency in the decolorization of selected dye and addition of 0.2 M 1-hydroxybenzotriazole (redox mediator) improved the efficiency of dye decolorization of fungal laccase.

5 Role of Enzymes

5.1 Laccase (E.C.1.10.3.2, p-benzenediol)

Laccase is an important enzyme that can catalyze many types of aromatic hydrocarbons such as phenolic compounds via oxidation–reduction mechanism and belong to the group oxidoreductive enzymes [26]. It can be found in bacteria, fungi as well as in plants too. Fungal laccase is broadly distributed in the species of the class Basidiomycetes, Ascomycetes, and Deuteromycetes and in white-rot fungal species, it played role in lignin degradation [58, 61, 93, 109]. Laccase molecule typically has four atoms of copper (Cu) and some time its structure also comprises three Cu atoms with a molecular mass of 50–100 kDa [51].

During dye degradation by fungi, the role of laccase is extensively reported in the studies. Laccase potentially contributed in the degradation of direct blue 1, malachite green, congo red when the fungus *Myrothecium roridum* and *Aspergillus terreus* were used [61, 109]. Abd El-Rahim et al. [1] investigated 17 fungal strains of the genus *Aspergillus* and *Lichtheimia* for the degradation of 20 dye viz. janus green B, direct blue 71, reactive orange, evans blue, fast green, crystal violet, methyl red, tartrazine, naphthol blue black, alura red AC, reactive blue 4, pararosaniline, safranin, alizarin yellow R, trypan blue, ponceau S, cibacron brilliant red 3B-A, brilliant green, direct violet 51 and direct red 80. After a critical evaluation, they found that the fungus *A. niger*, *A. terreus*, *A. oryzae*, *A. fumigatus* showed laccase activity in the degradation of direct violet while absent in the degradation of methyl red. Akpinar and Urek [4] employed Peach waste generated from fruit juice industry for fungal laccase production by *Pleurotus eryngii* and successfully used in methyl orange, reactive black, tartrazine and reactive red 2 dyes degradation. In another study, significant role of laccase was recorded in degradation of solar brilliant red 80 by *Schizophyllum commune*. These studies suggested the magnitude of laccase in the decolorization/degradation of dyes [8].

The mechanisms of laccase in the demolition/degradation of aromatic hydrocarbon (phenolic compounds) are variously suggested by the researchers. It oxidizes phenolic group to phenoxy radical via eliminating hydrogen (H) atom from the OH (hydroxyl) group [11]. Laccase show soaring affinity toward molecular oxygen and as proposed by Burke and Cairney [21] it reacts with type-1-copper reduction by the precursor, transfers electron as of type-1-Cu to type-2-Cu as well as Type-3-Cu in a trinuclear bunch, and reduction of molecular oxygen on type-2-Cu and Type-3-Cu sites. This similar pathway was proposed for the degradation of acid red 97 by fungal laccase isolated from *Peroneutypa scoparia* where hydrogen of phenyl group is broken by laccase and converted into phenoxy radical, subsequently it reacts with OH^- and breaks the azo bond (-N=N-) of acid red 97 and forms naphthalene 1,2 diene and benzenes sulfonic acid as end products [93]. Iark et al. [58] also reported successful degradation of congo red dye by the laccase produced by fungus *Oudemansiella canarii*. In mass spectrometry and Fourier transform infrared spectroscopy (FTIR) they found four types of compound as end products of congo red but no any dye degradative clear path was suggested. Shanmugam et al. [106] utilizes *Trichoderma asperellum* laccase for degradation of malachite green (MG). They suggested that the MG degradation happened by hydroxylation of central carbon by laccase and form Michler's ketone that further catalyzed into benzaldehyde and 4-aminobenzophenone as end products via some intermediates compound formation with laccase-oxidative cleavage and mediated deamination mechanisms. So, fungal laccase may react differentially to degrade the dyes and produced various kinds of less/non-toxic end products and has crucial role in fungal dye degradation.

5.2 Manganese Peroxidase (EC 1.11.1.13)

Manganese peroxidase (MnP), a ligninolytic enzyme first noticed in *Phanerochaete chrysosporium* in 1983, Mn^{2+} was used as substrate and oxidizes into Mn^{3+} [52]. The chelators stabilized to oxidized Mn^{3+} which is turn as very active low molecular mass and diffusible oxidoreductive intermediary. It attacks nonspecifically on hydrocarbon molecules. MnP can oxide different types of organic compounds as well recalcitrant xenobiotic organic pollutants such dyes, nitroaminotaluene, phenol derivates, etc., and can also depolymerize lignin into their simpler form [128]. In fungal dye degradation, its enhanced activity has been reported in many studies. MnP involved in MG, congo red, indigo carmine and reactive red 120 degradation by fungus *Myrothecium roridum, Aspergillus terreus* and *Phanerochaete chrysosporium, P. sordida* strain YK-624, respectively ([49, 61, 104, 81]).

MnP is highly reactive agent that can catalyze dye via oxidoreductase mechanism and diminish the dye toxicity potential. In an investigation, Jasinska et al. [61] used the crude MnP produced by fungus *Phanerochaete chrysosporium* in decolorization of indigo carmine which successfully reduces indigo carmine up to 90% within 6 h. After decolorization assay, they investigated the MnP degraded end products of indigo carmine. Isatin was found as end product of indigo carmine in gas chromatography mass spectroscopic (GCMS) analysis. They suggested that Mn^{3+} first attack on indigo carmine remove SO_2^{2-} then react with NH group of the ring to form intermediate compounds that react with H_2O and separated into isatin, end product of indigo carmine catalyzed by MnP. In another study, cDNA encoding MnP of *P. chrysosporium* was expressed in *Pichia pastoris* for MnP production. The produced MnP was utilized for degradation of MG. Where MnP mediated hydroxyl react with central carbon to mineralize MG into its N, N-dimethylaniline (N,Ndimethyl-benzenamine), 4-dimethylamino-benzophenone hydrate and methylbenzaldehyde compounds as its MnP catalyzed end products [104]. Fungal MnP have crucial role in degradation of dye.

5.3 Lignin Peroxidase (EC 1.11.1.14)

Major sources of lignin peroxidase are the white-rot basidiomycetes. Lignin peroxidase (LiP) is also called ligninase or diaryl propane peroxidase [18]. It is a globular glycoprotein extracellularly released by producers such as *Phanerochaete chrysosporium* as a secondary enzyme [42] and contains just about 343 amino acids having molecular size of 38–42 kDa [28, 60, 107]. Structurally, LiP is monomer of hemoprotein with a size of around 40 × 40 × 50 Å [95] and its folding motif have 8-alpha major and 8 minor helices and antiparallel short 3 beta-sheets [29, 40]. Its isoelectric point is in between 3.0 to 4.7 and is highly active at lower pH around 3.0 [46]. Its production can be affected by change in the ratio of source of carbon (C) and nitrogen (N) in the growth medium [54]. Its four disulfide bond upholds its overall structure and two Ca-binding sites may provide active site for reactivity of LiP toward substrates (lignin) [97]. Many of the species of fungi reported to produce LiP but strain *P. chrysosporium* contain multiple LiP-encoding gene and produce LiP extracellularly as abundant isozyme (H10, H8, H7, H6, H2 and H1) [42, 119]. LiP can catalyze both phenolic along with non-phenolic aromatic unit of lignin either in the existence or nonexistence of mediators which is ended by Trp171 residue through long-range electron transfer that is linked with the heme group [85, 96]. LiP is noticed in many studies for its role in degradation of synthetic dyes like Azo group of dye by fungi [61, 109]. LiP producing fungal species generally belong to the class ascomycetes and basidiomycetes and white-rot ligninolytic fungi [28, 61, 107].

In an investigation, the combined activity of LiP, MnP and laccase was reported in direct blue-1 degradation by *A. terreus* GS28. The direct blue-1 was mineralized into three types of compound namely 4,5-Diazotricyclo[4.3.0.0(3,7)non-4-in-2-one], 1,2-benzene dicarboxylic acid,3-nitro and phenol,2,6-bis(1,1-Diemethylethyl) [109]. Oliveira et al. [90] investigated the LiP production capacity of *Ganoderma lucidum* (GRM117) and *Pleurotus ostreatus* (PLO9). Obtained crude LiP was immobilized on carbon nano-tubes and successfully demonstrated as a biocatalyst for decolorization of RBBR. Jadhav et al. [59] also reported the participation of LiP in biotransformation of direct blue GLL into 3-(naphthalene-1-ylazo)-naphthlene-1, 5 disulfonic acid. Disperse yellow 3 degradation was investigated in a study by *Phanerochaete chrysosporium.* LiP, horseradish peroxidase and MnP was involved in degradation of disperse yellow 3 and it was converted into 4-methyl-1,2-benzoquinone (III) or 1,2-naphthoquinone (VI) and 4-acetamidophenyldiazene that after oxidation reaction converted to acetanilide as major product [112]. The combined action of LiP, MnP and laccase was also reported in the Scarlet RR degradation by *Peyronellaea prosopidis* [13].

5.4 Other Enzymes

Except Lip, MnP and Laccase many other enzymes have also been found for their active participation in degradation/decolorization of many dyes. These enzymes mostly belong to oxydoreductase enzymes group. Spadaro and Renganathan [112] reported that horseradish peroxidase accompanied the catalysis of disperse Yellow 3 into the acetanilide with MnP and LiP in *P. chrysosporium.* In *Diutina rugosa* the activity of NADH-DCIP (dichlorophenol indophenols) reductase was reported to involve in asymmetric cleavage and reduction of indigo dye escorted with laccase and LiP [14]. NADH-DCIP-mediated asymmetric cleavage, desulfonation and dehydroxylation of AR-88 were found in *Achaetomium strumarium.* In GCMS investigation, sodium naphthalene-1-sulfonate, naphthalen-2-ol, and 1,4-dihydronaphthalene were analyzed as intermediate products after action of fungal NADH-DCIP [12]. Enhanced NADH-DCIP activity was reported in *Pichia occidentalis* G1 in the degradation of acid red B [110]. In an investigation dye-decolorizing (DyP type) peroxidase enzyme was reported in *Geotrichum candidum.* DyP type enzyme vigorously associated in trypan blue, methyl orange, eriochrome black t, and congo red dyes degradation that belong to the group of azo dye. Its molecular mass was 63 kDa and was highly active at pH 6.0 and 35 °C in degradation of methyl orange [100]. Si et al. [108] investigated the dye decolorization ability of a white rot fungus *Trametes pubescens* Cui 7571 for congo red and successful decolorization was recorded. In the congo red decolorization, tyrosinase enzyme participated in the dye degradation including laccase, MnP and LiP and four by-products of dye namely naphthalene amine, naphthalene diazonium, biphenyl and biphenyl amine were found as final dye degraded metabolites. Bilirubin oxidase (BOX) is another oxidoreductase reported in degradation of RBBR by *Myrothecium* sp. IMER1, a non-ligninolytic fungus. BOX enzyme has also the potential to degrade congo red and indigo carmine under in vitro condition [131]. Cytochrome P-450 is another factor that holds big and various number of enzymes participated in dye biodegradation. Its role in MG decolorization by *Cunninghamella elegans* ATCC 36,112 was reported by Cha et al. [23] which reduces the MG into leucomalachite green by the action of cytochrome P-450 monooxygenases.

5.5 Role of Other Bioactive Compounds

Synthetic dyes degradation by fungi is mainly driven by oxidoreductase including laccase, MnP and LiP. But several reports have suggested the role of other bioactive molecules' low-molecular weight factors in the degradation of dye. In an investigation, Jasinska et al. [61] studied the malachite green decolorization potential of *Myrothecium roridum* IM 6482. They found that laccase participated in malachite green degradation but other active factor was also involved in degradation process. To analyze active factors, heat inactivation with adding sodium azide was done to neutralize the laccase activity and obtained non-enzymatic liquid coating low-molecular masses (less than 3 kDa) were used for the malachite green (MG) degradation. Non-enzymatic liquid also showed the decolorization potential of MG with efficiency of 8 to 11%. This might be due to the Fenton-like reaction driven by peroxides, which have been proven to oxidize chemical structure through oxidation by hydroxyl radical. To confirm this hypothesis, they conducted the decolorization experiment with catalase, thiourea, superoxide dismutase (SOD) (scavengers of reactive oxygen species, ROS) and did not found significant change in MG decolorization, confirming the presence of low-molecular-mass-factor. Fenton-like reaction, a non-enzymatic mechanism for the decolorization of dye has been also proposed by several workers [47, 64, 86]. In another investigation, Hu et al. [53] purified the low-molecular weight peptide from the culture of *Phanerochaete chrysosporium* that has the phenol peroxidase activity (Pc Factor). The molecular weight of Pc factor was 600 Da with high thermostability and similar to the observation of Jasinska et al. [61].

5.6 Role of Gene

Exploration of molecular mechanism associated with dye degradation by fungi is a crucial step for better understanding of fungal application in the field of bioremediation and has great scientific interest. Gene expressions have important role in metabolic process of the fungi which establishes the growth and development in addition to tolerance toward toxic dye under contaminated environment as well as dyes degradation by regulating multiple types of enzymes activity. Various stressors (such as dyes) can induce set of the gene in the organism in turn to connect the metabolic process with dye tolerance and degradation. There are two groups of genes expression are found in stressed conditions: *regulatory gene* and *functional gene*. Regulatory group of genes encodes various types of transcriptional genes while functional group of genes encode various types of compound and enzymes that are helpful in tolerance and degradation of dye and other toxic organic and inorganic compounds [68].

In an investigation, laccase-producing *Ganoderma lucidum* molecularly analyzed and showed 15 types of laccase isoenzyme genes. Out of them, *Glac1* was highly involved in laccase production for RBBR degradation [99]. Similarly, Değerli et al. [36] found *up* and down-regulation of 10 laccase-producing genes in lichen-forming fungi that were capable to degrade RBBR dye. In mid of this, *lac8* gene was highly up-regulated and possibly associated with enhanced dye degradation. A laccase encoding *lac-T* gene was highly *up*-regulated in *Trametes hirsuta* at the time of textile effluent decolorization and during the decolorization process significant increase in laccase activity was recorded [117]. Laccase encoding gene: *lacI* in *Pichia pastoris*, *Lac1* and *Lac2* in *Trametes hirsuta* MX2 and lac48424-1 in *Trametes* sp. 48,424 overexpression was reported in the studies in the presence of dyes such as crystal violet, malachite green, methyl orange, bromophenol blue, RBBR, acid red 1 and neutral red ([41, 55, 45]). In an exploration, Lee et al. [79] conducted RBBR degradation experiment using *Phlebia brevispora*. They found manganese peroxidase was stimulated in addition to laccase enzymes in the degradation of RBBR and GeneFishing technology confirmed the differential expression of two genes that possibly involved in the enzymes' regulation. In another study, halotolerant *Pichia occidentalis* A2 showed enhanced degradation of acid red B under Static Magnetic Field (SMF) of 206.3 mT. In transcriptomic investigation, it was found that 145 genes were up-regulated and 22 genes were down-regulated under these conditions that might be encoded by the enzymes or other functional proteins [127]. Conversely, Huy et al. [56] reported that synthetic dye can directly suppress the expression level of gene *Lacc110* in *Fusarium solani* HUIB02 while induces the Lacc42 expression level.

5.7 Role of Surface Functional Group in Dye Decolorization

Fungal cell wall is mainly made up of lipid, protein and carbohydrates that carry many types of surface functional group including –COOH, C=C, C=O, O–H, N–H and C-H [68, 75]. The surface functional groups do not participate in dye degradation but can increase the decolorization efficiency of fungi by adsorbing the dye molecules on the mycelia surface of fungi. Singh and Dwivedi [109] reported that almost 34.4% of total 98.2% decolorization of direct blue-1 dye happened via biosorption process by fungus *A. terreus* GS28 which showed that adsorption of dye on mycelia surface has crucial role in decolorization of dye by fungi. Further, in Fourier transform infrared spectroscopic investigation they found that O–H and C=C are involved in direct blue-1 adsorption via hydrogen bond and π-π interaction. Chakraborty et al. (2013 observed that morphology of *Alternaria alternata* CMERI F6 was more amorphous after decolorization of Congo red than control in scanning electron microscopic observation. Due to high amorphousness, the peak of x-ray diffraction analysis at 2 degree values of 28° was reduced which confirms the role of adsorption in congo red decolorization. In one more investigation, Asses et al. [9] examined the decolorization of congo red by fungus *Aspergillus niger* and they also found that surface adsorption of dye molecules occurred in decolorization of Congo red that was confirmed from changes FTIR spectra. Similar observations are reported in various studies in decolorization of synthetic dye by fungi [6, 62, 63, 65].

6 Fungal Genetic Engineering and Their Role in Dye Degradation

This era is dealing with heavy pollution load and the available tools and remediating agents are not sufficient to remove these pollutants. The bioremediation agents are considered as more environmental-friendly and sustainable to remove and degrade the pollutants from contaminated site. At present, several types of bioremediation agents including microorganisms and plants have been explored that can effectively degrade/remove the pollutants from the contaminated sites. These remediators cannot be used in every type of environment due to the fact that in the open environment there are multiple factors such as pH, temperature, presence of other contaminants that can affect the growth, development and pollutant removal/degradation efficiency of these bio-agents. Through the genetic engineering, functional genes of one organism that responsible for degradation and removal of pollutants can be expressed into another desired organism that contains pleasing features for its applicability in *on-site* or engineered systems. This approach can also enhance the applicability of fungi in the field of dyes degradation/decolorization. Many of the reports are available on the genetic engineering for fungi that showed enhanced dye degradation characteristic after gene manipulation ([32, 38, 129, 17, 82, 130]). In a recent study, through genetic engineering, Đurđić et al. [38] inserted LiP gene in pCTCON2 vector to find a saturation

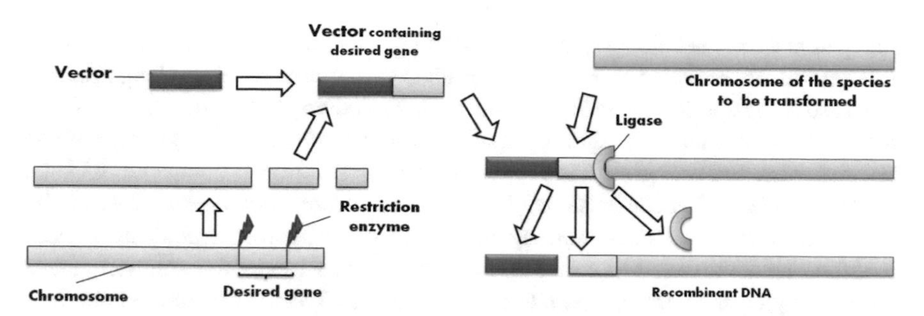

Fig. 2 The schematic diagram represents the genetic engineering process for obtaining genetically modified species for better productivity and performance of fungi

mutagenesis library. Further LiP variants genes and wild type gene were expressed in *S. cerevisiae* EBY100 and were used for degradation of structurally different azo dyes (Amido black 10B, Evans blue and Guinea green). The mutant ML3 and ML8 showed high catalytic activity toward Evans blue, ML2 and ML6 toward amido black 10B and ML3 and ML5 can catalyze Guinea green. In another investigation *Pleos-dyp1* gene of the fungus *Pleurotus ostreatus* that capable to degrade Acetyl Yellow G (AYG), RBBR and Acid Blue 129 (AB129) was successfully expressed in *Trichoderma atroviride*. Successfully genetically manipulated *T. atroviride* can degrade mono and di-azo dye, anthracenedione and anthraquinone dyes with its extracellularly produced DyP1 (dye-decolorizing peroxidase) activity. However, the recombinant *T. atroviride* was able to degrade AYG and AB129 at significant level [32]. In a study, Xu et al. [129] used cDNA library of 1092 bp full length from Manganese peroxidase producing fungus *Ganoderma lucidum* strain that designated as *GluMnP1* to construct eukaryotic expression vector pAO815::*GlMnP*. The vector pAO815::*GlMnP* was successfully transferred in *Pichia pastoris* SMD116. The recombinant *Pichia pastoris* SMD116 was capable to decolorize four dyes namely drimaren red K-4Bl, drimaren blue CL-BR, disperse navy blue HGL and drimaren yellow X8GN and phenol. Similarly, *mnp3* gene responsible for MnP production was cloned from *Cerrena unicolor* BBP6 and functionally expressed in *Pichia pastoris*. The resulting recombinant has great potential in decolorization of RBBR, methyl orange, bromophenol blue, crystal violet and brilliant blue R and two polyaromatic hydrocarbons (phenanthrene and fluorene) [130] (Fig. 2).

7 Toxicity of Fungal Degraded Products

Synthetic dyes have some level of toxicity to different types of organisms that can affect the normal physiological and metabolic process of the organism. As discussed above, fungi have huge potential for their application in bioremediation of dye contaminated wastewater and fungi can degrade the dye molecules into their smaller or constituents units. Mostly, these fungal degraded end products were less toxic than parent dye molecules (Table 3). In a study, Assess et al. [9] conducted an experiment using *Aspergillus niger* for the degradation of congo red dye. Congo red was degraded into naphthalene sulfonate and cycloheptadienylium and in the phytotoxicity assessment *Zea maize* and *Solanum leucopersicum* showed higher germinate rate and shoot and root growth in fungal treated dye solution as compared to Congo red dye solution. In another study, *Trichoderma tomentosum* degraded Acid Red 3R into 4-aminonapthalene-1-sulfonic acid, 8-amino-7-hydroxynapthelene-1 and 3-disulfonic acid. Further, the toxicity assessment of fungal treated dye solution showed less toxicity on the germination rate and growth of *Glycine max* and *Adenanthera microsperma* as compared to pure acid red 3R solution. Conversely, some studies were also reported that fugal treated some dye solution may more toxic than their parent dye molecule. In an investigation, Almeida and Corso [6] tested the procion red MX-5B dye degradation ability of *Aspergillus terreus*. In the end-product analysis, primary and secondary amines were found as fungal degraded metabolites of procion red MX-5B. In the toxicity analysis, fungal treated dye solution showed 10-folds increase in toxicity toward shoot and root growth of *Lactuca sativa* while cause 100% death of *Artemia salina* larvae as compared to parent dye solution. Therefore, the toxicity test should be taken into consideration after treatment of dye contaminated water which can reduce the environmental risk.

8 Conclusion

Fungi have enormous potential for their application in bioremediation of wastewater. This chapter discussed the application of fungi in dye decolorization process and the role of cell surface functional groups, enzyme, proteins and genes in decolorization process. Fungi in the decolorization of dye use biosorption and biodegradation mechanism. On the surface of fungi, many types of functional groups are present that are associated with biosorption of dye molecules. While, MnP, LiP and laccase extracellularly secreted by fungi that play crucial role in dye degradation. Some other enzymes including DCIP reductase, Azo-reductase and *Bilirubin oxidase* have also been reported for their role in dye degradation. Several studies have suggested that these enzymes are expressed by functional group of genes that play important role in dye degradation. Transgenic fungi have shown enhanced dye degradation potential that provide a new aspect in the field of dye decolorization by fungi and is needed more research. Genetically modified microorganisms are expected to have huge potential for future application.

Table 3 Toxicity of fungal degraded dye end products

Name of the Fungi	Dye	Degraded product	Toxicity test	Toxicity of the dye degraded products	References
Aspergillus terreus	Procion Red MX-5B	Primary and secondary amines	Phytotoxicity	Fungal degraded metabolites increased the percent inhibition of root growth (0.87 cm) around 5–50% of *Lactuca sativa* which denote a tenfold increase in toxicity as compare to pure dye. The pure dye were non-toxic for *Artemia salina* larvae while degraded product cause 100% death of larvae	Almeida and Corso [6]
Ceriporia lacerate	Congo Red	Napthylamine and benzidine		In the toxicity test, *Amaranthus mangostanus L.* and *Sesamum indicum* showed less germination rate (55.83 and 38.89%) in fungal degraded dye end products as compared to pure dye (62.50 and 60.00%)	Wang et al. [126]
Aspergillus niger	Congo Red	Napthalene sulfonate and Cycloheptadienylium	Phytotoxicity	*Zea mais and Solanum leucopersicum where degraded* showed maximum percent germination (82.30 and 81.27%), root length (3.16 and 3.20 cm) and shoot length (10.9 and 4.66 cm) in fungal degraded Congo red metabolite as compared to pure dye {germination (60.35 and 60.31%), root length (1.7 and 2.06 cm) and shoot length (6.5 and 1.8 cm)}	Assess et al. [9]

(continued)

Table 3 (continued)

Name of the Fungi	Dye	Degraded product	Toxicity test	Toxicity of the dye degraded products	References
Trichoderma tomentosum	Acid Red 3R	4-aminonapthalene-1-sulfonic acid, 8-amino-7-hydroxynapthelene-1 and 3-disulfonic acid	Phytotoxicity	After biodegradation, the toxicity test was done with two plant seed *Glycine max* and *Adenanthera microsperma*, which denote that degraded product was less toxic (percent germination-80 and 87%, root length-2.4 and 2.8 cm and shoot length-4.1 and 2.8 cm) as compared to pure dye (percent germination-73 and 83%, root length-0.9 and 1.3 cm and shoot length-3.0 and 2.6 cm)	He et al. [50]
Oudeman-siella canarii	Congo Red	Napthalene	Microtox test	The microtox assay detected that degraded product was less toxic than pure dye	Iark et al. [58]
Peyronellaea prosopidis	Scarlet RR	N-(1l3-chlorinin-2-yl)-2-{methyl[(4-oxo-3,4-dihydroquinolin-2-yl)methyl] amino}acetamide, N-ethyl-1l3-chlorinin-2-amine, N-(1l3-chlorinin-2-yl)-2-{[(4-oxo-3,4-dihydronaphthalen-2-yl)methyl]amino}acetamide, and 5-({[2-(1l3-chlorinin-2-ylamino) ethyl]amino}methyl)cyclohexa-2,4-dien-1-one	Phytotoxicity	Toxicity test was analyzed with *Triticum aestivum* and *Peyronellaea prosopidis* where degraded product showed higher percent germination, root length and shoot length as compared to pure dye	Bankole et al. [13]
Aspergillus Flavus	Malachite green	N-demethylated primary and Secondary arylamines	Phytotoxicity and Microbial toxicity	On the basis of toxicity assessment, degraded product was less toxic as compare to control	Barapatre et al. [15]

(continued)

Table 3 (continued)

Name of the Fungi	Dye	Degraded product	Toxicity test	Toxicity of the dye degraded products	References
Aspergillus terreus	Direct Blue-1	[4,5-Diazotricyclo[4.3.0.0(3,7) non-4-in-2-one], [(Phenol,2,6-bis(1,1-Diemethylethyl)-] and [1,2-Benzene Dicarboxylic Acid, 3-Nitro)]	Phytotoxicity test	Toxicity test was done with *Solanum leucopersicum* and *Triticum aestivum* seed where degraded product was less toxic as compared to pure dye	Singh and Dwivedi [109]
Aspergillus niger	Reactive red-120	2-aminobenzenesulfonic acid and 3-methanesulfinylbut-3-en-2-one	Phytotoxicity test	Mung beans seed used to check the toxicity of degraded metabolites where percent germination, root length and shoot length were less toxic in comparison with Reactive red	Su and Lin (2013)
Phomopsis sp.	Remazol Brilliant Blue R	Napthylamine and Benzidine	Phytotoxicity and Microbial test	Toxicity test was analyzed with *Phaseolus mungo* seed and *Azotobacter vinelandii* (MTCC 2460), *Azospirillum brasilense* (MTCC 4034) and *Bacillus cereus* (430) microbes where results showed that degraded product was less toxic as compared to pure dye	Navada et al. [87]

Acknowledgements V. Kumar and G. Singh thankful to University Grants Commission (UGC), Government of India for providing UGC-Non NET fellowship.

References

1. Abd El-Rahim WM, Moawad H, Azeiz AZA, Sadowsky MJ (2017) Optimization of conditions for decolorization of azo-based textile dyes by multiple fungal species. J Biotechnol 260:11–17
2. Abe FR, Machado AL, Soares AM, de Oliveira DP, Pestana JL (2019) Life history and behavior effects of synthetic and natural dyes on Daphnia magna. Chemosphere 236:124390
3. Agrawal K, Verma P (2019) Column bioreactor of immobilized Stropharia sp. ITCC 8422 on natural biomass support of L. cylindrica for biodegradation of anthraquinone violet R. Bioresour Technol Rep 8:100345
4. Akpinar M, Urek RO (2017) Induction of fungal laccase production under solid state bioprocessing of new agroindustrial waste and its application on dye decolorization. 3 Biotech 7(2):98
5. Alam R, Ardiati FC, Solihat NN, Alam MB, Lee SH, Yanto DHY, Kim S et al (2020) Biodegradation and metabolic pathway of anthraquinone dyes by Trametes hirsuta D7 immobilized in light expanded clay aggregate and cytotoxicity assessment. J Hazard Mater 124176
6. Almeida EJR, Corso CR (2014) Comparative study of toxicity of azo dye Procion Red MX-5B following biosorption and biodegradation treatments with the fungi Aspergillus niger and Aspergillus terreus. Chemosphere 112:317–322
7. Andleeb S, Atiq N, Robson GD, Ahmed S (2012) An investigation of anthraquinone dye biodegradation by immobilized Aspergillus flavus in fluidized bed bioreactor. Environ Sci Pollut Res 19(5):1728–1737
8. Asgher M, Yasmeen Q, Iqbal HMN (2013) Enhanced decolorization of Solar brilliant red 80 textile dye by an indigenous white rot fungus Schizophyllum commune IBL-06. Saudi J Biol Sci 20(4):347–352
9. Asses N, Ayed L, Hkiri N, Hamdi M (2018) Congo red decolorization and detoxification by Aspergillus niger: removal mechanisms and dye degradation pathway. BioMed Res Int 2018
10. Bae JS, Freeman HS (2007) Aquatic toxicity evaluation of new direct dyes to the Daphnia magna. Dyes Pigm 73(1):81–85
11. Baldrian P, Merhautová V, Gabriel J, Nerud F, Stopka P, Hrubý M, Beneš MJ (2006) Decolorization of synthetic dyes by hydrogen peroxide with heterogeneous catalysis by mixed iron oxides. Appl Catal B: Environ 66(3–4):258–264
12. Bankole PO, Adekunle AA, Govindwar SP (2018) Enhanced decolorization and biodegradation of acid red 88 dye by newly isolated fungus, Achaetomium strumarium. J Environ Chem Eng 6(2):1589–1600
13. Bankole PO, Adekunle AA, Obidi OF, Chandanshive VV, Govindwar SP (2018) Biodegradation and detoxification of Scarlet RR dye by a newly isolated filamentous fungus, Peyronellaea Prosopidis. Sustain Environ Res 28(5):214–222
14. Bankole PO, Adekunle AA, Obidi OF, Olukanni OD, Govindwar SP (2017) Degradation of indigo dye by a newly isolated yeast, Diutina rugosa from dye wastewater polluted soil. J Environ Chem Eng 5(5):4639–4648
15. Barapatre A, Aadil KR, Jha H (2017) Biodegradation of malachite green by the ligninolytic fungus Aspergillus flavus. CLEAN–Soil Air Water 45(4):1600045
16. Bazin I, Hassine AIH, Hamouda YH, Mnif W, Bartegi A, Lopez-Ferber M, Gonzalez C et al (2012) Estrogenic and anti-estrogenic activity of 23 commercial textile dyes. Ecotoxicol Environ Saf 85:131–136
17. Behrens CJ, Zelena K, Berger RG (2016) Comparative cold shock expression and characterization of fungal dye-decolorizing peroxidases. Appl Biochem Biotechnol 179(8):1404–1417

18. Biko OD, Viljoen-Bloom M, Van Zyl WH (2020) Microbial lignin peroxidases: applications, production challenges and future perspectives. Enzym Microb Technol 109669
19. Bosco F, Mollea C, Ruggeri B (2017) Decolorization of Congo Red by Phanerochaete chrysosporium: the role of biosorption and biodegradation. Environ Technol 38(20):2581–2588
20. Bouacem K, Rekik H, Jaouadi NZ, Zenati B, Kourdali S, El Hattab M, Bouanane-Darenfed A (2018) Purification and characterization of two novel peroxidases from the dye-decolorizing fungus Bjerkandera adusta strain CX-9. Int J Biol Macromol 106:636–646
21. Burke R, Cairney J (2002) Laccases and other polyphenol oxidases in ecto-and ericoid mycorrhizal fungi. Mycorrhiza 12(3):105–116
22. Casieri L, Anastasi A, Prigione V, Varese GC (2010) Survey of ectomycorrhizal, litter-degrading, and wood-degrading Basidiomycetes for dye decolorization and ligninolytic enzyme activity. Antonie Van Leeuwenhoek 98(4):483–504
23. Cha CJ, Doerge DR, Cerniglia CE (2001) Biotransformation of Malachite Green by the FungusCunninghamella elegans. Appl Environ Microbiol 67(9):4358–4360
24. Chairin T, Nitheranont T, Watanabe A, Asada Y, Khanongnuch C, Lumyong S (2013) Biodegradation of bisphenol A and decolorization of synthetic dyes by laccase from white-rot fungus, Trametes Polyzona. Appl Biochem Biotechnol 169(2):539–545
25. Chakraborty S, Basak B, Dutta S, Bhunia B, Dey A (2013) Decolorization and biodegradation of congo red dye by a novel white rot fungus Alternaria alternata CMERI F6. Biores Technol 147:662–666
26. Chawachart N, Khanongnuch C, Watanabe T, Lumyong S (2004) Rice bran as an efficient substrate for laccase production from thermotolerant basidiomycete Coriolus versicolor strain RC3. Fungal Divers 15:23–32
27. Chequer FMD, Angeli JPF, Ferraz ERA, Tsuboy MS, Marcarini JC, Mantovani MS, de Oliveira DP (2009) The azo dyes Disperse Red 1 and Disperse Orange 1 increase the micronuclei frequencies in human lymphocytes and in HepG2 cells. Mutat Res Genet Toxicol Environ Mutagen 676(1–2):83–86
28. Chio C, Sain M, Qin W (2019) Lignin utilization: a review of lignin depolymerization from various aspects. Renew Sustain Energy Rev 107:232–249
29. Choinowski T, Blodig W, Winterhalter KH, Piontek K (1999) The crystal structure of lignin peroxidase at 1.70 Å resolution reveals a hydroxy group on the Cβ of tryptophan 171: a novel radical site formed during the redox cycle. J Mol Biol 286:809–827
30. Ciullini I, Tilli S, Scozzafava A, Briganti F (2008) Fungal laccase, cellobiose dehydrogenase, and chemical mediators: combined actions for the decolorization of different classes of textile dyes. Biores Technol 99(15):7003–7010
31. Couto SR (2009) Dye removal by immobilised fungi. Biotechnol Adva 27(3):227–235
32. Cuamatzi-Flores J, Esquivel-Naranjo E, Nava-Galicia S, López-Munguía A, Arroyo-Becerra A, Villalobos-López MA, Bibbins-Martínez M (2019) Differential regulation of Pleurotus ostreatus dye peroxidases gene expression in response to dyes and potential application of recombinant Pleos-DyP1 in decolorization. PloS One 14(1):e0209711
33. Culp SJ, Mellick PW, Trotter RW, Greenlees KJ, Kodell RL, Beland FA (2006) Carcinogenicity of malachite green chloride and leucomalachite green in B6C3F1 mice and F344 rats. Food Chem Toxicol 44(8):1204–1212
34. de Lima ROA, Bazo AP, Salvadori DMF, Rech CM, de Palma Oliveira D, de Aragão Umbuzeiro G (2007) Mutagenic and carcinogenic potential of a textile azo dye processing plant effluent that impacts a drinking water source. Mutat Res Genet Toxicol Environ Mutagen 626(1–2):53–60
35. Deska M, Kończak B (2019) Immobilized fungal laccase as "green catalyst" for the decolourization process–state of the art. Process Biochem 84:112–123
36. Değerli E, Yangın S, Cansaran-Duman D (2019) Determination of the effect of RBBR on laccase activity and gene expression level of fungi in lichen structure. 3 Biotech 9(8):297
37. Ding Y, Sun C, Xu X (2009) Simultaneous identification of nine carcinogenic dyes from textiles by liquid chromatography/electrospray ionization mass spectrometry via negative/positive ion switching mode. Eur J Mass Spectrom 15(6):705–713

38. Đurđić KI, Ostafe R, Prodanović O, Đelmaš AĐ, Popović N, Fischer R, Prodanović R et al (2020) Improved degradation of azo dyes by lignin peroxidase following mutagenesis at two sites near the catalytic pocket and the application of peroxidase-coated yeast cell walls. Front Environ Sci Eng 15(2):1–10
39. Elmorsi TM, Riyad YM, Mohamed ZH, Abd El Bary HM (2010) Decolorization of Mordant red 73 azo dye in water using H2O2/UV and photo-Fenton treatment. J Hazard Mater 174(1–3):352–358
40. Falade AO, Nwodo UU, Iweriebor BC, Green E, Mabinya LV, Okoh AI (2017) Lignin peroxidase functionalities and prospective applications. MicrobiologyOpen 6(1):e00394
41. Fan F, Zhuo R, Sun S, Wan X, Jiang M, Zhang X, Yang Y (2011) Cloning and functional analysis of a new laccase gene from Trametes sp. 48424 which had the high yield of laccase and strong ability for decolorizing different dyes. Bioresource Technol 102(3):3126–3137
42. Farrell RL, Murtagh KE, Tien M, Mozuch MD, Kirk TK (1989) Physical and enzymatic properties of lignin peroxidase isoenzymes from Phanerochaete chrysosporium. Enzyme Microb Technol 11(6):322–328
43. Ferraz ER, Grando MD, Oliveira DP (2011) The azo dye Disperse Orange 1 induces DNA damage and cytotoxic effects but does not cause ecotoxic effects in Daphnia similis and Vibrio fischeri. J Hazard Mater 192(2):628–633
44. Flores JC, Esquivel-Naranjo E, Nava-Galicia S, López-Munguía A, Arroyo-Becerra A, Villalobos-López MA, Bibbins-Martínez M (2019) Differential regulation of Pleurotus ostreatus dye peroxidases gene expression in response to dyes and potential application of recombinant Pleos-DyP1 in decolorization. Plos one 14(1):e0209711
45. Fonseca MI, Molina MA, Winnik DL, Busi MV, Fariña JI, Villalba LL, Zapata PD (2018) Isolation of a laccase-coding gene from the lignin-degrading fungus Phlebia brevispora BAFC 633 and heterologous expression in Pichia pastoris. J Appl Microbiol 124(6):1454–1468
46. Furukawa T, Bello FO, Horsfall L (2014) Microbial enzyme systems for lignin degradation and their transcriptional regulation. Front Biol 9(6):448–471
47. Gomaa OM (2012) Ethanol induced response in Phanerochaete chrysosporium and its role in the decolorization of triarylmethane dye. Ann Microbiol 62(4):1403–1409
48. Güngördü A, Birhanli A, Ozmen M (2013) Biochemical response to exposure to six textile dyes in early developmental stages of Xenopus laevis. Environ Sci Pollut Res 20(1):452–460
49. Harazono K, Watanabe Y, Nakamura K (2003) Decolorization of azo dye by the white-rot basidiomycete Phanerochaete sordida and by its manganese peroxidase. J Biosci Bioeng 95(5):455–459
50. He XL, Song C, Li YY, Wang N, Xu L, Han X, Wei DS (2018) Efficient degradation of azo dyes by a newly isolated fungus Trichoderma tomentosum under non-sterile conditions. Ecotoxicol Environ Saf 150:232–239
51. Heinzkill M, Bech L, Halkier T, Schneider P, Anke T (1998) Characterization of laccases and peroxidases from wood-rotting fungi (family Coprinaceae). Appl Environ Microbiol 64(5):1601–1606
52. Hofrichter M (2002) lignin conversion by manganese peroxidase (MnP). Enzyme Microb Technol 30(4):454–466
53. Hu QH, Qiao SZ, Haghseresht F, Wilson MA, Lu GQ (2006) Adsorption study for removal of basic red dye using bentonite. Ind Eng Chem Res 45(2):733–738
54. Huang S, Huang D, Qitang WU, Meifang HOU, Xiaoyan TANG, Jian ZHOU (2020) Effect of environmental C/N ratio on activities of lignin-degrading enzymes produced by Phanerochaete chrysosporium. Pedosphere 30(2):285–292
55. Huang Q, Wang C, Zhu L, Zhang D, Pan C (2020) Purification, characterization, and gene cloning of two laccase isoenzymes (Lac1 and Lac2) from Trametes hirsuta MX2 and their potential in dye decolorization. Mol Biol Rep 47(1):477–488
56. Huy ND, Ha DTT, Khoo KS, Lan PTN, Quang HT, Loc NH, Show PL et al (2020) Synthetic dyes removal by Fusarium oxysporum HUIB02 and stimulation effect on laccase accumulation. Environ Technol Innov 19:101027

57. IARC (1993) Monographs on the evaluation of carcinogenic risks to humans, occupational exposures of hairdressers and barbersand personal use of hair colourants; some hair dyes, cosmetic colourants, industrial dyestuffs and aromatic amines, vol 57
58. Iark D, dos Reis Buzzo AJ, Garcia JAA, Côrrea VG, Helm CV, Corrêa RCG, Peralta RM et al (2019) Enzymatic degradation and detoxification of azo dye Congo red by a new laccase from Oudemansiella canarii. Bioresour Technol 289:121655
59. Jadhav UU, Dawkar VV, Telke AA, Govindwar SP (2009) Decolorization of Direct Blue GLL with enhanced lignin peroxidase enzyme production in Comamonas sp UVS. J Chem Technol Biotechnol Int Res Process Environ Clean Technol 84(1):126–132
60. Janusz G, Kucharzyk KH, Pawlik A, Staszczak M, Paszczynski AJ (2013) Fungal laccase, manganese peroxidase and lignin peroxidase: gene expression and regulation. Enzyme Microb Technol 52(1):1–12
61. Jasińska A, Paraszkiewicz K, Sip A, Długoński J (2015) Malachite green decolorization by the filamentous fungus Myrothecium roridum–mechanistic study and process optimization. Biores Technol 194:43–48
62. Jin X, Ning Y (2013) Laccase production optimization by response surface methodology with Aspergillus fumigatus AF1 in unique inexpensive medium and decolorization of different dyes with the crude enzyme or fungal pellets. J Hazard Mater 262:870–877
63. Kabbout R, Taha S (2014) Biodecolorization of textile dye effluent by biosorption on fungal biomass materials. Phys Procedia 55:437–444
64. Karimi A, Aghbolaghy M, Khataee A, Shoa Bargh S (2012) Use of enzymatic bio-Fenton as a new approach in decolorization of malachite green. Sci World J 2012
65. Kaushik P, Malik A (2009) Fungal dye decolourization: recent advances and future potential. Environ Int 35(1):127–141
66. Khan R, Bhawana P, Fulekar MH (2013) Microbial decolorization and degradation of synthetic dyes: a review. Rev Environ Sci Biotechnol 12(1):75–97
67. Krishnamoorthy R, Jose PA, Ranjith M, Anandham R, Suganya K, Prabhakaran J, Kumutha K (2018) Decolourisation and degradation of azo dyes by mixed fungal culture consisted of Dichotomomyces cejpii MRCH 1-2 and Phoma tropica MRCH 1-3. J Environ Chem Eng 6(1):588–595
68. Kumar V, Dwivedi SK (2021b) Mycoremediation of heavy metals: processes, mechanisms, and affecting factors. Environ Sci Pollut Res 28:10375–10412
69. Kumar V, Singh S, Singh G, Dwivedi SK (2019) Exploring the cadmium tolerance and removal capability of a filamentous fungus fusarium solani. Geomicrobiol J 36(9):782–791
70. Kumar CG, Mongolla P, Joseph J, Sarma VUM (2012) Decolorization and biodegradation of triphenylmethane dye, brilliant green, by Aspergillus sp. isolated from Ladakh, India. Process Biochem 47(9):1388–1394
71. Kumar V, Dwivedi SK (2019a) Hexavalent chromium stress response, reduction capability and bioremediation potential of *Trichoderma* sp. isolated from electroplating wastewater. Ecotoxicol Environ Saf 185:109734
72. Kumar V, Dwivedi SK (2019b) Hexavalent chromium reduction ability and bioremediation potential of *Aspergillus flavus* CR500 isolated from electroplating wastewater. Chemosphere 237:124567
73. Kumar V, Dwivedi SK (2020) Multimetal tolerant fungus *Aspergillus flavus* CR500 with remarkable stress response, simultaneous multiple metal/loid removal ability and bioremediation potential of wastewater. Environ Technol Innov 101075
74. Kumar V, Dwivedi SK (2021a) A review on accessible techniques for removal of hexavalent Chromium and divalent Nickel from industrial wastewater: recent research and future outlook. J Clean Prod 295:126229. https://doi.org/10.1016/j.jclepro.2021.126229
75. Kumar V, Dwivedi SK (2021c) Bioremediation mechanism and potential of copper by actively growing fungus Trichoderma lixii CR700 isolated from electroplating wastewater. J Environ Manag 277:111370
76. Kunjadia PD, Sanghvi GV, Kunjadia AP, Mukhopadhyay PN, Dave GS (2016) Role of ligninolytic enzymes of white rot fungi (Pleurotus spp.) grown with azo dyes. SpringerPlus 5(1):1487

77. Lade HS, Waghmode TR, Kadam AA, Govindwar SP (2012) Enhanced biodegradation and detoxification of disperse azo dye Rubine GFL and textile industry effluent by defined fungal-bacterial consortium. Int Biodeterior Biodegrad 72:94–107
78. Lallawmsanga AKP, Singh BP (2019) Exploration of macrofungi in sub-tropical semi-evergreen Indian forest ecosystems. Biol Macrofungi 1
79. Lee AH, Jang Y, Kim GH, Kim JJ, Lee SS, Ahn BJ (2017) Decolorizing an anthraquinone dye by Phlebia brevispora: Intra-species characterization. Eng Life Sci 17(2):125–131
80. Li WY, Chen FF, Wang SL (2010) Binding of reactive brilliant red to human serum albumin: insights into the molecular toxicity of sulfonic azo dyes. Protein Pept Lett 17(5):621–629
81. Li HX, Xu B, Tang L, Zhang JH, Mao ZG (2015) Reductive decolorization of indigo carmine dye with Bacillus sp. MZS10. Int Biodeterior Biodegradation 103:30–37
82. Liu S, Xu X, Kang Y, Xiao Y, Liu H (2020) Degradation and detoxification of azo dyes with recombinant ligninolytic enzymes from Aspergillus sp. with secretory overexpression in Pichia pastoris. R Soc Open Sci 7(9):200688
83. Maas R, Chaudhari S (2005) Adsorption and biological decolourization of azo dye Reactive Red 2 in semicontinuous anaerobic reactors. Process Biochem 40(2):699–705
84. Mansour HB, Ayed-Ajmi Y, Mosrati R, Corroler D, Ghedira K, Barillier D, Chekir-Ghedira L (2010) Acid violet 7 and its biodegradation products induce chromosome aberrations, lipid peroxidation, and cholinesterase inhibition in mouse bone marrow. Environ Sci Pollut Res 17(7):1371–1378
85. Miki Y, Pogni R, Acebes S, Lucas F, Fernandez-Fueyo E, Baratto MC, Basosi R et al (2013) Formation of a tyrosine adduct involved in lignin degradation by Trametopsis cervina lignin peroxidase: a novel peroxidase activation mechanism. Biochem J 452(3):575–584
86. Moldes D, Fernández-Fernández M, Sanromán M (2012) Role of laccase and low molecular weight metabolites from Trametes versicolor in dye decolorization. Sci World J 2012
87. Navada KK, Sanjeev G, Kulal A (2018) Enhanced biodegradation and kinetics of anthraquinone dye by laccase from an electron beam irradiated endophytic fungus. Int Biodeterior Biodegrad 132:241–250
88. Nestmann ER, Douglas GR, Matula TI, Grant CE, Kowbel DJ (1979) Mutagenic activity of rhodamine dyes and their impurities as detected by mutation induction in Salmonella and DNA damage in Chinese hamster ovary cells. Can Res 39(11):4412–4417
89. Novotný Č, Dias N, Kapanen A, Malachová K, Vándrovcová M, Itävaara M, Lima N (2006) Comparative use of bacterial, algal and protozoan tests to study toxicity of azo-and anthraquinone dyes. Chemosphere 63(9):1436–1442
90. Oliveira SF, da Luz JMR, Kasuya MCM, Ladeira LO, Junior AC (2018) Enzymatic extract containing lignin peroxidase immobilized on carbon nanotubes: potential biocatalyst in dye decolourization. Saudi J Biol Sci 25(4):651–659
91. Oliveira GAR, Ferraz ERA, Chequer FMD, Grando MD, Angeli JPF, Tsuboy MS, Zanoni MVB (2010) Chlorination treatment of aqueous samples reduces, but does not eliminate, the mutagenic effect of the azo dyes Disperse Red 1, Disperse Red 13 and Disperse Orange 1. Mutat Res Genet Toxicol Environ Mutagen 703(2):200–208
92. Pan H, Xu X, Wen Z, Kang Y, Wang X, Ren Y, Huang D (2017) Decolorization pathways of anthraquinone dye Disperse Blue 2BLN by Aspergillus sp. XJ-2 CGMCC12963. Bioengineered 8(5):630–641
93. Pandi A, Kuppuswami GM, Ramudu KN, Palanivel S (2019) A sustainable approach for degradation of leather dyes by a new fungal laccase. J Clean Prod 211:590–597
94. Parshetti GK, Kalme SD, Gomare SS, Govindwar SP (2007) Biodegradation of Reactive blue-25 by Aspergillus ochraceus NCIM-1146. Biores Technol 98(18):3638–3642
95. Piontek K, Glumoff T, Winterhalter K (1993) Low pH crystal structure of glycosylated lignin peroxidase from Phanerochaete chrysosporium at 2.5 Å resolution. FEBS Lett 315(2):119–124
96. Plácido J, Capareda S (2015) Ligninolytic enzymes: a biotechnological alternative for bioethanol production. Bioresour Bioprocess 2(1):23

97. Poulos TL, Edwards SL, Wariishi H, Gold MH (1993) Crystallographic refinement of lignin peroxidase at 2 A. J Biol Chem 268(6):4429–4440
98. Purnomo AS, Mauliddawati VT, Khoirudin M, Yonda AF, Nawfa R, Putra SR (2019) Bio-decolorization and novel bio-transformation of methyl orange by brown-rot fungi. Int J Environ Sci Technol 16(11):7555–7564
99. Qin P, Wu Y, Adil B, Wang J, Gu Y, Yu X, Chen X et al (2019) Optimization of Laccase from Ganoderma lucidum decolorizing Remazol Brilliant Blue R and Glac1 as main laccase-contributing gene. Molecules 24(21):3914
100. Rajhans G, Sen SK, Barik A, Raut S (2020) Elucidation of fungal dye-decolorizing peroxidase (DyP) and ligninolytic enzyme activities in decolorization and mineralization of azo dyes. J Appl Microbiol
101. Rawat AP, Kumar V, Singh DP (2020) A combined effect of adsorption and reduction potential of biochar derived from Mentha plant waste on removal of methylene blue dye from aqueous solution. Sep Sci Technol 55(5):907–921
102. Rizqi HD, Purnomo AS (2017) The ability of brown-rot fungus Daedalea dickinsii to decolorize and transform methylene blue dye. World J Microbiol Biotechnol 33(5):92
103. Robens JF, Dill GS, Ward JM, Joiner JR, Griesemer RA, Douglas JF (1980) Thirteen-week subchronic toxicity studies of Direct Blue 6, Direct Black 38, and Direct Brown 95 dyes. Toxicol Appl Pharmacol 54(3):431–442
104. Saravanakumar T, Palvannan T, Kim DH, Park SM (2013) Manganese peroxidase H4 isozyme mediated degradation and detoxification of triarylmethane dye malachite green: optimization of decolorization by response surface methodology. Appl Biochem Biotechnol 171(5):1178–1193
105. Sen SK, Raut S, Bandyopadhyay P, Raut S (2016) Fungal decolouration and degradation of azo dyes: a review. Fungal Biol Rev 30(3):112–133
106. Shanmugam S, Ulaganathan P, Swaminathan K, Sadhasivam S, Wu YR (2017) Enhanced biodegradation and detoxification of malachite green by Trichoderma asperellum laccase: degradation pathway and product analysis. Int Biodeterior Biodegrad 125:258–268
107. Sharma HK, Xu C, Qin W (2019) Biological pretreatment of lignocellulosic biomass for biofuels and bioproducts: an overview. Waste Biomass Valorization 10(2):235–251
108. Si J, Cui BK, Dai YC (2013) Decolorization of chemically different dyes by white-rot fungi in submerged cultures. Ann Microbiol 63(3):1099–1108
109. Singh G, Dwivedi SK (2020) Decolorization and degradation of Direct Blue-1 (Azo dye) by newly isolated fungus Aspergillus terreus GS28, from sludge of carpet industry. Environ Technol Innov 100751
110. Song L, Shao Y, Ning S, Tan L (2017) Performance of a newly isolated salt-tolerant yeast strain Pichia occidentalis G1 for degrading and detoxifying azo dyes. Biores Technol 233:21–29
111. Sosa-Martínez JD, Balagurusamy N, Montañez J, Peralta RA, Moreira RDFPM, Bracht A, Morales-Oyervides L et al (2020) Synthetic dyes biodegradation by fungal ligninolytic enzymes: process optimization, metabolites evaluation and toxicity assessment. J Hazard Mater 400:123254
112. Spadaro JT, Renganathan V (1994) Peroxidase-catalyzed oxidation of azo dyes: mechanism of Disperse Yellow 3 degradation. Arch Biochem Biophys 312(1):301–307
113. Stammati A, Nebbia C, De Angelis I, Albo AG, Carletti M, Rebecchi C, Dacasto M et al (2005) Effects of malachite green (MG) and its major metabolite, leucomalachite green (LMG), in two human cell lines. Toxicol in vitro 19(7):853–858
114. Stiborová M, Martínek V, Rýdlová H, Hodek P, Frei E (2002) Sudan I is a potential carcinogen for humans: evidence for its metabolic activation and detoxication by human recombinant cytochrome P450 1A1 and liver microsomes. Can Res 62(20):5678–5684
115. Sudha M, Saranya A, Selvakumar G, Sivakumar N (2014) Microbial degradation of azo dyes: a review. Int J Curr Microbiol App Sci 3(2):670–690
116. Tang W, Xu X, Ye BC, Cao P, Ali A (2019) Decolorization and degradation analysis of Disperse Red 3B by a consortium of the fungus Aspergillus sp. XJ-2 and the microalgae Chlorella sorokiniana XJK. RSC Adva 9(25):14558–14566

117. Tapia-Tussell R, Pereira-Patrón A, Alzate-Gaviria L, Lizama-Uc G, Pérez-Brito D, Solis-Pereira S (2020) Decolorization of textile effluent by Trametes hirsuta Bm-2 and lac-T as possible main laccase-contributing gene. Curr Microbiol 1–9
118. Teerapatsakul C, Parra R, Keshavarz T, Chitradon L (2017) Repeated batch for dye degradation in an airlift bioreactor by laccase entrapped in copper alginate. Int Biodeterior Biodegrad 120:52–57
119. Tien M, Kirk TK (1988) Lignin peroxidase of Phanerochaete chrysosporium. Methods Enzymol 161:238–249
120. Tochhawng L, Mishra VK, Passari AK, Singh BP (2019). Endophytic Fungi: role in dye Decolorization. Advances in endophytic fungal research. Springer, Cham, pp 1–15
121. Topaç Şağban FATMA, Dindar E, Uçaroğlu S, Baskaya H (2009) Effect of a sulfonated azo dye and sulfanilic acid on nitrogen transformation processes in soil
122. Tsuboy MS, Angeli JPF, Mantovani MS, Knasmüller S, Umbuzeiro GA, Ribeiro LR (2007) Genotoxic, mutagenic and cytotoxic effects of the commercial dye CI Disperse Blue 291 in the human hepatic cell line HepG2. Toxicol in Vitro 21(8):1650–1655
123. Varjani S, Rakholiya P, Ng HY, You S, Teixeira JA (2020) Microbial degradation of dyes: an overview. Bioresour Technol 123728
124. Venturini S, Tamaro M (1979) Mutagenicity of anthraquinone and azo dyes in Ames' Salmonella typhimurium test. Mutat Res Genet Toxicol 68(4):307–312
125. Voběrková S, Solčány V, Vršanská M, Adam V (2018) Immobilization of ligninolytic enzymes from white-rot fungi in cross-linked aggregates. Chemosphere 202:694–707
126. Wang N, Chu Y, Zhao Z, Xu X (2017) Decolorization and degradation of Congo red by a newly isolated white rot fungus, Ceriporia lacerata, from decayed mulberry branches. Int Biodeterior Biodegrad 117:236–244
127. Wang X, Wang Y, Ning S, Shi S, Tan L (2020) Improving Azo dye decolorization performance and halotolerance of Pichia occidentalis A2 by static magnetic field and possible mechanisms through comparative transcriptome analysis. Front Microbiol 11:712
128. Wesenberg D, Kyriakides I, Agathos SN (2003) White-rot fungi and their enzymes for the treatment of industrial dye effluents. Biotechnol Adv 22(1–2):161–187
129. Xu H, Guo MY, Gao YH, Bai XH, Zhou XW (2017) Expression and characteristics of manganese peroxidase from Ganoderma lucidum in Pichia pastoris and its application in the degradation of four dyes and phenol. BMC Biotechno 17(1):1–12
130. Zhang H, Zhang X, Geng A (2020) Expression of a novel manganese peroxidase from Cerrena unicolor BBP6 in Pichia pastoris and its application in dye decolorization and PAH degradation. Biochem Eng J 153:107402
131. Zhang X, Liu Y, Yan K, Wu H (2007) Decolorization of anthraquinone-type dye by bilirubin oxidase-producing nonligninolytic fungus Myrothecium sp. IMER1. J Biosci Bioeng 104(2):104–110

Enzyme Action for Dye Degradation

Bhautik Dave and Gaurav Sanghvi

Abstract Development in industries with new discoveries and applications are getting acknowledged over the globe. Industries from different segments viz: textile, leather, foods are using the recalcitrant compounds in manufacturing process which is hard to degrade. All such industries discharged the unprocessed waste in water bodies or nature and become serious environmental problem for ecological niche. These dyes cannot be degraded naturally by chemical and physical methods as these methods are not economically viable and environmental friendly. Alternative like biological methods which are effective, cheap and eco-friendly can be a substitute for chemical hazardous methods. Microorganisms like bacteria, fungi and others can degrade many synthetic dyes easily and this method is also applicable for big scale and most preferable for big industries. Different enzymes produced by this microorganism can cleave and break the bonds of complex dye structures like azo bonds and transformation of these hazardous dyes into simple and non-hazardous forms. Different bacterial, fungal and plant enzymes are discovered and isolated from different species of microorganisms and plants by using recombinant and different molecular techniques. Increasement in modification and specificity of enzyme for specific and high activity has also occurred and further research is still going on. There was a simple rule of nature "anything comes from the nature it has to be go back in the nature" so by using this biological method and use of microorganisms we have hope for saving our environment for future life.

Keywords Dye degradation · Fungi · Enzymes · Lacasse · Peroxidase

---

B. Dave · G. Sanghvi (✉)
Department of Microbiology, Marwadi University, Rajkot 360001, India
e-mail: gaurav.sanghvi@marwadieducation.edu.in

S. S. Muthu and A. Khadir (eds.), *Dye Biodegradation, Mechanisms and Techniques*,
Sustainable Textiles: Production, Processing, Manufacturing & Chemistry,
https://doi.org/10.1007/978-981-16-5932-4_6

1 General Introduction

1.1 Dyes

Dyes from natural origin used for painting and dyeing are known from primitive times. Usage of natural dyes started many years ago around 10,200 BC. During this time, the dyes were derived from diverse natural sources like vegetables, plants, and flowers. Dyes extracted and produced from natural sources include indigo, logwood (from plants), alizarin, tyrian purple, yellow, etc., which were all derived from different natural sources. Due to extensive use of natural dyes during 1850s the search for synthetic dyes started to replace/substitute the natural dyes. The synthetic-based colorant "Mauve" was the first synthetic dye reported by William Perkins from Great Britain. This invention has led to a way for better designing of development strategies for manufacturing of synthetic dyes on large scale.

With the expansion in science, coal tar was further revealed as a probable alternative to natural dyes. This discovery has led to development of more than 50 synthetic dyes from the coal tar using different synthetic pathways.

General classification of dyes is mentioned below (Fig. 1).

Synthetic dyes are broadly used in all segments of industries including textiles, leather, food and pharma. These industries discharge a large volume of colored effluent composed of various organic and inorganic chemicals, dyes, waste water, dust, solid particles which contain different and many types of pollutants. Pollution, specifically in the environment, is a common and important worldwide problem. The rapid climb of industries has encouraged the production of heavy tons of liquid waste. Environmental pollution, because of development and industrialization, has

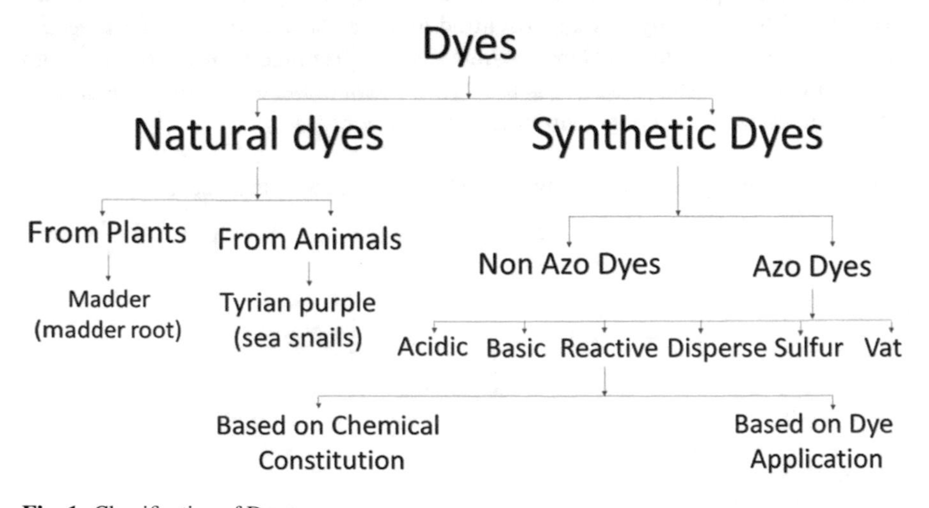

Fig. 1 Classification of Dyes

been identified as a remarkable issue throughout the world which has a detrimental and harmful effect on humans and all ecological niches.

1.2 Environmental Concern

Synthetic dyes fall under the xenobiotic compounds which are hard to degrade biologically, chemically and physically. Among the different class of dyes, azo dyes with the presence of one or more azo groups (–N=N–), are most versatile group of anthraquinone type of dyes. Synthetic dyes are considered as major pollutants and it is hard to remove it from the effluents by conventional wastewater treatments. They are human-caused pollutants causing worsening all the types of life forms viz. Quality of water and air, disturbance in food web and ecological cycles [73]. Moreover, due to its toxicity and carcinogenicity dyes have become a health hazard to all living things and a big problem for environment. Many of intermediate products of the dyes were also found toxic and potentially carcinogenic to all living systems [56].

Specifically, textile manufacturing industries consume enormous volumes of clean water for various processing stages and with that it also produces gallons of liquid effluents with contaminated water which contains different dyes and chemicals [2]. About 45–70 l of effluents are released per kilogram of clothing and fabric during the coloring process of different industries like clothing, leather and color manufacturing industries [53].

Textile effluents and all other color using industry's effluents with contaminated water containing different dyes and chemicals have high chemical and biological oxygen demand, and it also contains degradable and non-biodegradable chemicals such as dyes, dispersants, and detergent, surfactants, levelling agents, heavy metal ions, dissolved inorganics, acids, alkali, toxic and carcinogenic chemicals that are discharged into water bodies and from water bodies it is going spread to other living bodies like soil and air. Major pollutants and potentially hazardous compounds are discharged from the wet processing stages of textile industries.

In textile industries from applied reactive dyes about 30% of them are wasted because of the process of dye hydrolysis in the alkaline dye bath and because of that dye effluent contains 0.6–0.8 g/l of dye which is going to be released into environment without any pre- or post-treatment [8]. It is expectable that only from textile industries around 15% of the dyestuff is released into wastewater [68]. It had been assessed that in the textile industry, about 80% dyes used are azo dyes. These azo dyes on discharge emit toxic materials and create an imbalance in terms of biological oxygen demand. These dyes even change pH of the water and the chemical constituent of the water body. As every ecosystem is interrelated and connected with each other, the overall diversity is disturbed leading to erratic harmful effect on every life form. Also, some azo dyes have been found connected to many life threating diseases like cancer, hepatocarcinoma, and even nuclear anomalies which cause chromosomal aberration in mammalian cells [77].

Physical methods	Chemical methods	Biological methods
Precipitation/sedimentation	Catalytic degradation	Fungal degradation
Coagulation	Chemical precipitation	Algal treatment
Filtration	Reduction	Anaerobic digestion
Adsorption	Oxidation	Stabilization
Flotation	Neutralization	Trickling filters
Flocculation	Ion exchange	Activated sludge
Reverse osmosis	Electrolysis	Enzymatic processes
Membrane treatment	Advanced oxidation process	Combinatorial systems
Distilation	Ozonation	Aerated lagoons
Solvent extraction	Electrochemical oxidation	Surface immobilization
Photolytic degradation	Fenton reaction	

Fig. 2 Different methods for dye degradations

Due to the advancement and improvement in scientific development and technology in chemistry of dyes, a synthesized dye is more resolute and resistant to all forms of degradation in nature [64]. Different methods like coagulation, flocculation, photocatalysis, electrochemical oxidation, filtration and membrane filtration were attempted to degrade azo dye from colored textile waste and leather industry waste. However, they are much economically expensive and not affordable for industrial scale as they generate amine residues containing sludge after degradation which shows carcinogenesis in humans and other aquatic animals. With attempt of all chemical and physical methods, the biological treatment is hope for degradation textile dye effluent. Biological treatment of wastewater varies widely on type of fungal and bacterial cultures used [33].

The implementation of environment defense regulation, which increases awareness of the influence on environment by dyestuffs and effect of pollution on all types of biodiversities has become an attractive topic for researchers and scientists in recent years (Fig. 2).

2 Enzymes: An Important Tool

Enzymes are the active proteins having catalytic activity of biochemical reactions. Enzymes are true biological catalysts having the ability to convert a specific compound by using different methods (as substrate) into products at higher reaction rates and useful for many more. Industrial enzymes represent the core thrust area in field of biological sciences.

One of the highest sailing technical enzymes is in the leather market, followed by the bioethanol market. The food and beverage industry of enzymes sections are

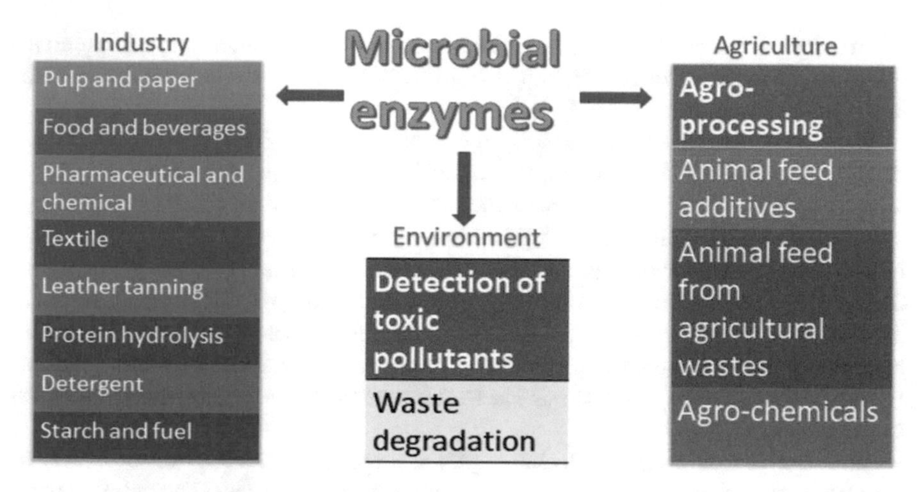

Fig. 3 Application of Microbial enzymes

expected to reach about 1.3 billion US dollar by 2015, from a value of 975 million US dollar in 2010, rising at a compound annual growth rate of 5.1% (Fig. 3).

Reports have established that microorganisms possess the ability to decolorize and degrade the chemically synthetic dyes and convert it to the detoxified compounds. Using microbes dyes are mineralize under specific environmental conditions [21, 79]. Biological methods are hope for saving environment in different ways. However, this method needs to be more reproducible and financially viable for large scale setup. The biological methods need to show high catalytic adaptability, and strength against the xenobiotic dyes and their products. These methods needs to be more resistant to salts, surfactants, detergents, metals like chromium, iron, zinc, different and extreme pHs or high temperatures [3].

It is confirmed that microorganisms can degrade and decolorize dyes. Even consortium of the microbes had led to better degradation results. Microbial enzymes viz: azoreductase, laccase, and peroxidase, are most studied for its capability of high rate of degrading azo dyes [80].

2.1 *Enzymology in Dye Degradation*

Microbes with capability to degrade synthetic dyes possess developed enzyme systems inside them to facilitate the catalytic activities of breaking down of azo bonds leading to dye degradation. These microorganisms produce many and wide types of reductive and oxidative enzymes with different mechanisms. Usually, these enzymes create free radicals that undergo a composite sequence of cleavage reaction ensuing in the decolorization and degradation of azo dyes [79].

Among the treatment of textile effluent, using the different reductive and oxidative enzymes produced by microbes, shows high decolorization and degradation of azo dyes with different mechanism [14].

Enzymatic Azo dyes have azo bridge (–N=N–) from that they are electron-deficient compounds because of the majority of the azo dyes have electron-withdrawing groups in it, which have the ability to generate an electron deficiency and gives them resistance from degradation by the enzymes produced by microbial system [1]. Under proper, specific and suitable conditions, these dyes can be degraded by microbial enzymes with using proper methods [90]. Mostly, for the treatment of industrial effluents containing dyes, enzymes are prerequisites as enzymes have ability to degrade the dye molecules. The characteristic specificity and isolation from different extreme habitats has led to the improvisation of enzymatic strength and toughness [39].

With rise of new recycling technologies and reduced consumption of water in process, has led to sustenance of textile industry in the environment. The few enzymes share common catalytic features which can catalyze the reactions, and display the broad substrate specificity.

2.2 *Classification of Dye Degrading Enzymes*

As per the international union of biochemistry and Molecular biology, enzymes are classified based on system and the type of reactions catalyze by them into seven categories:

1. oxidoreductases
2. transferases
3. hydrolases
4. lyases
5. isomerases
6. ligases
7. translocases.

Oxidoreductases, transferases, and hydrolases are the most abundant forms of enzymes. Oxidoreductase and laccase can degrade synthetic dyes, these are main classes and are further sub-classified according to the chemical name of the substrate and its reaction mechanism.

The most significant dye degrading enzymes which are well studied and widely used are: azoreductase, laccases and peroxidases so in this chapter we are going to know about their sources and applications, structure, mechanism, because they show great activity of degradation against many dyes in different ways [39].

3 Azoreductase

This enzyme has capability to decolorize azo dyes and convert the azo dyes to aromatic amines. For this breaking of azo bonds, Enzyme azoreductase have catalytic activity for reduction of azo groups by utilizing electron donor as NADH and/or NADPH for whole reaction.

Classification of azo reductase

(1) Oxygen sensitive: Microbes secrets this type of enzymes under anaerobic conditions or microaerophilic conditions.
(2) Oxygen insensitive: Obligate aerobic microbes generally secrets these enzymes under aerobic condition.

From the research of Rafii (1990) the presence of azoreductase in anaerobic bacteria is reported, which have the ability to decolorize sulfonated azo dyes in complex media conditions. *Clostridium* and *Eubacterium* are also reported species for sulphonate dye decolorization. Most of the Azoreductase enzymes originated and screened from these strains are oxygen-sensitive, extracellularly and formed constitutively. But far ahead research done by *C. perfringens* exposed that enzyme flavin adenine dinucleotide dehydrogenase, is responsible for azo dye reduction and have the ability to reduce nitroaromatic and other not easily biodegradable compounds [72].

Oxidoreductive Azoreductase enzyme are primarily showing its activity by breaking the bonds resulting in the degradation [39]. Azoreductase are characterized by presence of FMN or FAD-dependent enzymes that use NAD(P)H and cause reduction of azo dyes [19].

High activity of Azoreductase is the one of key factors for decolorization and degradation of azo dyes. These enzymes catalyzes the reductive breakdown of the azo bridge to aromatic amines with support of reducing counterparts such as NADH, NADPH, and $FADH_2$ [14]. At the end of reaction, colorless aromatic amines are produced which have some other activities because they all are original compounds those have their own action and reaction and maybe they are harmful so that they have to be treated again before releasing it into environment. With aid of advanced research and recombinant technology, different types of azoreductases were isolated, screened and characterized from different microbes possessing degradation ability of azo dyes [74].

3.1 *Structure and Mechanism*

By knowing usefulness of azoreductases, many studies and research are underway and different types of azoreductase have been characterized from diversity of microorganisms. These azoreductase enzymes have specific mode of action against a particular dye or can even catalyze diminution of many dyes. Flavin-dependent

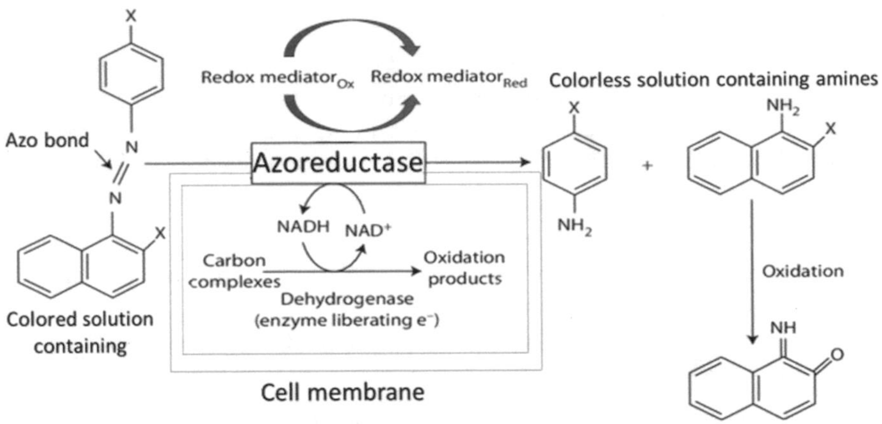

Fig. 4 Mechanism of Azoreductase enzyme

azoreductases are used to solve the structure by crystallization and other techniques. Revealing exact structure of azoreductase can become a crucial step in the path of the rationalization molecular mechanism and its utility [10]. The first exploration of FMN-dependent azoreductase (AzoA) structure is done by [45] from *E. faecalis.* The determination of FMN-dependent NADH-azoreductase (AzoR) crystal structure from *E. coli* has opened the gate for better understanding of the reaction mechanism [38]. Based on research done so far, two-cycle ping-pong mechanism is established which is responsible for reduction reaction [65]. Determination and characterization of the structure of azoreductase which was obtained from species *P. aeruginosa* (paAzoR), was done by using two similar companion models (1) paAzoR2 and (2) paAzoR3 [93]. The study reported the role of NADPH in azoreductases and its mechanism for reduction of FMN (Fig. 4).

At molecular level, FMN is reduced at the active site by NADH. In this reduction process, reduced FMN generates one or two hydrides from N atoms. The reductant equal, like FMN, NADH, and NADPH are always found to be involved in mechanism of azo dyes reduction. In the initial phase of reaction, azo compounds are reduced to hydrazine and added hydrazines are altered into two amines [59].

3.2 *Classification*

Classification is needful for better understanding of azoreductase enzyme. Because of low amino acid homology, it's hard to classify them based on their primary amino acid level. However, it can be categorized in two classes based on its function.

1. Flavin dependent azoreductases
2. Flavin independent azoreductases.

Flavin independent or flavin-dependent reductase enzymes are utilized as an electron donor and shows catalytic activity. Due to this cleavage activity, aromatic amine products are formed in both aerobic as well as anaerobic situation. The flavin-dependent reductases make use of the reducing equivalents NADH and NADPH electron donors [46]. The mechanisms of breakage of azo association by azoreductases engross a four-electron, four proton reduction reaction. The mechanism works with two step reaction.

Step 1: In initial step with aid of reducing equivalents transformation of two electrons happens to the azo dye and they form a hydrazo intermediate. This hydrazo intermediate further breaks into aromatic amines.

Step 2: transfer of two electrons (Singh et al. 2019) [19].

Flavin reliant azoreductases showcases strong similarity with respect to the structural, and mechanism point of view with bigger family of Flavin-dependent quinone reductases [19]. These enzymes with reducing capability not only reduce the azo dyes but also reduce quinonimines, quinones, and nitro groups.

These enzymes also have different applications and as per recent research it is stated that the azoreductase obtained from *B. subtilis* and *E. coli* were fretful to thiol-specific stress in cellular responses [44]. Enzyme Lot6p obtained from *S. cerevisiae*, homolog of azoreductase was also found to be in increased levels during oxidative stress [71]. Moreover, like this azoreductases, other family members of these enzymes are revealed, and still the recombinant substrate remains to develop with it rapidly.

There is another mechanism by which azoreductase enzymes transferring electrons to the dye molecules includes:

(1) Direct enzymatic reduction.

In the mechanism of the direct bacterial enzymatic reduction, a direct physical contact or interaction of azoreductase is going to happen in which azo dye is to transfer electrons from reducing equivalents which is produced by the oxidation of the substrate or coenzyme [11].

(2) Indirect enzymatic reduction.

Indirect enzymatic reduction is completed by a redox mediator which acts as an electron transporter between the dye and the enzyme azoreductase [14]. The reduced electron passage can devolve electrons to electron-deficient azo dyes [66]. The non-sulfonated azo dyes can penetrate and pass through the cell membrane due to their ability to decolorize and also catalyze flavin-dependent cytoplasmic azoreductase. Most of the azo dyes are high molecular weight polar compounds. These dyes cannot strew in cell membranes, hence diminution of such azo dye is not reliant on their intracellular uptake. On the contrary, most of the bacterial cell membranes are hard and it is not easy to penetrate and pass-through this bacterial cell for flavin-containing cofactors and by that they limit the transfer of reducing counterparts by flavins between cytosol to the sulfonated azo dyes [78].

With continuation in search of dye degrading enzyme, recent studies have proved that there are other important enzymes like laccases, veratrine alcohol oxidase, and DCIP(2,6-dichlorophenol-indophenol) reductase which shows high activity for dye degradation. Even though, azoreductases have wide range of activity but apart from it, other recently found enzymes also plays important role in the dye degradation mechanism [67, 79, 81].

3.3 Sources and Application

Azoreductase is identified and derived from various sources and have different uses in different areas. Normally, they are monomeric and dimeric in nature. Tetrameric form of NADPH-dependent azoreductases is an exception reported from *S. aureus* [46, 65].

Microorganisms like bacteria, fungi and yeast showed proficient degradation of azo dyes and quinones under different experimental conditions. The efficiency of the oxidoreductase enzyme system was first verified by. They observed highest decolorization and detoxification of 18 different azo dyes in three model wastewaters. Around 80% decolorization and detoxification of azo dyes it is proved that greater than 50% of them is done by oxidoreductase–enzyme system and this shows potential of this enzyme that it has potential to be used to take action against the toxic azo-dye-containing effluent discharged by textile, leather and color manufacturing industries.

3.4 Laccase

Laccases are one of the biggest and profoundly studied groups of enzyme for dye degradation. They are copper-containing enzymes belonging to family of copper-containing polyphenol oxidases (Singh et al. 2015b). Laccases are the oxidoreductases commonly found and obtained from higher plants, fungi, bacteria, and insects. These enzymes show a wide range of substrate specificity and have the capability to degrade different xenobiotics with diverse and complex chemical structures. Normally laccases merge to one of the electrons of four substrate molecules and form four-electron reductive cleavage. As they are multi-copper oxidases, they are also identified as multi-copper phenol oxidases [28].

Laccases have displayed great potential in the bioremediation process, because of the specific characteristics such as capacity of nonspecific oxidation, lack of necessity for cofactors, and utilization of accessible oxygen as an electron acceptor.

These enzymes have great applications in various biotechnological processes [35, 58, 60, 61, 89]. Due to their high rate of activity, Laccases have been studied in some depth on account of their capacity to decolorize and degrade azo dyes present in textile effluent. They are also used for phenolic compound degradation, in the presence of oxygen as an electron acceptor [83]. Phenolic compound's oxidation starts with the formation of unstable phenoxy radicals as the deprotonation process of the hydroxyl group of phenols leads to the establishment of quinine. Usually, laccases cannot oxidise non-phenolic contents due to high redox potential and its complex structure [28]. On the contrary, the compounds with low molecular weight can supress the oxidation of these compounds. Mostly low molecular weight compounds are redox mediators which can oxidize laccase into radicals which is stable and also can oxidize the high redox potential substrate approximately.

The redox mediators can be natural, as an example, 4-hydroxybenzoic acid, 3-hydroxyanthranilic acid, 4-hydroxy benzyl alcohol, 2,2′-azinobis(3-ethylthiazoline-6-sulphonic acid) (ABTS), 1-hydroxy benzotriazole which are produced by ligninolytic fungi or synthetic [40]. Few reported laccases have a ability to degrade azo dyes with or without using mediators (redox) [63]. These mediators have a stimulatory influence on the degradation of textile dye and improve the adequacy and tolerability of laccases.

Mostly laccase is better and preferable for the detoxification of complex chemicals [17] discharged from the textile, leather, petrochemical, paper and pulp industries. They are used in bioremediation, biodegradation and detoxification for treating the recalcitrant compounds present in soil [35, 57, 60, 61]. Moreover, laccases can also generate a polymeric product useful in synthesis of organic mixture [51, 75]. Additionally, they are also potential enzyme for manufacturing of biosensors and biofuel cells.

Most of the known laccases are discovered from fungi (e.g., white-rot fungi) and other plant origins. Many reports in last few years have indicated presence of laccases in bacteria also [16, 28]. Bacterial laccases have advantages for biotechnological processes due to the absence of post-translational modifications and ease of manipulation using protein engineering approach.

3.4.1 Sources and Applications

Laccases are the most important enzyme for dye degradation because it can oxidize both toxic and nontoxic substrates. It is applied widely in many types of industries like textile, food processing, wood processing, leather manufacturing, color manufacturing, pharmaceutical and chemical. This enzyme is ecologically sustainable and very specific toward its mechanism, and a capable catalyst for many reactions.

There are hundreds of applications of laccase but specifically for dye degradation they can degrade many structurally complex dyes. *Trametes versicolor* (TRV) in the form of pellets are used for the treatment a black liquor discharged from textile industries. The pellets were able to decolorize dyes and showed significant reduction in chemical oxygen demand. After this experiment, they concluded that *T. versicolor* can produce laccases having the ability for complete decolorization of dyes like Congo red, Reactive Blue 15, Amaranth, with no dye sorption. They also found that after decolorization process, toxicity of few dyes changed and final product became less or nontoxic.

Laccase is normally founded from higher plants and fungi but they are also found in some bacterial species such as Streptomyces cyaneus, Streptomyces lavendulae and Marinomonas Mediterranea. Romero et al. [26] founded that bacteria *S. maltophilia* which can decolorize some synthetic dyes like Congo red, methylene blue, toluidine blue, methyl orange, methyl green and methyl pink and along with the industrial effluent with complex and different chemical structures.

Basidiomycetes such as *Theiophora terrestris*, *Lenzites, Betulina* and *Phanerochaete chrysosporium* and white-rot fungi such as *Trametes versicolor*, *Phlebia radiate* and Pleurotus ostreatus, also produce laccase. Many Trichoderma species such as *T. longibrachiatum*, *T. harzianum* and *T. atroviride* are also sources of laccases. With different activities many other species also produce laccase, like *Monocillium indicum* from Laccase family is the first laccase to be considered from Ascomycetes with having peroxidase activity [23]. Ligninolytic laccase is produced by *Pycnoporus cinnabarinus*, while *Pycnoporus sanguineus* produces laccase as phenoloxidase. In plants, laccase plays a role for significations oppositely in fungi it has been concerned in delignification, and for fruiting, pigment production, sporulation, plant pathogenesis and body formation [69].

3.4.2 Fungal Laccases

The fungal laccases were identified and reported for the first time by Bertrand [7]. He is the first person who noticed that in mushroom boletus. This mushroom produces laccase enzyme and when they come in contact with air, the color is changed. Several fungal species are reported for laccase production and among them white-rot fungi are the most prominent producers and widely used for dye degradation. Species like *Trametes Versicolor* (TRV), *Agaricus bisporus*, *Phanerochaete chrysosporium*, *Pleurotus ostreatus* and *Coprinus cinereus* shows many activities like pigment production, plant pathogenesis, fruiting body formation, lignin degradation and sporulation and stress define fungal laccases are involved.

Diversity of structures with different laccases present in nature displays its complexity; but most of them are in monomeric form. Few white rot fungi were reported to have homo and hetero and multimeric forms of laccases. Molecular weight in the range of 50–140 kDa is reported depending upon the organism, although a range of distinctive fungal laccases consists a range of 60–70 kDa around pH 4.0 [5].

Usually fungal laccases are glycosylated, with increase in mass of 10–25%. Characteristics of laccases like conformational stability and protection of the enzyme against proteolysis are due to presence of carbohydrates present in laccases [60, 61].

Advancement in molecular biological techniques and use of them in recombinant fungal laccases, obtained with better applications in bioremediation and biodegradation resolutions because of high specificity, stability and strength of enzyme. For example, proteins Lcc1 and Lcc3 the recombinant one from *T. trogii* and produced in *P. pastoris* showed itself as a biocatalyst of numerous polluting dyes for oxidative degradation. Dye like indigo carmine, which is the most important and widely used dye for manufacturing blue jeans [18]. Furthermore, a recombinant laccase (Lcc IIIb) was the most effective in degradation from *T. versicolor* in *Y. lipolytic*, has shown the ability of decolorization of five phenolic azo dyes with more than 40% effectiveness.

Wang et al. [92] expressed an ectomycorrhizal fungal laccase from the *Laccaria bicolour* in which showed capability of decolorizing more than 80% of the crystal violet one of the mostly used dye for laboratory staining purpose. Balcazar-Lopez [4] reported a recombinant lacasses from *P. Sanguineus*. The expressed laccases illustrated similar features to those of the innate enzyme. The recombinant lacasses showed more than 90% removal of the benzopyrene present in discharged wastewater [4].

3.4.3 Bacterial Laccase

Laccase founded in bacteria also has great potential in dye degradation and firstly it is discovered in *A. lipoferum* in year 1993 [29]. Later on laccases were reported from different species of bacteria from different genera, like *Pseudomonas, Proteobacterium, Bacillus* and *Marinomonas*. Moreover, some of the laccases have also been isolated and reported from the Archaea such as *Haloferax volcanii* [91]. Generally, under natural conditions bacterial laccases are involved in toxin oxidation, pigmentation and morphogenesis processes (Singh et al. 2010) [55]. Structurally, 50–70 kDa is the range of the molecular weight of these enzymes, with major monomeric intracellular protein [32, 54]. For example, the extracellular enzyme produced by *Bacillus tequilensis SN4*.

3.4.4 Bacterial Enzyme: CotA-Laccase

Mostly studied bacterial laccase is the external endospore coat protein CotA-laccase derived from *Bacillus subtilis*. Structurally it is composed of three cupredoxin domains [24]. Similar bacterial multicopper enzymes (MCOs) consist of the copper homeostasis protein (CueO) isolated from *E. coli* [76].

Bacterial laccases have three-dimensional structure but some are also founded with two-domain laccases in Streptomyces, Nitrosomonas, and Amycolatopsis, and they are famous as small laccases (SLACs). The first study, on synthetic dyes decolorization was done by using bacterial laccases combined with recombinant CotA-laccase extracted from *Bacillus subtilis*. This enzyme is thermally active and essentially thermally stable with half-life of 2 h at 80 °C (Pereira et al. 2009a, b).

Bacterial laccases have been widely used in bioremediation and degradation of synthetic dyes. Liu et al. reported pH and thermostable lacasses from *Klebsiella pneumoniae* with dye degradation ability. used in many different industrial processes as an example a reactive dye reactive dark blue M-2GE, brilliant blue X-BR, bromophenol blue, congo red, and malachite green, only in 90 min short reaction time under varied pH values at 70 °C. Another example CotA-laccase mutant WLF from *B. pumilus* [50] has been verified for the dye degradation with higher decolorization yields for complex synthetic dyes. Azide-resistant laccases from halotolerant *Bacillus safensis*, also displayed siginificant decolorization of la dyes like malachite green and toluidine blue [88].

Recombinant laccases have shown more efficient activity compared to enzyme isolated from original isolated strain [47]. The recombinant laccase reported from *Streptomyces ipomoea* was expressed in *E. coli,* in presence of mediators enhanced the detoxification and decolorization of many dyes, like indigo carmine, orange II and reactive black 5. It also decreases the toxicity of dyes and its products used in textile industries [9].

Two major differences were found in comparison of bacterial and fungal laccases

(1) The no prerequisite of redox mediators
(2) A maximum activity at the neutral to an alkaline range of pH.

By technological viewpoint, it is a significant advantage that bacterial CotA-laccase exhibited the absence of strict requirement for redox mediators compared to fungal enzymes. Additionally, low-molecular-weight compounds are pricey with some limitations for big scale production [34, 43, 52].

3.4.5 Structure of Laccase

Laccases are associated to the superfamily cupredoxin, this family is also called multi-domain cupredoxins. This family has unique characteristics that they are composed of cupredoxin fold also known as Greek key motif.

Laccases are multicopper oxidative enzyme (MCOs); remarkably in special oxidation state they have four copper atoms which all are forming their catalytic sites.

1. Type-1
2. Type-2
3. Type-3

All the types create Greek key motifs with β-strands in head-to-head space but not in sequence [34]. Further, copper atoms classification is constructed and depends on specific ion binding and analytical characteristics. The three copper ions T1, T2 and T3 and the further characterized as following segments.

(1) T1: paramagnetic 'blue' copper,
(2) T2: paramagnetic 'non-blue' copper,
(3) T3: a diamagnetic spin-coupled copper–copper pair [62].

T1 copper has an optimum absorbance at 610 nm. The position of the ions is trigonal where the cysteine and histidine molecules are in variable nature bonded with axial ligand. Type-T2 and the two type-T3 coppers jointly generate a cluster, where reduction of oxygen takes place with liberation of water. Maximum absorbance of that cluster is 330 nm. In T3 copper atoms, the antiferromagnetic coupling is maintained by a hydroxyl bridge [41].

Commonly, laccases are made up of three cupredoxin domains which are homologous in nature. In domain 3, single nuclear copper site and formation of trinuclear cluster are going to happen at edge between domain 1 and 3, allowing trinuclear cluster formation. With this feature and topology, domain 2 main function is to join domains 1 and 3 in position and allowing the trinuclear cluster formation.

Most of the domain 2 laccases, are obtained from bacteria and because of that they are also considered as small laccases. In domain 1 or 2 mononuclear copper site occurs, but for trinuclear cluster formation oligomerize as homotrimers are essentially required. These homotrimers are catalytic site at the edge between the domain 1 and the domain 2 of monomer [41].

In both situations, the relative position and distance between the copper sites are preserved and the cluster in majority of laccases is about 12 Å [34]. Laccases with three-domain are mostly originated in fungi but in some bacteria, archaea, plants and insects also it observed [55]. In bioinformatics there was a database called "BioCat-Net" with all the information on different types of laccases and MCOs [87]. Due to their numerous biological activities in different areas, Laccases are also well-thought-out as "moonlighting" proteins [76]. Bioinformatic tool, protein data bank (PDB) has data of more than 70 fungal laccases and a few bacterial laccases structures, and also crystallized in their different forms like mutant, wild-type, and other derivative forms, as well as complex diversity of substrate-like ligands and oxygen reactive species [48].

3.4.6 Mechanism

Because of, lots of uses and potential applications in different areas of laccases with different biocatalytic processes for different industries and environmental solutions has spontaneously augmented the interest in understanding their mechanism of action, from it we can better understand that how is it works and also, we can improve its potential and specificity by using different recombinant techniques.

General, laccases have ability to oxidize an extensive range of substrates; characteristically substituted aromatic amines and phenols, which are transformed into free radicals. Unstable chemical products mainly produce free radicals which starts domino reactions that leads to form complex chemical transformations of biological significance like lignin synthesis and degradation [49].

In general laccase reaction, includes one electron ($1e^-$), sequential oxidations of four molecules of reducing substrates, concurrently with two double electron ($2 \times 2e^-$) reductions of oxygen atoms into their individual water molecules. There is catalytic exchange of $4H^+$ equivalents from this all reaction are attended [70]. From the mechanistic, structural, kinetical points of view, a laccase reaction is moved toward as two half-reactions connected by step interior electron transfer (IET), which is strengthened by the catalytic copper ions placed at the sites of trinuclear cluster, T1 Cu and T2 Cu/T3 Cuα/T3 Cuβ domains [25]. Chiefly, the eleven (one Cys and ten His) residues with fully well-maintained nature which establishing the sites of TNC laccase and T1 copper, and generally for all MCOs, clarify their crucial role and its importance in the catalytic action. This association has been proven experimentally by using different methods and techniques like sequence evaluation and mutagenic approaches in many research and studies [36].

Similarly, other different catalytic steps having to play an important role which is involved in mechanism of laccase with highly conserved residues. In terms of structure and function of laccase, all advancements for better understanding are carried out, but a complete vision of their mechanisms and molecular properties with their kinetic performance is not clear yet. As the usefulness of laccase there was deep research are going through it for finding better product and from that by using advanced recombinant and molecular techniques different forms of laccase with different application have been discovered from years.

3.5 Peroxidase

Peroxidases are heme-containing proteins by use of hydrogen peroxide (H_2O_2) or organic hydroperoxides (R–OOH) as electron-accepting co-substrates they show reaction of oxidizing a different type of compounds. Peroxidases are specific attention of different industries for different processes because of their catalytic adaptability and enzymatic stability for redox conversion [37]. Plants, microorganisms, and animals are common resources from where oxidoreductive peroxidases are found [22]. They precisely catalyze the chemical reactions in the presence of hydrogen peroxide as a terminal electron acceptor as an alternative of oxygen. For their role in the degradation of azo dye Peroxidases are famous. Extracellular and intracellular both the peroxidases show dye degradation and also involved in it with different mechanism. It has ability to catalyze the degradation of dyes without the formation of hazardous aromatic amines and that is the main feature of this enzyme because from this feature any kind of pre- or post-treatment doesn't need.

3.5.1 Sources and Applications

For the treatment of azo dyes, fungal systems presented and proved themselves as one of the best for reducing azo dyes. Fungi capability is related to the creation and producing exoenzymes and from this peroxidases and phenol oxidases are best examples and widely usable. Peroxidases are hemoproteins and in the presence of hydrogen peroxide they catalyze the reactions. For the decolorization of textile dyes, mostly for azo dyes, peroxidase enzyme derived from the plants *Saccharum spontaneum* and *Ipomea palmitate* (1.003 IU/g of a leaf) is an alternative for as commercial source of soybean and horseradish peroxidase enzyme.

3.5.2 Lignin Peroxidase

First ligninolytic peroxidases are discovered from *P. chrysosporium* and named lignin peroxidase (LiP) because it is very specific to lignin degradation, but it is also showing activity against dyes. This enzyme also has some other names like diarylpropane oxygenase, ligninase I, diarylpropane peroxidase, hydrogen-peroxide oxidoreductase. For the production of this enzyme Neem hull waste is used, which contains a high amount of lignin and other phenolic compounds with species Phanerochaete chrysosporum by using solid-state fermenter and also by some other physical methods. All-out decolorization attained by partially purified lignin peroxidase was 80% for Porocion Brilliant Blue HGR, 83% for Ranocid Fast Blue, 70% for Acid Red 119.

This enzyme can oxidize non-phenolic aromatic compounds like veratryl alcohol, and the effects of different concentrations of veratryl alcohol, hydrogen peroxide, enzyme and dye on the competence of decolorization have been examined.

3.5.3 Mechanism

In (LiP) enzyme facilitated catalytic mechanism of degradation of azo dyes includes the oxidation of the phenolic ring by H_2O_2 which oxidize LiP into two successive monoelectronic oxidation steps. By this oxidation, radical cation (carbonium ion) at the carbon is produced which bears the azo bridge. By water the nucleophilic attack happens at the phenolic carbon which leads to the formation of quinone and the consistent phenyldiazene. This phenyldiazene can be further oxidized by O_2 by a monoelectronic oxidation reaction to a phenyl radical and the azo linkage is eliminated as N_2 [15].

3.5.4 Manganese Peroxidase

Manganese peroxidase (MnP) is first discovered by the research groups of Michael H. Gold and Ronald in 1985 from the fungus *Phanerochaete chrysosporium*. And protein was genetically sequenced in *P. chrysoporium* in 1989. The enzyme is supposed to be unique to Basidiomycota as no bacterium, yeast, or mold species has yet been found which naturally produces it.

3.5.5 Mechanism

The MnP mediated mechanism of degradation of dye includes the oxidation of Mn^{2+} to Mn^{3+} by H2O2-oxidized MnP as same as Lip. Mn^{3+} is then stabilized by organic acids and forms an Mn^{3+}-organic acid complex. This complex acts as an active oxidant which have ability to oxidize the dyes mostly textile dyes.

So, the Mn^{3+} is responsible for the oxidation of phenolic compounds but Mn^{2+} is necessary for the accomplishment of the catalytic cycle of MnP from Pleurotus species and Bjerkandera adusta isolated MnP isoenzymes is different from the isoenzymes isolated from *P. chrysosporium* because they can oxidize 2,6-dimethoxyphenol (DMP) and veratryl alcohol in an Mn^{2+}-independent reaction so it acts differently and specifically [30]. LiPs and MnPs both the enzymes have a similar reaction mechanism during their catalytic cycle which begins with the oxidation of the enzyme by H_2O_2 to an oxidized state and then from the next step their activity is different.

Some research and studies have shown that LiP or MnP both the enzymes from *P. chrysosporium* is directly showing activity against the degradation of various xenobiotic compounds and dyes. In many studies, better degradation of different sulfonated azo dyes either by MnP and Mn^{2+} or by LiP was verified [7].

4 Future Perspectives

In this chapter we saw the enzymology behind the dye degradation, microbial decolorization of textile dyes can occur by either biosorption or enzymatic degradation or by a combination of both. The microorganisms can produce enzymes (oxidative and reductive) which catalyze the breakdown of textile dyes. Treatment systems based on microbial consortia are more effective as compared to a single pure microorganism because of the cooperative metabolic activities of the microbial community in different ways.

Several researchers are also focusing on the development of consortia to remove the textile dyes and tolerate the extreme conditions of textile effluent. They are trying to identify group of microbes which can work together and increase range of degradation activity.

Various advanced methods such as GMMs and their enzymes, immobilized cells or enzymes, biofilms, and MFCs have great and specific potential for the treatment of textile effluents. However, the potential of microbes and other advanced methods have been tested using dye solution under lab conditions only in most cases. The ability of these methods to treat real textile effluent has not been fully clarified. Moreover, textile industries consume large amounts of potable water and, therefore, recycling and reuse of textile effluent are also necessary to limit the amount of wastewater to protect the environment. Treatment systems having the capability to complete the mineralization of textile dyes and other pollutants present in textile effluent as well as to recycle and reuse the treated water are going to be necessary for the future.

By the advancement of biology and technology, molecular biological methodologies, like cloning, gene recombination techniques, site-directed mutagenesis, random mutagenesis, directed evolution, rational design, heterologous expression, and metagenomics, and also from bioinformatics tools is used to accelerate the evolution processes in such a way that the bioremediation process is improved. But by advances in molecular genetics and genetic engineering have made it possible to clone and express almost any gene in a suitable microbial host.

5 Conclusion

Effluent released by the textile industry is a complex mixture with many mixed dyes, mixed chemicals of various pollutants. The discharge of dye-containing textile effluent into the environment has caused serious concern because of its aesthetic value, toxicity, mutagenicity, and carcinogenicity to water bodies. Various Physico-chemical methods have been used for the treatment of textile effluent with dyes and other chemical, but these methods have several limitations like not affordable for big scale. Additionally, use of synthetic dyes leads to environmental and health issues for which proper environmental solutions are perquisites. The alternative for solving this challenge would be use of biological methods and among that use of enzymes would be the best optimal solution. Enzymes like laccases, peroxidases, reductases, from diverse microbes can serve as an ideal tool to convert the dyes to simple eco-friendly degradable compounds.

References

1. Anburaj J, Kuberan T, Sundaravadivelan C, Kumar P (2011) Biodegradation of azo dye by Listeria sp. Int J Environ Sci 1(7):1760
2. Andleeb SAADIA et al (2010) Biological treatment of textile effluent in stirred tank bioreactor. Int J Agric Biol 12(2):256–260
3. Anjaneyulu Y, Sreedhara Chary N, Samuel Suman Raj D (2005) Decolourization of industrial effluents—available methods and emerging technologies—a review. Rev Environ Sci Biotechnol 4(4):245–273
4. Balcázar-López E et al (2016) Xenobiotic compounds degradation by heterologous expression of a Trametes sanguineus laccase in Trichoderma atroviride. PLoS One 11(2):e0147997
5. Baldrian P (2006) Fungal laccases—occurrence and properties. FEMS Microbiol Rev 30:215–242
6. Banat IM et al (1996) Microbial decolorization of textile-dyecontaining effluents: a review. Bioresour Technol 58(3):217–227
7. Bertrand la GS (1896) Presence simultanee de la laccase et de la tyrosinase dans le suc de quelques champignons. CR Hebd Seances Acad Sci 123:463–465
8. Beydilli MI, Pavlostathis SG, Tincher WC (1998) Decolorization and toxicity screening of selected reactive azo dyes under methanogenic conditions. Water Sci Technol 38(4–5):225–232
9. Blánquez A et al (2019) Decolorization and detoxification of textile dyes using a versatile Streptomyces laccase-natural mediator system. Saudi J Biol Sci 26(5):913–920
10. Blumel S, Stolz A (2003) Cloning and characterization of the gene coding for the aerobic azoreductase from Pigmentiphaga kullae K24. Appl Microbiol Biotechnol 62:186–190
11. Bragger JL et al (1997) Investigations into the azo reducing activity of a common colonic microorganism. Int J Pharm 157(1):61–71
12. Bürger S, Stolz A (2010) Characterisation of the flavin-free oxygen-tolerant azoreductase from Xenophilus azovorans KF46F in comparison to flavin-containing azoreductases. Appl Microbiol Biotechnol 87(6):2067–2076
13. Casas N et al (2013) Laccase production by Trametes versicolor under limited-growth conditions using dyes as inducers. Environ Technol 34(1):113–119
14. Chacko JT, Subramaniam K (2011) Enzymatic degradation of azo dyes-a review. Int J Environ Sci 1(6):1250

15. Chivukula M, Spadaro JT, Renganathan V (1995) Lignin peroxidase-catalyzed oxidation of sulfonated azo dyes generates novel sulfophenyl hydroperoxides. Biochemistry 34(23):7765–7772
16. Claus H (2003) Laccases and their occurrence in prokaryotes. Arch Microbiol 179(3):145–150
17. Couto SR, Herrera JLT (2006) Industrial and biotechnological applications of laccases: a review. Biotechnol Adv 24(5):500–513
18. Darvishi F et al (2018) Laccase production from sucrose by recombinant Yarrowia lipolytica and its application to decolorization of environmental pollutant dyes. Ecotoxicol Environ Saf 165:278–283
19. Deller S et al (2006) Characterization of a thermostable NADPH: FMN oxidoreductase from the mesophilic bacterium Bacillus subtilis. Biochemistry 45(23):7083–7091
20. Díaz-Cruz MS et al (2014) Fungal-mediated biodegradation of ingredients in personal care products. In: Personal care products in the aquatic environment. Springer, Cham, pp 295–317
21. Dos Santos AB, Cervantes FJ, Van Lier JB (2007) Review paper on current technologies for decolourisation of textile wastewaters: perspectives for anaerobic biotechnology. Bioresour Technol 98(12):2369–2385
22. Duarte-Vázquez MA et al (2003) Removal of aqueous phenolic compounds from a model system by oxidative polymerization with turnip (Brassica napus L var purple top white globe) peroxidase. J Chem Technol Biotechnol Int Res Process Environ Clean Technol 78(1):42–47
23. Eggert C, Temp U, Eriksson K-EL (1996) Laccase-producing white-rot fungus lacking lignin peroxidase and manganese peroxidase: role of laccase in lignin biodegradation. 130–150. https://doi.org/10.1021/bk-1996-0655.ch010
24. Enguita FJ et al (2003) Crystal structure of a bacterial endospore coat component: a laccase with enhanced thermostability properties. J Biol Chem 278(21):19416–19425
25. Gabdulkhakov AG et al (2018) Incorporation of copper ions into T2/T3 centers of two-domain laccases. Mol Biol 52(1):23–29
26. Gago-Ferrero P et al (2012) Evaluation of fungal-and photo-degradation as potential treatments for the removal of sunscreens BP3 and BP1. Sci Total Environ 427:355–363
27. Gedikli S et al (2010) Enhancement with inducers of lacasse production by some strains and application of enzyme to dechlorination of 2,4,5-trichlorophenol. Electron J Biotech 13(6):6–7
28. Giardina P et al (2010) Laccases: a never-ending story. Cell Mol Life Sci 67(3):369–385
29. Givaudan A et al (1993) Polyphenol oxidase in Azospirillum lipoferum isolated from rice rhizosphere: evidence for laccase activity in non-motile strains of Azospirillum lipoferum. FEMS Microbiol Lett 108(2):205–210
30. Glenn JK, Akileswaran L, Gold MH (1986) Mn (II) oxidation is the principal function of the extracellular Mn-peroxidase from Phanerochaete chrysosporium. Arch Biochem Biophys 251:688–696
31. Grönqvist S et al (2005) Oxidation of milled wood lignin with laccase, tyrosinase and horseradish peroxidase. Appl Microbiol Biotechnol 67(4):489–494
32. Guan Z-B et al (2018) Bacterial laccases: promising biological green tools for industrial applications. Cell Mol Life Sci 75(19):3569–3592
33. Hadibarata T et al (2012) Decolorization of azo, triphenylmethane and anthraquinone dyes by laccase of a newly isolated Armillaria sp. F022 Water Air Soil Pollut 223(3):1045–1054
34. Hakulinen N, Rouvinen J (2015) Three-dimensional structures of laccases. Cell Mol Life Sci 72(5):857–868
35. Haritash AK, Kaushik CP (2009) Biodegradation aspects of polycyclic aromatic hydrocarbons (PAHs): a review. J Hazard Mater 169(1–3):1–15
36. Hoegger PJ et al (2006) Phylogenetic comparison and classification of laccase and related multicopper oxidase protein sequences. FEBS J 273(10):2308–2326
37. Hofrichter M et al (2010) New and classic families of secreted fungal heme peroxidases. Appl Microbiol Biotechnol 87(3):871–897
38. Ito K, Nakanishi M, Lee WC, Zhi Y, Sasaki H, Zenno S, Saigo K, Kitade Y, Tanokura M (2008) Expansion of substrate specificity and catalytic mechanism of azoreductase by X-ray crystallography and site-directed mutagenesis. J Biol Chem 283:13889–13896

39. Kandelbauer A, Guebitz GM (2005) Bioremediation for the decolorization of textile dyes—a review. Environ Chem 269–288
40. Khlifi R et al (2010) Decolourization and detoxification of textile industry wastewater by the laccase-mediator system. J Hazard Mater 175(1–3):802–808
41. Komori H, Miyazaki K, Higuchi Y (2009) X-ray structure of a two-domain type laccase: a missing link in the evolution of multi-copper proteins. FEBS Lett 583(7):1189–1195
42. Kunamneni A et al (2007) Fungal laccase—a versatile enzyme for biotechnological applications. Communicating current research and educational topics and trends in applied microbiology 1:233–245
43. Lawton TJ et al (2009) Crystal structure of a two-domain multicopper oxidase. J Biol Chem 284(15):10174–10180
44. Leelakriangsak M, Borisut S (2012) Characterization of the decolorizing activity of azo dyes by Bacillus subtilis azoreductase AzoR1. Songklanakarin J Sci Technol 34(5)
45. Liu ZJ, Chen H, Shaw N, Hopper SL, Chen L, Chen S, Cerniglia CE, Wang BC (2007) Crystal structure of an aerobic FMN-dependent azoreductase (AzoA) from Enterococcus faecalis. Arch Biochem Biophys 463:68–77
46. Liu G et al (2009) Acceleration of azo dye decolorization by using quinone reductase activity of azoreductase and quinone redox mediator
47. Liu H et al (2015) Overexpression of a novel thermostable and chloride-tolerant laccase from Thermus thermophilus SG0. 5JP17-16 in Pichia pastoris and its application in synthetic dye decolorization. PLoS One 10(3):e0119833
48. Liu Z et al (2016) Crystal structure of CotA laccase complexed with 2,2-azinobis-(3-ethylbenzothiazoline-6-sulfonate) at a novel binding site. Acta Crystallogr F 72(4):328–335
49. Llevot A et al (2016) Selective laccase-catalyzed dimerization of phenolic compounds derived from lignin: towards original symmetrical bio-based (bis) aromatic monomers. J Mol Catal 12:34–41
50. Luo Q et al (2018) Functional expression enhancement of Bacillus pumilus CotA-laccase mutant WLF through site-directed mutagenesis. Enzyme Microb Technol 109:11–19
51. Madhavi V, Lele SS (2009) Laccase: properties and applications. BioResources 4:1694–1717
52. Majumdar S et al (2014) Roles of small laccases from Streptomyces in lignin degradation. Biochemistry 53(24):4047–4058
53. Manu B, Chaudhari S (2002) Anaerobic decolorisation of simulated textile wastewater containing azo dyes. Bioresour Technol 82(3):225–231
54. Martins LO et al (2015) Laccases of prokaryotic origin: enzymes at the interface of protein science and protein technology. Cell Mol Life Sci 72(5):911–922
55. Mate DM, Alcalde M (2017) Laccase: a multi-purpose biocatalyst at the forefront of biotechnology. Microb Biotechnol 10(6):1457–1467
56. Mendes S et al (2011) Synergistic action of azoreductase and laccase leads to maximal decolourization and detoxification of model dye-containing wastewaters. Bioresour Technol 102(21):9852–9859
57. Mendes S, Robalo MP, Martins LO (2015) Bacterial enzymes and multi-enzymatic systems for cleaning-up dyes from the environment. In: Microbial degradation of synthetic dyes in wastewaters. Springer Cham, pp 27–55
58. Mikolasch A, Schauer F (2009) Fungal laccases as tools for the synthesis of new hybrid molecules and biomaterials. Appl Microbiol Biotechnol 82(4):605–624
59. Misal SA, Gawai KR (2018) Azoreductase: a key player of xenobiotic metabolism. Bioresour Bioprocess 5(1):1–9
60. Morozova O, Shumakovich G, Gorbacheva M, Shleev S, Yaropolov A (2007a) Blue laccases. Biochemistry (Mosc) 72:1136–1150
61. Morozova OV et al (2007b) Laccase-mediator systems and their applications: a review. Appl Biochem Microbiol 43(5):523–535
62. Mot AC, Silaghi-Dumitrescu R (2012) Laccases: complex architectures for one-electron oxidations. Biochemistry (Moscow) 77(12):1395–1407

63. Moya R et al (2010) Contributions to a better comprehension of redox-mediated decolouration and detoxification of azo dyes by a laccase produced by Streptomyces cyaneus CECT 3335. Bioresour Technol 101(7):2224–2229
64. Muhd Julkapli N, Bagheri S, and Bee Abd Hamid S (2014) Recent advances in heterogeneous photocatalytic decolorization of synthetic dyes. Sci World J 692307.https://doi.org/10.1155/2014/692307
65. Nakanishi M, Yatome C, Ishida N, Kitade Y (2001) Putative ACP phosphodiesterase gene (acpD) encodes an azoreductase. J Biol Chem 276:46394–46399
66. Ogata D et al (2010) Crystallization and preliminary X-ray studies of azoreductases from Bacillus sp. B29. Acta Crystallogr Sect F Acta Crystallogr F 66(5):503–505
67. Olukanni OD et al (2010) Decolorization and biodegradation of Reactive Blue 13 by Proteus mirabilis LAG. J Hazard Mater 184(1–3):290–298
68. Pearce CI, Lloyd JR, Guthrie JT (2003) The removal of colour from textile wastewater using whole bacterial cells: a review. Dyes Pigm 58(3):179–196
69. Pointing SB, Vrijmoed LLP (2000) Decolorization of azo and triphenylmethane dyes by Pycnoporus sanguineus producing laccase as the sole phenoloxidase. World J Microb Biot 16(3):317–318
70. Polyakov KM et al (2019) The subatomic resolution study of laccase inhibition by chloride and fluoride anions using single-crystal serial crystallography: insights into the enzymatic reaction mechanism. Acta Crystallogr F Sect D Struct Biol 75(9):804–816
71. Pricelius S et al (2007) Enzymatic reduction and oxidation of fibre-bound azo-dyes. Enzyme Microb Technol 40(7):1732–1738
72. Rafii F, Cerniglia CE (1995) Reduction of azo dyes and nitroaromatic compounds by bacterial enzymes from the human intestinal tract. Environmental Health Perspectives 103 (suppl 5): 17–19
73. Rai HS et al (2005) Removal of dyes from the effluent of textile and dyestuff manufacturing industry: a review of emerging techniques with reference to biological treatment. Crit Rev Environ Sci Technol 35(3):219–238
74. Ramalho PA et al (2005) Azo reductase activity of intact Saccharomyces cerevisiae cells is dependent on the Fre1p component of plasma membrane ferric reductase. Appl Environ Microbiol 71(7):3882–3888
75. Riva S (2006) Laccases: blue enzymes for green chemistry. Trends Biotechnol 24(5):219–226
76. Roberts SA et al (2002) Crystal structure and electron transfer kinetics of CueO, a multicopper oxidase required for copper homeostasis in Escherichia coli. PNAS 99(5):2766–2771
77. Rodríguez-Couto S (2015) Degradation of azo dyes by white-rot fungi Microbial degradation of synthetic dyes in wastewaters. Springer, Cham, pp 315–331
78. Russ R, Rau J, Stolz A (2000) The function of cytoplasmic flavin reductases in the reduction of azo dyes by bacteria. AEM 66(4):1429–1434
79. Saratale RG et al (2010) Decolorization and biodegradation of reactive dyes and dye wastewater by a developed bacterial consortium. Biodegradation 21(6):999–1015
80. Sarkar S et al (2017) Degradation of synthetic azo dyes of textile industry: a sustainable approach using microbial enzymes. Water Conserv Sci Eng 2(4):121–131
81. Senan RC, Abraham TE (2004) Bioremediation of textile azo dyes by aerobic bacterial consortium. Biodegradation 15(4):275–280
82. Shaffiqu TS et al (2002) Degradation of textile dyes mediated by plant peroxidases. Appl Biochem Biotechnol 102(1):315–326
83. Sharma P, Goel R, Capalash N (2007) Bacterial laccases. World J Microb Biot 23(6):823–832
84. Shekher R et al (2011) Laccase: microbial sources, production, purification, and potential biotechnological applications. SAGE-Hindawi. https://doi.org/10.4061/2011/217861
85. Singh G et al (2011) Laccase from prokaryotes: a new source for an old enzyme. Rev Environ Sci Biotechnol 10(4):309–326
86. Singh RL, Singh PK, Singh RP (2015) Enzymatic decolorization and degradation of azo dyes—a review. Int Biodeterior Biodegrad 104:21–31

87. Sirim D et al (2011) The Laccase engineering database: a classification and analysis system for laccases and related multicopper oxidases. Database bar006. https://doi.org/10.1093/database/bar006
88. Siroosi M et al (2018) Decolorization of dyes by a novel sodium azide-resistant spore laccase from a halotolerant bacterium, Bacillus safensis sp. strain S31. Water Sci Technol 77(12):2867–2875
89. Stoj C, Kosman D (2005) Copper proteins: oxidases. In: King R (ed) Encyclopedia of inorganic chemistry, vol II. Wiley, Hoboken
90. Sugumar S, Thangam B (2012) BiodEnz: a database of biodegrading enzymes. Bioinformation 8(1):40
91. Uthandi S et al (2010) LccA, an archaeal laccase secreted as a highly stable glycoprotein into the extracellular medium by Haloferax volcanii. AEM 76(3):733–743
92. Wang B, Yan Y, Xu J, Fu X, Han H, Gao J, Li Z, Wang L, Tian Y, Peng R (2018) Heterologous expression and characterization of a laccase from Laccaria bicolor in Pichia pastoris and Arabidopsis thaliana. J Microbiol Biotechnol 28:2057–2063
93. Wang CJ, Hagemeier C, Rahman N, Lowe E, Noble M, Coughtrie M, Sim E, Westwood I (2007) Molecular cloning, characterisation and ligand-bound structure of an azoreductase from Pseudomonas aeruginosa. J Mol Biol 373:1213–1228

Interaction of Dye Molecules with Fungi: Operational Parameters and Mechanisms

Moises Bustamante-Torres, David Romero-Fierro, Jocelyne Estrella-Nuñez, Samantha Pardo, and Emilio Bucio

Abstract Dyes can cause considerable pollution that directly affects the environment. Diverse industries produce a significant amount of dyes every year. Therefore, various technologies have been developed to face this problem. Compared with traditional methods, the fungal mechanism has significant advantages. Fungal selection is based on the pollutants present in the medium and all the treatments used before fungal applications. Fungal biosorption and enzymatic degradation are relevant mechanisms of dye degradation. Some fungal enzymes and external factors (redox mediator, pH, temperature, oxygen concentration, carbon source) are essential in this process. We reviewed the importance of fungal technology as an alternative mechanism.

Keywords Dyes · Environment · Industries · Fungal · Enzymatic degradation · Pollutants · External factors

M. Bustamante-Torres (✉)
Biomedical Engineering Department, School of Biological and Engineering, Yachay Tech University, Urcuqui 170522, Ecuador
e-mail: moises.bustamante@yachaytech.edu.ec

D. Romero-Fierro · J. Estrella-Nuñez
Chemistry Department, School of Chemical and Engineering, Yachay Tech University, Urcuqui 170522, Ecuador
e-mail: david.romero@yachaytech.edu.ec

J. Estrella-Nuñez
e-mail: jocelyne.estrella@yachaytech.edu.ec

S. Pardo
Environmental Engineering Faculty, Salesian Polytechnic University, Quito 170702, Ecuador

M. Bustamante-Torres · D. Romero-Fierro · E. Bucio (✉)
Department of Radiation Chemistry and Radiochemistry, Institute of Nuclear Sciences, National Autonomous University of Mexico, 04510 Mexico City, Mexico
e-mail: ebucio@nucleares.unam.mx

S. S. Muthu and A. Khadir (eds.), *Dye Biodegradation, Mechanisms and Techniques*, Sustainable Textiles: Production, Processing, Manufacturing & Chemistry,
https://doi.org/10.1007/978-981-16-5932-4_7

1 Introduction

1.1 Overview of Dye Fungi

Dye can be defined as a molecule that encompasses two chemical groups: chromophore and auxochrome, which act by providing color and fixing the dye molecules on the tissue, respectively [37]. Dyes are highly used in some industries and in some cases, they are reported as harmful residues, which are dangerous for the environment. Dyes affect the photosynthetic activity in the aquatic ecosystem, through the reduction of light penetration and by the presence of toxic compounds [89]. Azo dyes are reactive synthetic dyes with a vast variety of colors and structures. These kinds of dyes are the most significant water-soluble synthetic dyes.

Microorganisms have been studied to degrade the accumulation of several organic compounds [99]. A feasible solution seems to be the employment of fungi because of their versatility for solving these kinds of problems with a wide number of applications. Fungi act as scavengers in order to break down compounds from dead animals and vegetable materials into simpler compounds. Due to their importance, fungi are used in food, pharmaceutical, and textile industries. Research on fungi continues to grow to find new solutions to a variety of existing problems.

Fungi are essential members of the environment [111]. All fungi are obligate heterotrophs, so, they utilize fixed (organic) C sources as substrate [88]. The fungal kingdom is very diverse, and its species grow in the form of unicellular yeast and/or branched hyphae, producing a large number of spores and other reproductive structures. The cell wall structure of fungi corresponds to a multilayer microfiber: external layer composed of glucan, mannan or galactan, while the internal layer is a microfibril layer [93]. Its crystalline characteristic consists of chitin chains (sometimes cellulose chains), and these yeasts are arranged in parallel.

Fungal technology has shown a promising use in biodegradation processes. Fungi and bacteria are known for their degrading potential, but in the case of fungi, it takes advantage of its superiority in enzyme production. There is a traditional classification of fungal species as white, brown, and soft-rot, which is based on their technical decay and the metabolic pathway involved in the breakdown of lignins and carbohydrates [40, 98].

Extracellular enzymes as phenoloxidases and peroxidases are related to the decomposition of lignin. Over the years, white-rot fungi have been studied by their capacity to degrade diverse contaminants, which can be applied to bioremediation systems [49, 92]. Fungal species have the advantage to survive in extreme environmental conditions. It does not need previous conditioning to a particular pollutant and the biodegradation process is easy for upscaling [107].

1.2 Environmental Issues

Due to the development of industrialization, environmental problems have appeared in recent years. Waste residue deposits from multiple industries such as papermaking, cosmetics, textiles, pharmaceuticals, and food have caused severe damage to the entire ecosystem. The residues made by the industries are one of the leading causes of contamination. As a consequence of residues from industries, dye contamination affects the soil, water, and air. Fungal treatment is produced by adsorption and enzymatic degradation to decolorize and detoxify the dye-contaminated effluents.

2 Dye Degradation Mechanism by Fungi

2.1 Biosorption

Biosorption is a mechanism related to physical and chemical interactions like adsorption, deposition, ion exchange, electrostatic interactions, metal ion chelation or complexation [79, 100] that can occur between pollutant ions and certain biological materials. In biotechnology, fungal biosorption consist on the concentration of

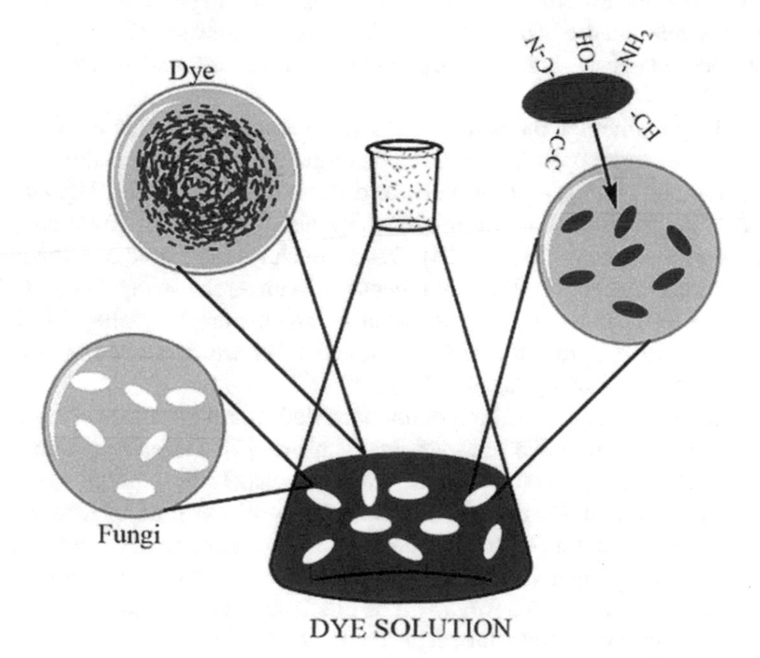

Fig. 1 Representation of fungal biosorption

pollutants on fungal cell wall, as is shown in Fig. 1 [11, 64]. In other words, biological materials are used as adsorbents in the biosorption process.

Biosorption is associated with the absorption of toxicants (adsorbate) using biomaterials (biosorbent) [79]. The adsorption of adsorbate will depend on the chemical composition of the cell wall. Functional groups like carboxylate, hydroxyl, amino, and phosphoryl groups are commonly a part of fungal cell wall [23]. The binding process requires minimal energy (21 kJ mol) to be activated and it occurs in a short period and it could be reversible [29, 56]. It means possible adsorbent desorption, regeneration, and reuse of the biosorbent in a new cycle.

Biosorption is significant in dye decolorization. This technique has simplicity, flexibility, easy operation, high efficiency, low cost, and the possibility of resource recovery [120]. Fermentation process generates a large amount of fungal biomass which is highly adaptable using genetic and morphological manipulation fungal biosorption [99]. Due to the existence of different organic pollutants with characteristic structures, the relationship between the chemical structure of organic compounds and microbial sorption requires special research.

Adsorption reaction occurs on the surface. Biosorption potential depends on the surface area, polarity, and ion state of the biomass. Therefore, due to the presence of a high percentage of cell wall materials, fungal biomass is considered a biosorbent [29, 45], thereby enhancing the chelating ability of fungal adsorbents.

The mechanism of biosorption involves the attachment of azo dye to the adsorbent, which results in a clear effluent [75]. The development of synthetic dyes, especially azo dyes has relevant use in diverse industrial areas. Its main structural characteristic is the presence of –N=N– bond, which provokes a kind of resistance to biodegradation [75].

Natural, agricultural or industrial wastes have been used as resources for biosorption. The most common could be seaweeds, fermentation waste, or activated sludge [4]. The microbial biomass used can be in different life stages (lag, log, stationary, death). Surface properties could be improved by taking advantage of different components present in the dead biomass [64]. Dead cells have been used because they can collect harmful substances. Water treatment has been a relevant application of dead biomass due to different advantages. Dead organisms are not vulnerable to toxic substances, do not require nutritional supply, live under less controlled environments, and can be used in many cycles [4, 67].

Biosorption developed in natural or uncontrolled environments involves a combination of passive and active transport mechanisms [6]. Different parameters are relevant in the process, such as initial biomass concentration, initial dye concentration, culture conditions, the effect of salinity on biosorption, pH, temperature, biomass pretreatment, kinetic studies, desorption, isotherm models, and biosorbent properties. These parameters are related to the surface functional groups of the cells responsible for dye adsorption [67, 75]. The efficiency of adsorption capacity can be improved by appropriate pretreatment processes, for example, using selected organic and inorganic chemicals [43, 44], and physical methods (e.g., high temperature).

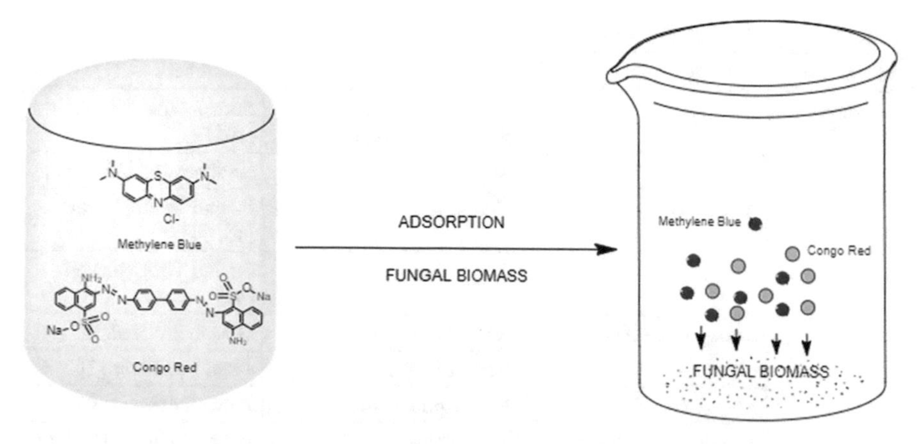

Fig. 2 Diagram of decolorization of methylene blue and Congo red dyes using fungal biomass

2.1.1 Biosorption Mechanism

Park et al. [80] reported the discoloration degree by biosorption mechanism. Fungal strains were combined with NaOH, vortexed, sonicated, and finally the absorbance was measured through the UV/Vis spectrophotometer [80]. The cell wall and the adsorbate ions fall in contact during the biosorption process, providing the actives sites for these adsorbate ions. Relevant changes in the type and quantity of adsorbent ions bounded are associated with the cell wall composition. The adsorption of adsorbate ions on the cell surface occurs through the physical and chemical interaction (physical adsorption, ion exchange, and chemical sorption) between the sorbent and the functional groups present on the microbial cell [29]. An intracellular accumulation can take place, when the adsorbate moves across the cell membrane, depending on the cell's metabolism.

Aspergillus (Asp.) Niger has been studied by Fu and Viraraghavan [43, 44] due to its biosorbent potential against four types of dyes. Figure 2 shows the chemical structure of Basic Blue and Congo Red, where carboxyl and amino groups, or amino, carboxylic acid, phosphate groups, and lipid moieties are the main binding sites, respectively. Electrostatic attraction is the main mechanism, while the chemical adsorption technique was studied in Disperse Red 1 [80].

2.1.2 Biosorption Isotherms

Equilibrium isotherms are useful to represent the relationship between the amount of adsorbate retained on the biosorbent and the concentration of adsorbate in the liquid phase. The design and optimization of the adsorption system requires equilibrium isotherms where it can be stablished the most appropriate correlation of the balance curve [67]. There are several models that predict the equilibrium.

Table 1 Fungi absorption examples

Mechanism	Culture	Dye	References
Biosorption	*Asp. foetidus*	Remazole, lignin, Alkali liquor, black liquor	Sumathi and Phatak [105]
	Asp. fumigatus	Methylene blue	Kabbout and Taha [55]
	Asp. niger	Congo red	Fu and Viraraghavan [43, 44]
		Basic blue 9	Fu and Viraraghavan [41]
	Rhizopus arrhizus	Reactive orange	O'Mahony et al. [77]
		Remazol black	Aksu and Tezer [7]
	Rhizopus Stolonifer	Bromophenol blue	Zeroual et al. [119]
	Trametes versicolor	Direct blue Direct red	Bayramoğlu and Yakup Arıca [14]
		Congo red	Binupriya et al. [18]
		Grey Lanaset	Blánquez et al. [19]
	Saracladium sp.	Remazol black	Nouri et al. [75]
	Penicillium oxalicum	Reactive red 19 Reactive red 241 Reactive yellow	Zhang et al. [120]
	Fomes fomentarius	Methylene blue Rhodamine B	Maurya et al. [67]
	Aspergillus foetidus	Remazole red Remazole dark blue HR Remazole brown gk	Sumathi and Phatak [105]
	Aspergillus foetidus	Reactive black 5	Patel and Suresh [81]
	Lentinus concinnus	Reactive yellow 86	Przystaś et al. [84]
	Aspergillus carbonarius	Methylene blue	Bouras et al. [21]
	Penicillium glabrum	Methylene blue	Bouras et al. [21]
	Trametes sp.	Acid blue	Puchana-Rosero et al. [85]
	Cladonia convoluta	Acid red	Bayazit et al. [12]

For non-lignin-decomposing fungi, such as *Asp. niger*, biosorption is mainly observed [42]. Several types of fungi are used to decolorize dyes in live or dead form. Table 1 details several species with potential biosorption activity.

2.2 Fungal Enzymes

Biosorption has several advantages and becomes a technical solution for treating water contaminated by dyes. However, removing the microbial biomass obtained

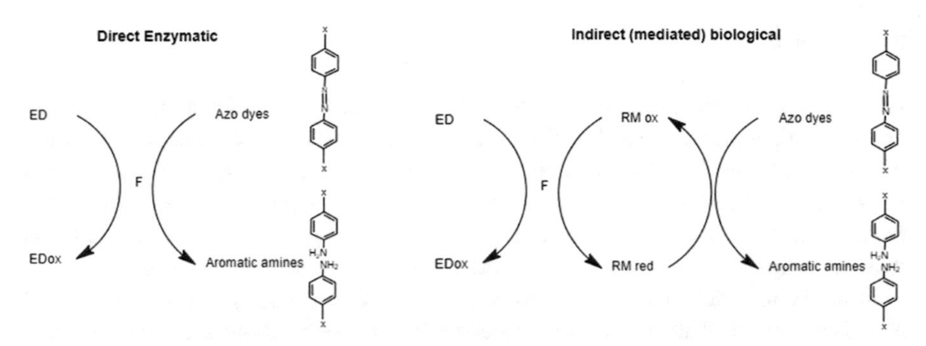

Fig. 3 Mechanisms of azo dye biodegradation. RM: Redox mediator; ED: electron donor; F: fungal (enzymes)

after treatment contains adsorbed dyes, which becomes a significant obstacle to its proposed function in the biological cleaning of colored water [25]. This has led people to seek other purification methods without ignoring the advantages of using fungi. This is the importance of enzymatic biodegradation as a purification mechanism.

When dyes are biodegraded by fungal action, lignin-modifying enzymes such as laccase, manganese peroxidase (MnP), and lignin peroxidase (LiP) are produced [86]. For each fungus, the relative enzyme contribution to the discoloration of the dye may be different. Different mechanical pathways are described in the literature, such as direct enzymatic and indirect biological (Fig. 3). For example, azo reductase degrades azo dyes, and azo reductase is almost exclusively anaerobic in nature [26, 94]. Compared with dye molecules, oxygen is able to inhibit the reduction mechanism and give priority to redox mediators [83, 94] using an indirect biological way. The study of dye biodegradation by fungi started with white-rot fungus *Phanerochaete chrysosporium*. The purified ligninase catalyzes the N-demethylation of crystal violet. The study proved that the lignin-degradation system is the cause of the biodegradation of crystal violet [9]. In addition, this fungus has been shown to be more capable of degrading other triphenylmethane dyes than crystal violet.

Fungi are considered to be suitable organisms for dye degradation. Fungal mycelium has the advantage to produce extracellular enzymes and dissolve insoluble substrates. Due to the increased cell-to-surface ratio, the physical and enzymatic contact between the fungus and the environment has increased [58]. Extracellular enzymes from fungi can convert aromatic substances such as lignin, polycyclic aromatic hydrocarbons, or pesticides. Fungal biomass has been used as an enzyme generator, which is related to the biotransformation process but not to biosorption [84].

Fungi are known to degrade the lignocellulosic substrates, which are materials composed of lignin, hemicelluloses, and cellulose. Each kind of enzyme has different pathways of action [49]. A wide range of species from white-rot fungi studied are *Phnerochaete chrisosporium, Bjerkandera sp., Trametes versicolor, Pleurotus ostreatus, Irpex lacteus*, among others, have the characteristic of enzyme production [84].

2.2.1 Laccases

Laccase is part of the multiple copper oxidase enzymes (MCOs). In its structure, there are four copper atoms located in different binding sites of the molecule. Each atom is different and is essential in the catalytic mechanism. Laccases are MCOs with a low substrate specificity and can oxidize organic molecules to promote processes such as lignin formation [85]. According to reports, the redox potential of laccase is lower than other ligninolytic peroxidases. It was initially reported that laccases would only oxidize phenolic substrates [12]. However, its oxidative capacity can be enhanced by mediators involved in the reaction mechanism. These mediators increase laccase activity and have low molecular weight, instability, and high reactivity in some cases [13]. The addition of compound like ammonium chloride enhances laccase production, which was reported in previous studies [12].

These enzymes can oxidize many organic compounds (mainly phenols and aromatic amines) using molecular oxygen [94]. The main feature is the ability to oxidize multiple substrates while oxidizing four-electron (from O_2 to H_2O) [21]. Laccase works by single-electron abstraction from electron-rich aromatic substrates where a stable entity could be obtained when generated free radicals can delocalize unpaired electrons [122].

Park et al. [80] analyzed laccase activities by the oxidation of syringaldazine in 0.15 M McIlvaine buffer (pH 4.6). The increase in absorbance was measured to calculate laccase activity. On the other hand, UV/Vis was used to determine laccase activity by oxidation of ABTS [87]. In addition, they studied the joint use of fungal biomass and laccase concluding the equally efficiency of laccase as the treatment with fungal biomass included. However, fewer metabolites are formed in fungal biomass-laccase treatment.

2.2.2 Peroxidades: Lignin Peroxidase (LiP) and Manganese-Dependent Peroxidase (MnP)

Peroxidases are hemoproteins (blood proteins) that catalyze a reaction using H_2O_2. Lignin and manganese peroxidases have similar reactions [13]. Afterward, there are two successive electron transfers in which the azo dyes substrates reduce the enzyme to its original form [53]. In the case of peroxidases, the dye decolorization is enhanced as a result of extracellular H_2O_2 production when strains were grown in glucose media [78]. Compared with lignin-modified peroxidase, laccase has a lower redox potential and only allows the direct oxidation of phenolic lignin subunits. However, synthetic mediators can oxidize substrates with high redox potentials by laccase.

Lignin peroxidases are characterized by the oxidation of aromatic compounds with high redox potential. Lignin peroxidases can directly oxidize phenolic and non-phenolic compounds, but manganese peroxidases must convert Mn^{2+} into Mn^{3+} chelates to oxidize phenolic compounds [68]. Manganese peroxidases are common enzymes produced by 96% of the species. Lignin peroxidase, manganese peroxidase, and laccase can degrade various aromatic compounds due to their non-specific enzymatic activity. There are different promising sources for fungal biomass, and the usual method of application is pellets. Compared with the dispersion medium, it has some advantages such as media with low viscosity, easy biomass separation, and higher resistance under severe conditions. However, the use of pellets has some defects related to the obstruction of the internal transportation of nutrients, results in the inability to form inanimate areas inside the pellets [30]. Table 2 shows some examples of fungal species and their respective enzyme activity for dye treatment.

Table 2 Enzyme examples for dye treatment

Fungi	Dye	Enzyme	References
Irpex lacteus	Remazol brilliant blue R	Laccase, MnP	dos Santos et al. [31]
	Reactive Orange 16 Remazol brilliant blut R	Laccase	Stolz [103]
Trametes versicolor	Amaranth Remazol brilliant blue	Laccase	McMullan et al. [69]
Phanerochaete chrysosporium	Direct blue	MnP	Glenn et al. [47]
	Methylene blue	LiP	Kinnunen et al. [60]
Lentinula edades	Azure B	Laccase, MnP	Espinosa-Ortiz et al. [36]
Funalia trogii	Azure B	Laccase	Novotný et al. [76]
Pycnoporus sanguineus	Acid blue	Laccase	Champagne and Ramsay [24]
A. ochraceus	Reactive blue	Laccase, LiP	Pazarlioglu et al. [82]
Trametes hirsutaa	Indigo carmine Phenol red	Laccase	Bayazit et al. [12]
Phanerochaete chrysosporium	Poly R481 Poly B411	MnP	McGuirl and Dooley [68]
Trametes versicolor	Reactive Black CKF	MnP	Ferreiraleitao et al. [39]
Exidia saccharina	–	MnP, Laccase	Kersten et al. [59]
Phlebia centrifuga	–	MnP, Laccase	Kersten et al. [59]
Hydrodontia sp.	–	MnP, Laccase, LiP	Kersten et al. [59]
Bjerkandera sp.	–	MnP, Lip	Kersten et al. [59]

3 Factor Affecting Fungal Degradation of Dyes

3.1 Redox Mediators (RM)

RM can accept and donate electrons, and in this way modify the transfer capacity between a redox pair by changing the speed of the process. According to several studies, laccases obtained from several basidiomycetes and ascomycetes are tested for several dyes. The results showed that adding 1-hydroxybenzotriazole (1-HOBt) as a redox mediator can promote azo dyes discoloration (see these dyes in Fig. 4), both individually and in color mixtures. Therefore, laccase/mediators systems can effectively treat effluents from the textile, dye, and printing industries [27].

Other studies of dye removal were focused on two redox using laccases obtained from fungus (*Peniophora sp*) and certain metal ions. The metal ions used were divalent ions, such as Cd^{2+}, Mn^{2+}, Co^{2+}, Ca^{2+}, and Cu^{2+}, and the redox mediators consist of 2,2'-Azino-bis (3-ethylbenzthiazoline-6-sulphonic acid) and 1-HOBt (Fig. 5). This study showed that the crystal violet dye has the highest degree of discoloration when 1 mM ABTS (96.3%) was used and HBT (86.0%) is present. Finally, the laccases can remove the color of crystal violet dye even in presence of metal ions [96].

In another study, it was found that certain laccases can decolour dyes without redox mediators up to a certain time limit. Moilanen et al. [72] used crude laccase from *Cerrena unicolor* and *Trametes hirsuta* fungi to discolor Remazol Brilliant blue, Congo red, Lanaset Gray, and Poly R-478 (Fig. 6). *C. unicolor* was able to discolor dyes without the need for a mediator for 19.5 h. On the other hand, *T. hirsuta* can reduce the efficiency of discoloration of all dyes, but in the same period of time.

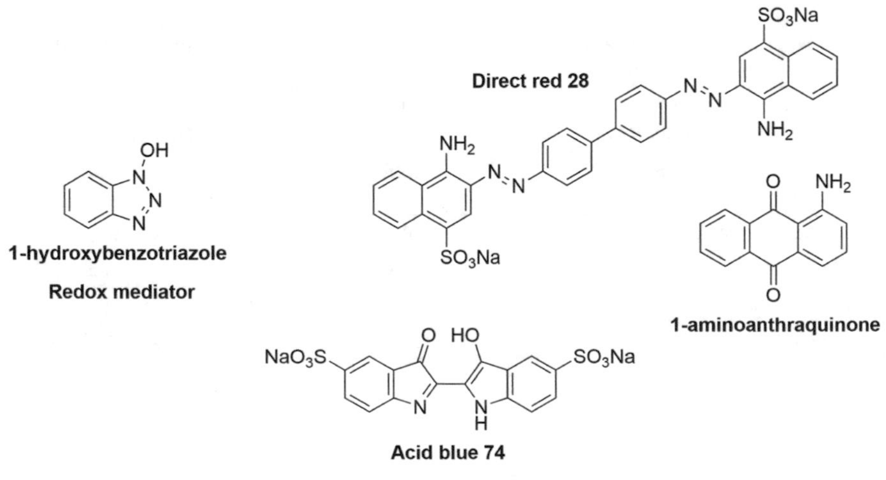

1-hydroxybenzotriazole
Redox mediator

Direct red 28

1-aminoanthraquinone

Acid blue 74

Fig. 4 Chemical structures: 1-hydroxybenzotriazole redox mediator and acid blue 74, direct red 28, and 1-aminoanthraquinone which are dyes to be degraded

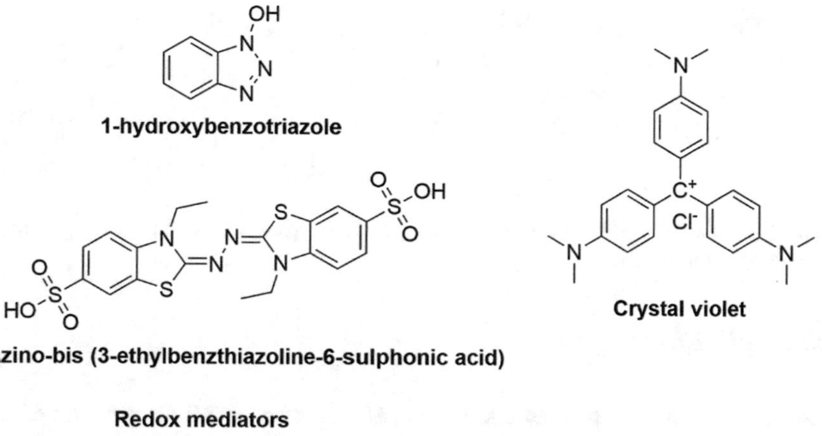

1-hydroxybenzotriazole

2,2'-Azino-bis (3-ethylbenzthiazoline-6-sulphonic acid)

Crystal violet

Redox mediators

Fig. 5 Chemical structures of redox mediator (HBT and ABTS) and crystal violet dye

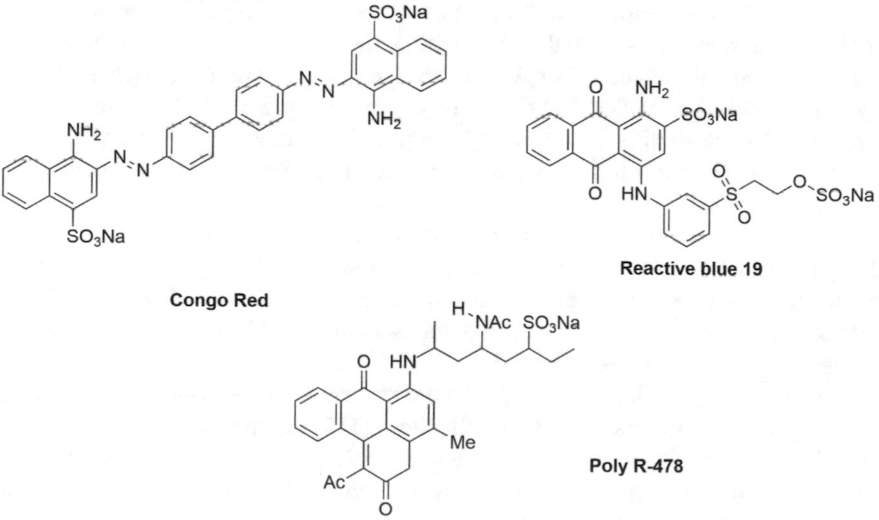

Congo Red

Reactive blue 19

Poly R-478

Fig. 6 Chemical structures of dyes: Congo red, reactive blue 19, Poly R-478 degraded by Cerrena unicolor and Trametes hirsuta fungi

In this case, the effect of the redox mediator was evaluated and its effectiveness in increasing discoloration was verified [72].

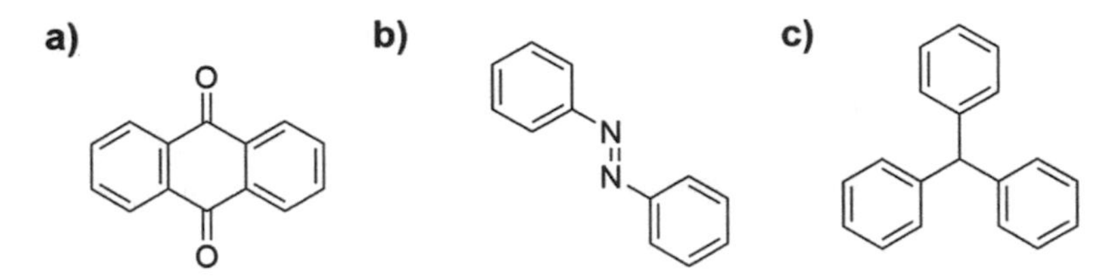

Fig. 7 Chemical structures of dyes belonging to chromophores groups: **a** anthraquinone, **b** azo, and **c** triphenylmethane

3.2 pH Effect

Several studies have tried to clarify the pH effect on fungal degradation of dyes in industrial effluents. The results are related to the types of fungi used and the dyes that need to be decolorized. The bioremediation ability of *Aspergillus niger* on direct violet textiles under alkaline and acid conditions was studied. Abd El-Rahim et al. [2] found that the greatest removal occurs under strong acid (pH 2) or weakly basic (pH 9) conditions for 24 h, and that it increases over time. Besides, Asses and his colleagues [8] determined a Congo red azo dye discoloration (200 mg/L), of 97% by *Aspergillus niger* under slightly acidic conditions (pH 5), at 28 °C for 6 days. The degraded metabolites were studied by liquid chromatography with tandem mass spectrometry. The study has shown that azo compounds require acidic pH conditions to be degraded [8].

Bettin et al. [15] used laccases from the fungus *Pleurotus sajor-caju* to decolor 22 dyes belonging to chromophores groups as show in Fig. 7. It was found that 19 of the tested dyes exhibited the greatest color-changing ability under acidic conditions (pH 3.2). An interesting fact is that two-way shaking of 100 rpm was found to have a negative effect on discoloration [15].

On the other hand, it has also been found that certain fungi can degrade dyes at several pH conditions. Yanto and his colleagues [115] experimented with three dyes as show in Fig. 8. These dyes were tested using *Pestalotiopsis sp.*, showing that it significantly decolorizes the dyes under acidic and basic conditions (pH 3–12) with great efficacy, but showing greater ability at pH 8 [115].

3.3 Oxygen Effect

Oxygen phenomenon is studied in a greater amount of color removed in cultures with greater oxygen transfer (aerated cultures). The reason for this is because oxygen acts as a redox mediator due to the presence of free electron pairs in its structure, allowing them to easily bind to fungi and perform a catalytic role [95]. Kirby et al. [61] verified this premise. They proved that *Phlebia tremellosa* can decolorize 8 synthetic dyes (especially Remazol Black B) (Fig. 9) with an efficiency higher than 94% using

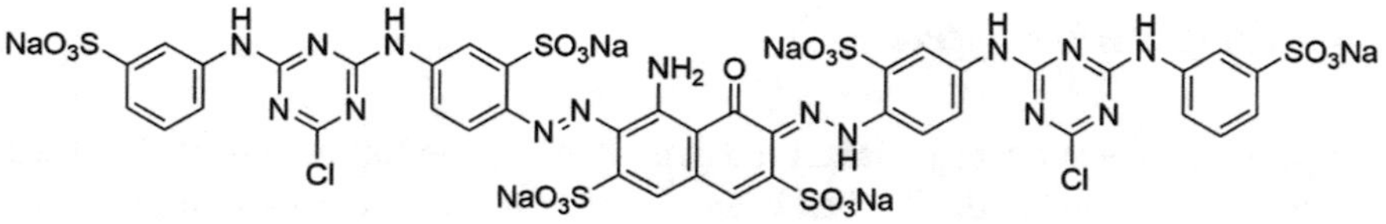

Reactive Green 19

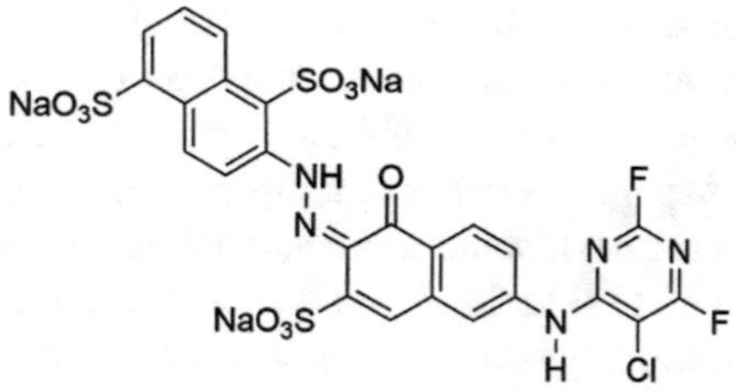

Reactive Orange 64

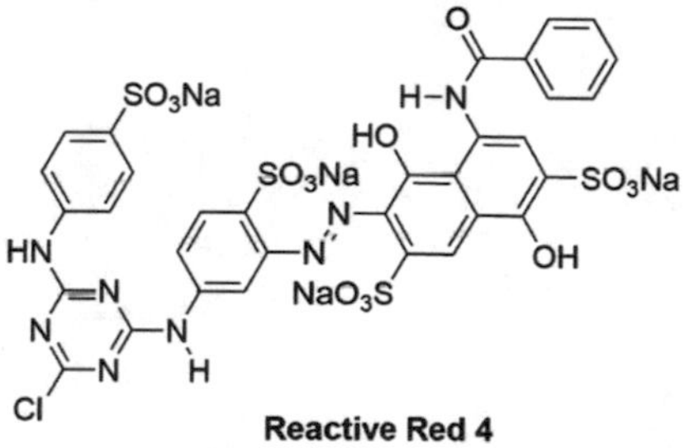

Reactive Red 4

Fig. 8 Chemical structures of dyes degraded experimented by Yanto et al. [115]

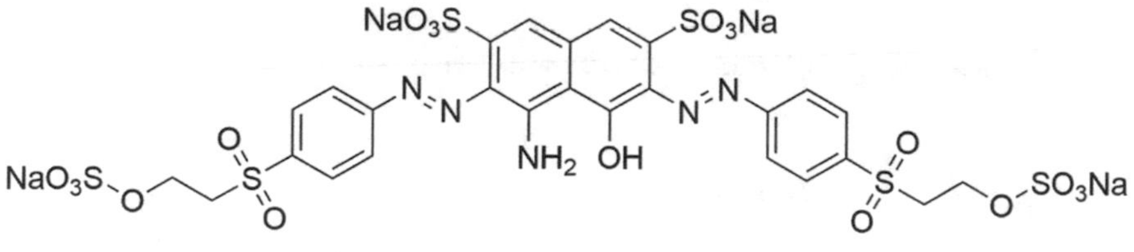

Remazol Black B

Fig. 9 Chemical structure of Remazol black B

oxygen in the culture medium. Similarly, through HPLC studies, it was found that it can completely degrade it [61].

3.4 Temperature Effect

Temperature affects the degradation medium, because the full growth of fungi will depend on this condition. The better the fungus develops; the more efficient the degradation of the dye will be. For example, under acidic conditions (pH 4) and close to standard temperature (T = 28 °C), the color of *Aspergillus (Asp) niger* directly and reactive dyes was significantly reduced as show Fig. 10 [51].

Erum and Ahmed [35] also used *Asp. flavus* and *Asp. terreus* for research. Fungal strains showing the highest decolorizing ability was produced at 30 °C. This result indicates that the genus Asp. grows better at temperatures range of 27–30 °C.

The same case occurs for the discoloration and biodegradation of dyes shown in Fig. 11. Mounguengui et al. [73] verified that the ideal temperature of 30 °C allows adequate growth of the fungi (*Perreniporia tephropora*) in the culture medium. The biodegradation of the dyes is controlled by means of UV–Vis spectrophotometer and HPLC, which confirms the strong degradation ability of the two dyes by fungi.

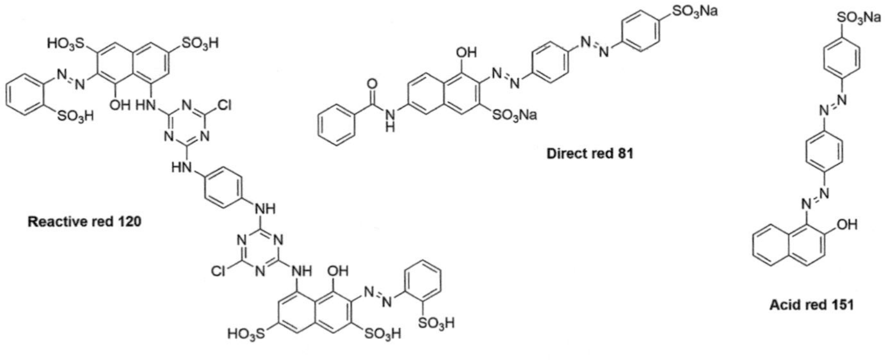

Fig. 10 Chemical structures of reactive red 120 and direct red 81

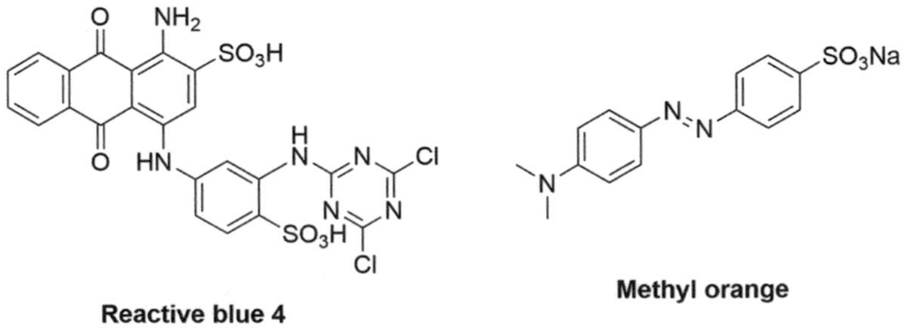

Fig. 11 Chemical structures of reactive blue 4 and methyl orange

4 Fungal Systems Involved in Dye Degradation

4.1 Yeast

Yeasts are single-celled organisms. They have various advantages (such as rapid growth), and they can resist adverse environments. Yeast presents a beneficial function for treating high-concentration organic wastewater. Among the most commonly used genera, we have ascomycetes that can decolorize reactive dyes by biodegradation, but only a few reports involve treating dyes with Basidiomycetes [66].

The use of yeast for the biodegradation of dyes can be carried out in different ways. Studies have shown the use of it in biosorption, where the yeast *Saccharomyces cerevisiae, Schizosaccharomyces pombe, Kluyveromyces marxianus, and Candida lipolytica* were analyzed for Remazol Blue reactive dye.

The pH and initial concentration of the pollutant were modified to 2 and 100 mg/l. The investigation reported an excellent absorption of all yeast for initial concentration, where Candida lipolytica reported the highest dye uptake capacity (250 mg/g) [5]. Also, the research demonstrated the biosorption kinetics of dye to dried yeast cells obeyed the second-order adsorption kinetics, concluding that the biosorption process can be used as a cost-effective method. On the other hand, the yeast-like *Rhodotorula mucilaginosa* has been investigated in wastewater polluted with heavy metals and reactive dye where the optimal pH for removing them varied for each contaminant and dye being pH 4 for copper (II), pH 6 for chromium (VI), and pH 5 for nickel(II), the concentration were 50 mg/l for all [34]. In the medium with copper(II), the maximum dye bioaccumulation was 96.1%. The dye bioaccumulation in medium with chromium (VI) was 76.1%, and 94.1% for the medium with nickel (II), respectively [34].

Bioprospecting procedures for yeasts that color dyes generally use carbon and nitrogen sources that are easily absorbed in the culture medium, this added to that the samples prepared in a laboratory many times present issues at the apply moment also the points pollution must be adapted to the characteristic nutrients of the yeast used in a laboratory. The investigation has studied yeasts capable of using dyes as the sole source of energy to isolate and select from the environment, increasing the selectivity of the medium. Yeasts of the basidiomycetous genus Trichosporon and the ascomycetous genus Cyberlindnera, Barnettozyma, and Candida were selected to isolate Candida sp MM 4035 was chosen for future studies [66].

4.2 Filamentous Fungi

Filamentous fungi are fungi formed by a series of tubular branches called hyphae, all of which form the mycelium, the vegetative organ. For the decontamination of organic matter, the mycelium secretes extracellular enzymes and acids that break down lignin,

Table 3 Filamentous fungi applications

Point pollution	Filamentous fungi	References
Sewage sludge	Penicillium corylophilum, Penicillium waskmanii, Penicillium citrinum, Aspergillus terries, Aspergillus flavus, Trichoderma Harzianum, Spicaria, and Hyaloflorae	Fakhru'l-Razi et al. [38],
Wastewater	*Monascus purpureus, Isaria spp., Emericella spp., Fusarium spp. and Penicillium spp*	Velmurugan et al. [110]
Sewage sludge	*Pleurotus ostreatus*	Bhattacharya et al. [16]

cellulose, and different types of compounds present in different environments [3]. For this reason, the place where they are found must have organic and energy-rich nutrients.

Filamentous fungal bioremediation is a cost-effective and ecologically essential option because the enzymes produced can degrade synthetic contaminants found in water and soil. Furthermore, fungal biomass also has biosorption properties, effectively removing toxic metals found in wastewater such as chromium, copper, mercury, nickel, cadmium, and lead [97].

Several alternative methods can be used to treat wastewater containing synthetic dyes. The physical, chemical, and biological treatments, where the latter are attractive for their low cost and low impact on the environment. Lignin degrading filamentous fungi are the most efficient microorganisms in the degradation of synthetic dyes [113]. Among its degrading enzymes are laccase, lignin peroxidase, and manganese peroxidase [113]. Besides, it has been reported that filamentous fungal biofilms produce many enzymes, being efficient both in consortia of several strains and individually [50].

Wastewater sludge includes various microorganisms, such as viruses, bacteria, fungi, algae, and yeast. For these reasons, the survey has analyzed wastewater sludge and wastewater, just like the precise environment in which the fungus grows [38], and they have found much variety of fungi on them (Table 3).

4.3 Genetically Modified Microorganisms or Enzymes

To obtain enzymes that can biodegrade dyes, it is necessary first to select the microorganisms that can produce enzymes for the biodegradation of lignins due to the structure of the lignin monomer that is very similar to the structures of dyes. Using fungi in wastewater, decontamination has been widely investigated [9]. The fungus has been shown to be a valuable organism for treating textile effluents and removing dyes. Fungi have superiority over unicellular organisms since fungi feed through their mycelia, solubilizing insoluble substrates with their enzymes. Due to the higher cell surface ratio, the physical and enzymatic contact between the fungus and the environment is better [90]. Another advantage to tolerate a high concentration of dyes is the extracellular nature of fungal enzymes.

Facts have shown that enzymatic dye degradation technology can provide satisfactory results in the shortest possible time without affecting the environment and with low energy consumption [70]. Besides being functional in a broader pH and temperature range and adapting in certain circumstances to media with organic solvents [108]. Some beneficial properties also stand out, such as convenient operation, improved steadiness, reuse, impeding interaction between phases, and hardening by multipoint covalent bonds [71]. The activity improved selectivity, specificity, resistance to inhibition, and purity [17]. On the other hand, correct isolation and enzymatic selection are fundamental for accomplishing an effective discoloration of the dye [104]. Additionally, a large number of investigations have verified the application of enzymes isolated from microorganism as fungi, yeast, and plants for the degradation of dye in wastewater [71, 106, 114].

The capacity of fungal cells is based on the creation of enzymes like oxidoreductase and peroxidase, which can effectively catalyze the degradation of the dye molecules. The oxidoreductase enzymes are involved in the oxidation process by degrading the phenolic ring of the dye into radicals and later into ions [71]. Subsequently, the peroxidase enzyme intervenes as an oxidation mediator in the process.

Enzymatic catalysis in dyes has been investigated using a batch reactor and the *Asp.* laccase enzyme [28]. The experiment was carried out for 10 min under optimal conditions. The enzyme concentration of 432 U/L was able to fade more than 84% of the mixed dye. Therefore, fungal enzymes based on laccase are a suitable prospect for actual industrial wastewater decontamination.

The use of Trametes hirsuta laccase in the degradation of azo and triarylmethane dyes has also been studied [1]. The results showed an 80% reduction of the dye seen at an initial concentration of 0.25 mM, and the enzyme reduced 50% of its activity during the experiment. Furthermore, at 60 °C and pH 4.5, the immobilized enzyme has a greater lifetime and more stability than in a free environment. This indicates that the use of these enzymes may be more efficient than that of soluble enzymes. That way, Zille et al. [121] investigated the discoloration of Reactive Black 5 in wastewater caused by immobilized Trametes villosa laccase. The authors report that even after the fourth cycle of repeated use, the enzyme can still show a 90% discoloration of the initial dye at 1 g/L in 24 h. Yu et al. [118] reported that under optimized conditions, the lignin-degrading enzyme Phanerochaete chrysosporium could degrade approximately 60% of the original 60 mg/L K-2BP reactive bright red dye in 15 min, which is more efficient than manganese peroxidase.

The production of mutated strains of microorganisms generate ligninase by UV radiation [62], one of the significant impediments is the difficult overexpression of these enzymes, which are very beneficial in the discoloration of dyes and are from white-rot fungi [54]. These fungi have mechanisms for stabilizing at high temperatures, which prevents the overexpression of the enzyme of interest [117]. Bailey et al. [10] In Trichoderma reesei, a thermostable laccase was produced from Trichoderma nigra fungus by intermittent, batch feeding, and continuous culture. Larrondo et al. [63] expressed two laccases cDNA clones corresponding to the Ceriporiopsis subvermispora and Coprinus coprinus. The two enzymes had similar behavior, which can be seen from the pH activity curve, which can oxidize 2,29-azidobis (3-ethylbenzothiazolin-6-sulfonate) (ABTS) or syringezine (SGZ) [58]. Yeasts have proven to be suitable as hosts during the generation of heterologous proteins because of their high growth capacity, and being unicellular, they are easy to manipulate [58].

4.4 *Advanced Oxidation Processes Combined with Fungal Processes*

Dye-containing wastewater is a powerful threat to the environment. Even if the concentration is deficient, these dyes will also affect aquatic life and the food chain. Conventional treatments such as adsorption, coagulation, filtration, electrocoagulation, photolysis, ozonation, photocatalysis, membrane filtration, etc., cannot eliminate the persistent organic contaminants [74]. The removal of various organic pollutants through the oxidation of electrochemically generated hydroxyl groups has attracted widespread attention.

The effectiveness of the advanced oxidation process (AOPs) is related to the production and availability of hydroxyl group (OH). Six advanced oxidation mechanisms can form OH and degrade pollutants. They are radiation, photolysis and photocatalysis, sonolysis, electrochemical oxidation technology, Fenton-based reaction, and ozone-based process [112]. These different mechanisms are selected of pollutant to be treated and the characteristic of sources pollution.

Research on textile industry wastewater usually focuses on the development of new treatment strategies for synthetic wastewater. The process mainly contains azo dyes and surfactants responsible for forming foam on the surface of rivers, abnormal growth of algae, and toxicity to certain aquatic organisms. In this sense, as shown in Table 4, AOPs have been widely proven to have the most outstanding prospects for treating textile wastewater. Although chemical oxidation is generally expensive for complete mineralization, it is widely reported that combining it with biological treatment can reduce operating costs and is a technically feasible options.

Using biological systems hand in hand with advanced oxidation processes has attracted much attention for the treatment of wastewater generated from industries related to dye production or application [48]. It has been determined that coupling AOPs with biological treatment can be organized in different ways: biological treatment as pretreatment, as a post-treatment, or such as mediator in the OH production [109].

Ganzenko et al. [46] deduced that due to the low rates of energy requirement and respect for the environment, the use of Fenton is a treatment with more significant advantages. Considering that its by-products could be easily degradable with

Table 4 AOPs characteristics

AOPs	Highlights	References
UV-O^3, UV-H_2O_2, UV-TiO_2 and the photo-Fenton reaction	Complete decolourization in 20 min with ozone, 75% after 90 min with the photo-Fenton reaction and after 150 min with UV/ozone	Ruppert et al. [91]
Fenton's oxidation (Fe^{2+}/H_2O_2) and ozonation	Ozonation increased to 99% remotion when 5 mg/l hydrogen peroxide was added	Solmaz et al. [101]
Homogeneous photochemical oxidation processes	O_3/H_2O_2/UV process showed 99% removal for colour	Yonar et al. [116]
Homogeneous and heterogeneous Fenton systems	Mesoporous Activated Carbon was used with Fenton's reagent for remotion	Karthikeyan et al. [57]

a biological post-treatment to eliminate all organic matter. Elias et al. [33] studied the degradation of Methyl Red, Biebrich Scarlet, and Orange II by electronic Fenton before the biological step. The fading rate of methyl red and Biebrich scarlet exceeds 90%, while orange II hardly degrades 80% after 4 h in electro-Fenton solution. Additional research on the methyl red dye validated the whole decolorization in 1-h, improved biodegradability, and eliminated toxicity trends. These tendencies leading the subsequent biological treatment to remove residual pollution [46].

Vasiliadou et al. [109] investigated white-rot fungi to produce OH through a system similar to Fenton, which increased the percentages of biodegradation in wastewater containing dyes. This method is called the advanced bio-oxidation process (ABOP). Lignolytic fungi and quinone substances were used to determine the efficiency of the system. The results show that the most promising media were 2,6-dimethoxy-1,4-benzoquinone and gallic acid [109].

On the other hand, Lucas et al. [65] studied the decolorization of Reactive Black 5 solution by way of combining a Fenton followed by an aerobic biological process (mediated by the yeast Candida oleophila). The results showed that with Fenton's reagent process as the primary treatment and the grown yeast cells as the secondary treatment, about 91% of the decolorization was achieved for the initial RB5 concentration of 500 mg/L, and the significant decolorization ability obtained from each individual was Concentrations of up to 200 mg/l have been successful. The use of fungi processes combined with advanced oxidation processes can be a viable and eco-friendly option. Also, the selection of the appropriate hybrid technology is the key to a feasible system.

4.5 Side Effects of the Enzyme Degradation Process

In some cases, microbial metabolism produces more toxic metabolites than the original compound [20]. Toxic products, mainly amines are the main problem from azo dyes [32]. The toxicity of aromatic amines (AAs) is related to the metabolic activation of the amino group, which can generate the reactive intermediate hydroxylamine known to damage DNA and proteins [22]. Furthermore, various azo dyes and amines in their degradation have shown mutagenic responses. The toxicity arises from the source and structure of the molecule. For instance, Acid Red 26 dye is carcinogenic due to the presence of a methyl group and the difference in the position of the sodium sulfonate [102]. Duarte Baumer et al. [32] investigated biodegradation of Reactive Blue 21, Reactive Blue 19, Reactive Black, and Reactive Red dyes. The results showed an increase in the toxicity after enzymatic treatment only in Reactive Blue 29. In conclusion, any bioremediation technology needs to evaluate the toxicity of the contaminants and metabolites formed after the degradation [52]. Therefore, toxicological tests should be applied to track the level of toxicity resulting from this transformation. Different methods can be used to assess toxicity, including phytotoxicity, ecotoxicity, genetic toxicity, mutagenicity, acute toxicity, microbial toxicity, and invertebrate toxicity [102]. Due to their low cost and easy operation, phytotoxicity methods have become more common.

5 Conclusion

The bioremediation process is essential to face environmental problems such as water pollution. In recent years, fungal technology has received more attention due to its promising performance in treating dye-contaminated effluents generated by different industries. The principal mechanisms of interest are biosorption and enzymatic degradation. In this process, the use of dead biomass is more advantageous because it is not vulnerable to toxic substances, does not require nutrients supply or a strictly controlled environment, and can be used in many cycles. In the case of enzyme degradation, many extracellular biomass enzymes (such as laccases and peroxidases) are used as biosorbents or enzyme generators to degrade different pollutants found in sewage.

Most reported studies are based on white-rot fungi, which produce diverse extracellular enzymes useful in dye industrial effluents. The simplicity, flexible operation, high efficiency, cost, and the possibility of resource recovery make these methods suitable. Besides, some external factors are essential in dye degradation. For instance, pH can degrade dyes under alkaline or acidic conditions. Likewise, oxygen produces a catalytic activity in the medium, affecting the dye concentration. Finally, the temperature is essential for fungi' growth and the degradation of certain dyes within a specific range.

Yeast and fungi present outstanding advantages in treating dyes. Ascomycetes show significant dye decolorization effects, and they are the primary yeast used for decontamination. Furthermore, it has been proved that the use of carbon and nitrogen sources can improve the selectivity and efficiency of the medium. On the other hand, filamentous fungi seem to be a cost-effective and ecologically essential option because the enzymes produced on their mycelium can degrade synthetic pollutants. Fungi mechanisms appear as feasible method of bioremediation. They must be thoroughly analyzed before use to reduce or avoid the production of toxic metabolites. Toxicity control should be carried out when treating fungal enzymes for bioremediation.

References

1. Abadulla E, Tzanov T, Costa S, Robra K-H, Cavaco-Paulo A, Gübitz G (2000) Decolorization and detoxification of textile dyes with a laccase from Trametes hirsuta. Appl Environ Microbiol 66:3357–3362
2. Abd El-Rahim W, El-Ardy O, Mohammad F (2009) The effect of pH on bioremediation potential for the removal of direct violet textile dye by Aspergillus niger. Desalination 249:1206–1211. https://doi.org/10.1016/j.desal.2009.06.037
3. Adenipekun CO, Lawal R (2012) Uses of mushrooms in bioremediation: a review. Biotechnol Mol Biol Rev 7:62–68. https://doi.org/10.5897/BMBR12.006
4. Aksu Z (2005) Application of biosorption for the removal of organic pollutants: a review. Process Biochem 40:997–1026. https://doi.org/10.1016/j.procbio.2004.04.008
5. Aksu Z, Dönmez G (2003) A comparative study on the biosorption characteristics of some yeasts for Remazol Blue reactive dye. Chemosphere 50:1075–1083. https://doi.org/10.1016/S0045-6535(02)00623-9
6. Aksu Z, Karabayır G (2008) Comparison of biosorption properties of different kinds of fungi for the removal of Gryfalan Black RL metal-complex dye. Biores Technol 99:7730–7741. https://doi.org/10.1016/j.biortech.2008.01.056
7. Aksu Z, Tezer S (2000) Equilibrium and kinetic modelling of biosorption of Remazol Black B by Rhizopus arrhizus in a batch system: effect of temperature. Process Biochem 36:431–439. https://doi.org/10.1016/s0032-9592(00)00233-8
8. Asses N, Ayed L, Hkiri N, Hamdi M (2018) Congo red decolorization and detoxification by Aspergillus niger: removal mechanisms and dye degradation pathway. Biomed Res Int 2018:1–9. https://doi.org/10.1155/2018/3049686
9. Azmi W, Sani RK, Banerjee UC (1998) Biodegradation of triphenylmethane dyes. Enzyme Microb Technol 22:185–191. https://doi.org/10.1016/S0141-0229(97)00159-2
10. Bailey MJ, Adamitsch B, Rautio J, von Weymarn N, Saloheimo M (2007) Use of a growth-associated control algorithm for efficient production of a heterologous laccase in Trichoderma reesei in fed-batch and continuous cultivation. Enzyme Microb Technol 41:484–491. https://doi.org/10.1016/j.enzmictec.2007.04.002
11. Banat I, Nigam P, Singh D, Marchant R (1996) Microbial decolorization of textile-dye containing effluents: a review. Biores Technol 58:217–227. https://doi.org/10.1016/s0960-8524(96)00113-7
12. Bayazıt G, Gül Ü, Ünal D (2018) Biosorption of Acid Red P-2BX by lichens as low-cost biosorbents. Int J Environ Stud 76:608–615. https://doi.org/10.1080/00207233.2018.1502959
13. Bayramoglu G, Yilmaz M (2018) Azo dye removal using free and immobilized fungal biomasses: isotherms, kinetics and thermodynamic studies. Fib Polym 19:877–886. https://doi.org/10.1007/s12221-018-7875-y

14. Bayramoğlu G, Yakup Arıca M (2007) Biosorption of benzidine based textile dyes "Direct Blue 1 and Direct Red 128" using native and heat-treated biomass of Trametes versicolor. J Hazard Mater 143:135–143. https://doi.org/10.1016/j.jhazmat.2006.09.002
15. Bettin F, Cousseau F., Martins K, Zaccaria S, Girardi V, da Silveira MM, Dillon AJP (2019) Effects of pH, temperature and agitation on the decolourisation of dyes by laccase-containing enzyme preparation from Pleurotus sajor-caju. Brazil Arch Biol Technol 62. https://doi.org/10.1590/1678-4324-2019180338
16. Bhattacharya S, Das A, Mangai G, Vignesh K, Sangeetha J (2011) Mycoremediation of Congo red dye by filamentous fungi. Braz J Microbiol 42:1526–1536. https://doi.org/10.1590/S1517-83822011000400040
17. Bilal M, Rasheed T, Zhao Y, Iqbal HMN (2019) Agarose-chitosan hydrogel-immobilized horseradish peroxidase with sustainable bio-catalytic and dye degradation properties. Int J Biol Macromol 124:742–749. https://doi.org/10.1016/j.ijbiomac.2018.11.220
18. Binupriya A, Sathishkumar M, Swaminathan K, Kuz C, Yun S (2008) Comparative studies on removal of Congo red by native and modified mycelial pellets of Trametes versicolor in various reactor modes. Biores Technol 99:1080–1088. https://doi.org/10.1016/j.biortech.2007.02.022
19. Blánquez P, Caminal G, Sarrà M, Vicent T (2007) The effect of HRT on the decolourisation of the Grey Lanaset G textile dye by Trametes versicolor. Chem Eng J 126:163–169. https://doi.org/10.1016/j.cej.2006.09.007
20. Boopathy R (2000) Factors limiting bioremediation technologies. Biores Technol 74:63–67. https://doi.org/10.1016/S0960-8524(99)00144-3
21. Bouras H, Isik Z, Arikan E, Yeddou A, Bouras N, Chergui A et al (2020) Biosorption characteristics of methylene blue dye by two fungal biomasses. Int J Environ Stud 1–17. https://doi.org/10.1080/00207233.2020.1745573
22. Brüschweiler BJ, Merlot C (2017) Azo dyes in clothing textiles can be cleaved into a series of mutagenic aromatic amines which are not regulated yet. Regul Toxicol Pharmacol 88:214–226. https://doi.org/10.1016/j.yrtph.2017.06.012
23. Bziwulska U, Bajguz A, Godlewska-Żyłkiewicz B (2004) The use of algae Chlorella vulgaris immobilized on cellex-T support for separation/preconcentration of trace amounts of platinum and palladium before GFAAS determination. Anal Lett 37:2189–2203. https://doi.org/10.1081/AL-200026696
24. Champagne P, Ramsay J (2005) Contribution of manganese peroxidase and laccase to dye decoloration by Trametes versicolor. Appl Microbiol Biotechnol 69:276–285. https://doi.org/10.1007/s00253-005-1964-8
25. Chander M, Arora DS (2007) Evaluation of some white-rot fungi for their potential to decolourise industrial dyes. Dyes Pigm 72:192–198. https://doi.org/10.1016/j.dyepig.2005.08.023
26. Chen H, Hopper SL, Cerniglia CE (2005) Biochemical and molecular characterization of an azoreductase from Staphylococcus aureus, a tetrameric NADPH-dependent flavoprotein. Microbiology 151:1433–1441. https://doi.org/10.1099/MIC.0.27805-0
27. Claus H, Faber G, König H (2002) Redox-mediated decolorization of synthetic dyes by fungal laccases. Appl Microbiol Biotechnol 59:672–678. https://doi.org/10.1007/s00253-002-1047-z
28. Cristóvão RO, Tavares APM, Brígida AI, Loureiro JM, Boaventura RAR, Macedo EA, Coelho MAZ (2011) Immobilization of commercial laccase onto green coconut fiber by adsorption and its application for reactive textile dyes degradation. J Mol Catal B Enzym 72:6–12. https://doi.org/10.1016/j.molcatb.2011.04.014
29. Dhankhar R, Hooda A (2011) Fungal biosorption—an alternative to meet the challenges of heavy metal pollution in aqueous solutions. Environ Technol 32:467–491. https://doi.org/10.1080/09593330.2011.572922
30. Domínguez A, Couto S, Sanromán M (2005) Dye decolorization by Trametes hirsuta immobilized into alginate beads. World J Microbiol Biotechnol 21:405–409. https://doi.org/10.1007/s11274-004-1763-x

31. Dos Santos A, Cervantes F, van Lier J (2007) Review paper on current technologies for decolourisation of textile wastewaters: Perspectives for anaerobic biotechnology. Biores Technol 98:2369–2385. https://doi.org/10.1016/j.biortech.2006.11.013
32. Duarte Baumer J, Valério A, de Souza SMGU, Erzinger GS, Furigo A Jr, de Souza Ulson AA (2018) Toxicity of enzymatically decolored textile dyes solution by horseradish peroxidase. J Hazard Mater 360:82–88. https://doi.org/10.1016/j.jhazmat.2018.07.102
33. Elias B, Guihard L, Nicolas S, Fourcade F, Amrane A (2011) Effect of electro-Fenton application on azo dyes biodegradability. Environ Prog Sustain Energy 30:160–167. https://doi.org/10.1002/ep.10457
34. Ertuğrul S, San NO, Dönmez G (2009) Treatment of dye (Remazol Blue) and heavy metals using yeast cells with the purpose of managing polluted textile wastewaters. Ecol Eng 35:128–134. https://doi.org/10.1016/j.ecoleng.2008.09.015
35. Erum S, Ahmed S (2011) Comparison of dye decolorization efficiencies of indigenous fungal isolates. Afri J Biotechnol 10:3399–3411. https://doi.org/10.5897/AJB10.1098
36. Espinosa-Ortiz E, Rene E, Pakshirajan K, van Hullebusch E, Lens P (2016) Fungal pelleted reactors in wastewater treatment: applications and perspectives. Chem Eng J 283:553–571. https://doi.org/10.1016/j.cej.2015.07.068
37. Exbrayat JM (2016) Microscopy: light microscopy and histochemical methods. In: Encyclopedia of food and health, pp 715–723. https://doi.org/10.1016/B978-0-12-384947-2.00460-8
38. Fakhru'l-Razi A, Zahangir Alam M, Idris A, Abd-Aziz S, Molla AH (2002) Filamentous fungi in Indah Water Konsortium (IWK) sewage treatment plant for biological treatment of domestic wastewater sludge. J Environ Sci Health—Part A Toxic/Hazard Subst Environ Eng 37:309–320. https://doi.org/10.1081/ESE-120002830
39. Ferreiraleitao V, Decarvalho M, Bon E (2007) Lignin peroxidase efficiency for methylene blue decolouration: comparison to reported methods. Dyes Pigm 74:230–236. https://doi.org/10.1016/j.dyepig.2006.02.002
40. Freitag M, Morrell J (1992) Decolorization of the polymeric dye Poly R-478 by wood-inhabiting fungi. Can J Microbiol 38:811–822. https://doi.org/10.1139/m92-133
41. Fu Y, Viraraghavan T (2000) Removal of a dye from an aqueous solution by the fungus Aspergillus niger. Water Qual Res J 35:95–112. https://doi.org/10.2166/wqrj.2000.006
42. Fu Y, Viraraghavan T (2001) Fungal decolorization of dye wastewaters: a review. Biores Technol 79:251–262. https://doi.org/10.1016/s0960-8524(01)00028-1
43. Fu Y, Viraraghavan T (2002) Dye biosorption sites in Aspergillus niger. Biores Technol 82:139–145. https://doi.org/10.1016/S0960-8524(01)00172-9
44. Fu Y, Viraraghavan T (2002) Removal of Congo Red from an aqueous solution by fungus Aspergillus niger. Adv Environ Res 7:239–247. https://doi.org/10.1016/s1093-0191(01)00123-x
45. Gadd GM (1994) Interactions of fungi with toxic metals. In: The genus Aspergillus. Springer US, pp 361–374. https://doi.org/10.1007/978-1-4899-0981-7_28
46. Ganzenko O, Huguenot D, van Hullebusch ED, Esposito G, Oturan MA (2014). Electrochemical advanced oxidation and biological processes for wastewater treatment: a review of the combined approaches. In: Environmental science and pollution research. Springer. https://doi.org/10.1007/s11356-014-2770-6
47. Glenn J, Akileswaran L, Gold M (1986) Mn(II) oxidation is the principal function of the extracellular Mn-peroxidase from Kaneswaran chrysosporium. Arch Biochem Biophys 251:688–696. https://doi.org/10.1016/0003-9861(86)90378-4
48. Guivarch E, Trevin S, Lahitte C, Oturan MA (2003) Degradation of azo dyes in water by Electro-Fenton process. Environ Chem Lett 1:38–44. https://doi.org/10.1007/s10311-002-0017-0
49. Gupta S, Annepu S, Summuna B, Gupta M, Nair S (2018) Role of mushroom fungi in decolourization of industrial dyes and degradation of agrochemicals. Fungal Biol 177–190. https://doi.org/10.1007/978-3-030-02622-6_8

50. Gutiérrez-Correa M, Ludeña Y, Ramage G, Villena GK (2012) Recent advances on filamentous fungal biofilms for industrial uses. In: Applied biochemistry and biotechnology, vol 167. Humana Press Inc., pp 1235–1253. https://doi.org/10.1007/s12010-012-9555-5
51. Husseiny SM (2008) Biodegradation of the reactive and direct dyes using Egyptian isolates. J Appl Sci Res 4:599–606
52. Jadhav SB, Phugare SS, Patil PS, Jadhav JP (2011) Biochemical degradation pathway of textile dye Remazol red and subsequent toxicological evaluation by cytotoxicity, genotoxicity and oxidative stress studies. Int Biodeterior Biodegrad 65:733–743. https://doi.org/10.1016/j.ibiod.2011.04.003
53. Jones S, Solomon E (2015) Electron transfer and reaction mechanism of laccases. Cell Mol Life Sci 72:869–883
54. Jönsson LJ, Saloheimo M, Penttilä M (1997) Laccase from the white-rot fungus Trametes versicolor: cDNA cloning of Icc1 and expression in Pichia pastoris. Curr Genet 32:425–430. https://doi.org/10.1007/s002940050298
55. Kabbout R, Taha S (2014) Biodecolorization of textile dye effluent by biosorption on fungal biomass materials. Phys Procedia 55:437–444. https://doi.org/10.1016/j.phpro.2014.07.063
56. Kaduková J, Virčíková E (2005) Comparison of differences between copper bioaccumulation and biosorption. In: Environment international, vol 31. Elsevier Ltd., pp 227–232. https://doi.org/10.1016/j.envint.2004.09.020
57. Karthikeyan S, Titus A, Gnanamani A, Mandal AB, Sekaran G (2011) Treatment of textile wastewater by homogeneous and heterogeneous Fenton oxidation processes. Desalination 281:438–445. https://doi.org/10.1016/j.desal.2011.08.019
58. Kaushik P, Malik A (2009) Fungal dye decolourization: recent advances and future potential. Environ Int 35:127–141. Elsevier Ltd. https://doi.org/10.1016/j.envint.2008.05.010
59. Kersten P, Kalyanaraman B, Hammel K, Reinhammar B, Kirk T (1990) Comparison of lignin peroxidase, horseradish peroxidase and laccase in the oxidation of methoxybenzenes. Biochemical J 268:475–480. https://doi.org/10.1042/bj2680475
60. Kinnunen A, Maijala P, Jarvinen P, Hatakka A (2017) Improved efficiency in screening for lignin-modifying peroxidases and laccases of basidiomycetes. Curr Biotechnol 6:105–115. https://doi.org/10.2174/2211550105666160330205138
61. Kirby N, Marchant R, McMullan G (2000) Decolourisation of synthetic textile dyes by Phlebia tremellosa. FEMS Microbiol Lett 188:93–96. https://doi.org/10.1111/j.1574-6968.2000.tb09174.x
62. Kirk TK, Croan S, Tien M, Murtagh KE, Farrell RL (1986) Production of multiple ligninases by Phanerochaete chrysosporium: effect of selected growth conditions and use of a mutant strain. Enzyme Microb Technol 8:27–32. https://doi.org/10.1016/0141-0229(86)90006-2
63. Larrondo LF, Avila M, Salas L, Cullen D, Vicuña R (2003) Heterologous expression of laccase cDNA from Ceriporiopsis subvermispora yields copper-activated apoprotein and complex isoform patterns. Microbiology (reading). https://doi.org/10.1099/mic.0.26147-0
64. Legorreta-Castañeda A, Lucho-Contastino C, Beltrán-Hernández R, Coronel-Olivares C, Vázquez-Rodriguez G (2020) Biosorption of water pollutants by fungal pellets. Water 12:1–38. https://doi.org/10.3390/w12041155
65. Lucas MS, Dias AA, Sampaio A, Amaral C, Peres JA (2007) Degradation of a textile reactive Azo dye by a combined chemical-biological process: Fenton's reagent-yeast. Water Res 41:1103–1109. https://doi.org/10.1016/j.watres.2006.12.013
66. Martorell MM, Pajot HF, de Figueroa LIC (2012) Dye-decolourizing yeasts isolated from Las Yungas rainforest. Dye assimilation and removal used as selection criteria. Int Biodeterior Biodegrad 66:25–32. https://doi.org/10.1016/j.ibiod.2011.10.005
67. Maurya N, Mittal A, Cornel P, Rother E (2006) Biosorption of dyes using dead macro fungi: Effect of dye structure, ionic strength and pH. Biores Technol 97:512–521. https://doi.org/10.1016/j.biortech.2005.02.045
68. McGuirl M, Dooley D (1999) Copper-containing oxidases. Curr Opin Chem Biol 3:138–144. https://doi.org/10.1016/s1367-5931(99)80025-8

69. McMullan G, Meehan C, Conneely A, Kirby N, Robinson T, Nigam P et al (2001) Microbial decolourisation and degradation of textile dyes. Appl Microbiol Biotechnol 56:81–87. https://doi.org/10.1007/s002530000587
70. Michniewicz A, Ledakowicz S, Ullrich R, Hofrichter M (2008) Kinetics of the enzymatic decolorization of textile dyes by laccase from Cerrena unicolor. Dyes Pigm 77:295–302. https://doi.org/10.1016/j.dyepig.2007.05.015
71. Mishra S, Maiti A (2019) Applicability of enzymes produced from different biotic species for biodegradation of textile dyes. In: Clean technologies and environmental policy. Springer. https://doi.org/10.1007/s10098-019-01681-5
72. Moilanen U, Osma JF, Winquist E, Leisola M, Couto SR (2010) Decolorization of simulated textile dye baths by crude laccases from Trametes hirsuta and Cerrena unicolor. Eng Life Sci 10:242–247. https://doi.org/10.1002/elsc.200900095
73. Mounguengui S, Attéké C, Saha-Tchinda JB, Ndikontar MK, Dumarcay S, Gérardin P (2014) Discoloration and biodegradation of two dyes by white-rot fungi Perreniporia tephropora MUCL 47500 isolated in Gabon. Int J Curr Microbiol App Sci 3:731–741
74. Nidheesh PV, Zhou M, Oturan MA (2018) An overview on the removal of synthetic dyes from water by electrochemical advanced oxidation processes. Chemosphere 197:210–227. https://doi.org/10.1016/j.chemosphere.2017.12.195
75. Nouri H, Azin E, Kamyabi A, Moghimi H (2020) Biosorption performance and cell surface properties of a fungal-based sorbent in azo dye removal coupled with textile wastewater. Int J Environ Sci Technol. https://doi.org/10.1007/s13762-020-03011-5
76. Novotný Č, Svobodová K, Erbanová P, Cajthaml T, Kasinath A, Lang E, Šašek V (2004) Ligninolytic fungi in bioremediation: extracellular enzyme production and degradation rate. Soil Biol Biochem 36:1545–1551. https://doi.org/10.1016/j.soilbio.2004.07.019
77. O'Mahony T, Guibal E, Tobin J (2002) Reactive dye biosorption by Rhizopus arrhizus biomass. Enzyme Microb Technol 31:456–463
78. O'Malley D, Whetten R, Bao W, Chen C, Sederoff R (1993) The role of laccase in lignification. Plant J 4:751–757
79. Özer A, Özer D, Ekiz HI (2005) The equilibrium and kinetic modelling of the biosorption of copper(II) ions on cladophora crispata. Adsorption 10:317–326. https://doi.org/10.1007/s10450-005-4817-y
80. Park C, Lee M, Lee B, Kim SW, Chase HA, Lee J, Kim S (2007) Biodegradation and biosorption for decolorization of synthetic dyes by Funalia trogii. Biochem Eng J 36:59–65. https://doi.org/10.1016/j.bej.2006.06.007
81. Patel R, Suresh S (2008) Kinetic and equilibrium studies on the biosorption of reactive black 5 dye by Aspergillus foetidus. Biores Technol 99:51–58. https://doi.org/10.1016/j.biortech.2006.12.003
82. Pazarlioglu N, Akkaya A, Akdogan H, Gungor B (2010) Biodegradation of direct blue 15 by free and immobilized trametes versicolor. Water Environ Res 82:579–585. http://www.jstor.org/stable/27870348. Accessed 25 Jan 2021
83. Pricelius S, Held C, Sollner S, Deller S, Murkovic M, Ullrich R et al (2007) Enzymatic reduction and oxidation of fibre-bound azo-dyes. Enzyme Microb Technol 40:1732–1738. https://doi.org/10.1016/j.enzmictec.2006.11.004
84. Przystaś W, Zabłocka-Godlewska E, Grabińska-Sota E (2018) Efficiency of decolorization of different dyes using fungal biomass immobilized on different solid supports. Braz J Microbiol 49:285–295. https://doi.org/10.1016/j.bjm.2017.06.010
85. Puchana-Rosero M, Lima E, Ortiz-Monsalve S, Mella B, da Costa D, Poll E, Gutterres M (2016) Fungal biomass as biosorbent for the removal of Acid Blue 161 dye in aqueous solution. Environ Sci Pollut Res 24:4200–4209. https://doi.org/10.1007/s11356-016-8153-4
86. Raghukumar C, Chandramohan D, Michel FC, Reddy CA (1996) Degradation of lignin and decolorization of paper mill bleach plant effluent (BPE) by marine fungi. Biotech Lett 18:105–106. https://doi.org/10.1007/BF00137820
87. Riegas-Villalobos A, Martínez-Morales F, Tinoco-Valencia R, Serrano-Carreón L, Bertrand B, Trejo-Hernández MR (2020) Efficient removal of azo-dye Orange II by fungal biomass

absorption and laccase enzymatic treatment. 3 Biotech 10:1–10. https://doi.org/10.1007/s13205-020-2150-5
88. Ritz K (2005) Fungi. Encycl Soils Environ 110–119. https://doi.org/10.1016/b0-12-348530-4/00147-8
89. Robinson T, McMullan G, Marchant R, Nigam P (2001) Remediation of dyes in textile effluent: a critical review on current treatment technologies with a proposed alternative. Biores Technol 77:247–255. https://doi.org/10.1016/s0960-8524(00)00080-8
90. Roy U, Manna S, Sengupta S, Das P, Datta S, Mukhopadhyay A, Bhowal A (2018) Dye removal using microbial biosorbents. Springer, Cham, pp 253–280. https://doi.org/10.1007/978-3-319-92162-4_8
91. Ruppert G, Bauer R, Heisler G (1994) UV-O_3, UV-H_2O_2, UV-TiO_2 and the photo-Fenton reaction—comparison of advanced oxidation processes for wastewater treatment. Chemosphere 28:1447–1454. https://doi.org/10.1016/0045-6535(94)90239-9
92. Sasek V, Cajthaml T (2005) Mycoremediation: current state and perspectives. Int J Medi Mush 7:360–361. https://doi.org/10.1615/intjmedmushr.v7.i3.200
93. Sağ Y (2001) Biosorption of heavy metals by fungal biomass and modeling of fungal biosorption: a review. Sep Purif Methods 30:1–48. https://doi.org/10.1081/SPM-100102984
94. Sen SK, Raut S, Bandyopadhyay P, Raut S (2016) Fungal decolouration and degradation of azo dyes: a review. Fungal Biol Rev. Elsevier Ltd. https://doi.org/10.1016/j.fbr.2016.06.003
95. Senthilkumar S, Perumalsamy M, Janardhana Prabhu H (2014) Decolourization potential of white-rot fungus Phanerochaete chrysosporium on synthetic dye bath effluent containing Amido black 10B. J Saudi Chem Soc 18:845–853. https://doi.org/10.1016/j.jscs.2011.10.010
96. Shankar S, Nill S (2015) effect of metal ions and redox mediators on decolorization of synthetic dyes by crude laccase from a novel white rot fungus Peniophora sp. (NFCCI-2131). App Biochem Biotechnol 175:635–647. https://doi.org/10.1007/s12010-014-1279-2
97. Sharma C, Hagen F, Moroti R, Meis JF, Chowdhary A (2015) Triazole-resistant Aspergillus fumigatus harbouring G54 mutation: is it de novo or environmentally acquired? J Global Antimicrob Resist 3:69–74. https://doi.org/10.1016/j.jgar.2015.01.005
98. Singh L (2017) Biodegradation of synthetic dyes: a mycoremediation approach for degradation/decolourization of textile dyes and effluents. J Appl Biotechnol Bioeng 3. https://doi.org/10.15406/jabb.2017.03.00081
99. Singh L, Singh V (2014) Textile dyes degradation: a microbial approach for biodegradation of pollutants. Microb Degradat Synth Dyes Wastewaters 187–204. https://doi.org/10.1007/978-3-319-10942-8_9
100. Singh H (2006) Mycoremediation. Fungal bioremediation. Wiley, New York. https://doi.org/10.1002/0470050594
101. Solmaz SKA, Birgul A, Ustun GE, Yonar T (2006) Colour and COD removal from textile effluent by coagulation and advanced oxidation processes. Color Technol 122:102–109. https://doi.org/10.1111/j.1478-4408.2006.00016.x
102. Solís M, Solís A, Pérez HI, Manjarrez N, Flores M (2012) Microbial decolouration of azo dyes: a review. Process Biochem. Elsevier. https://doi.org/10.1016/j.procbio.2012.08.014
103. Stolz A (2001) Basic and applied aspects in the microbial degradation of azo dyes. Appl Microbiol Biotechnol 56:69–80. https://doi.org/10.1007/s002530100686
104. Sudha M, Saranya A, Selvakumar G, Sivakumar N (2014) Microbial degradation of Azo Dyes: a review. Int J Curr Microbiol App Sci 3:70–690
105. Sumathi S, Phatak V (1999) Fungal treatment of bagasse based pulp and paper mill wastes. Environ Technol 20:93–98. https://doi.org/10.1080/09593332008616797
106. Sun H, Yang H, Huang W, Zhang S (2015) Immobilization of laccase in a sponge-like hydrogel for enhanced durability in enzymatic degradation of dye pollutants. J Colloid Interface Sci 450:353–360. https://doi.org/10.1016/j.jcis.2015.03.037
107. Svobodová K, Novotný Č (2017) Bioreactors based on immobilized fungi: bioremediation under non-sterile conditions. Appl Microbiol Biotechnol 102:39–46. https://doi.org/10.1007/s00253-017-8575-z

108. Torres E, Bustos-Jaimes I, Le Borgne S (2003) Potential use of oxidative enzymes for the detoxification of organic pollutants. Appl Cataly B Environ. Elsevier. https://doi.org/10.1016/S0926-3373(03)00228-5
109. Vasiliadou IA, Molina R, Pariente MI, Christoforidis KC, Martinez F, Melero JA (2019) Understanding the role of mediators in the efficiency of advanced oxidation processes using white-rot fungi. Chem Eng J 359:1427–1435. https://doi.org/10.1016/j.cej.2018.11.035
110. Velmurugan P, Kamala-Kannan S, Balachandar V, Lakshmanaperumalsamy P, Chae JC, Oh BT (2010) Natural pigment extraction from five filamentous fungi for industrial applications and dyeing of leather. Carbohyd Polym 79:262–268. https://doi.org/10.1016/j.carbpol.2009.07.058
111. Volk TJ (2013) Fungi. Encycl Biodiver 2:624–640
112. Wang JL, Xu LJ (2012) Advanced oxidation processes for wastewater treatment: formation of hydroxyl radical and application. Crit Rev Environ Sci Technol 42:251–325. https://doi.org/10.1080/10643389.2010.507698
113. Wesenberg D, Kyriakides I, Agathos SN (2003) White-rot fungi and their enzymes for the treatment of industrial dye effluents. In: Biotechnology advances, vol 22. Elsevier Inc., pp 161–187. https://doi.org/10.1016/j.biotechadv.2003.08.011
114. Yang J, Gunn J, Dave SR, Zhang M, Wang YA, Gao X (2008) Ultrasensitive detection and molecular imaging with magnetic nanoparticles. Analyst 133:154–160. https://doi.org/10.1039/b700091j
115. Yanto DHY, Tachibana S, Itoh K (2014) Biodecolorization and biodegradation of textile dyes by the newly isolated saline-pH tolerant fungus Pestalotiopsis sp. J Environ Sci Technol 7:44–55. https://doi.org/10.3923/jest.2014.44.55
116. Yonar T, Yonar GK, Kestioglu K, Azbar N (2005) Decolorisation of textile effluent using homogeneous photochemical oxidation processes. Color Technol 121:258–264. https://doi.org/10.1111/j.1478-4408.2005.tb00283.x
117. Yoshitake A, Katayama Y, Nakamura M, Iimura Y, Kawai S, Morohoshi N (1993) N-linked carbohydrate chains protect laccase III from proteolysis in Coriolus versicolor. J Gen Microbiol 139:179–185. https://doi.org/10.1099/00221287-139-1-179
118. Yu G, Wen X, Li R, Qian Y (2006) In vitro degradation of a reactive azo dye by crude ligninolytic enzymes from nonimmersed liquid culture of Phanerochaete chrysosporium. Process Biochem 41:1987–1993. https://doi.org/10.1016/j.procbio.2006.04.008
119. Zeroual Y, Kim B, Kim C, Blaghen M, Lee K (2006) Biosorption of bromophenol blue from aqueous solutions by Rhizopus stolonifer biomass. Water Air Soil Pollut 177:135–146. https://doi.org/10.1007/s11270-006-9112-3
120. Zhang Y, Wei D, Huang R, Yang M, Zhang S, Dou X et al (2011) Binding mechanisms and QSAR modeling of aromatic pollutant biosorption on Penicillium oxalicum biomass. Chem Eng J 166:624–630. https://doi.org/10.1016/j.cej.2010.11.034
121. Zille A, Tzanov T, Gübitz GM, Cavaco-Paulo A (2003) Immobilized laccase for decolourization of reactive Black 5 dyeing effluent. Biotech Lett 25:1473–1477. https://doi.org/10.1023/A:1025032323517
122. Zucca P, Cocco G, Sollai F, Sanjust E (2016) Fungal laccases as tools for biodegradation of industrial dyes. Biocatalysis 1:82–108. https://doi.org/10.1515/boca-2015-0007

Effect of Environmental and Operational Parameters on Sequential Batch Reactor Systems in Dye Degradation

Ahmad Hussaini Jagaba, Shamsul Rahman Mohamed Kutty, Mohamed Hasnain Isa, Augustine Chioma Affam, Nasiru Aminu, Sule Abubakar, Azmatullah Noor, Ibrahim Mohammed Lawal, Ibrahim Umaru, and Ibrahim Hassan

Abstract **Issues**: Dye-containing wastewater not effectively managed is among the major contributors to water contamination. It is considered a major threat to public health and the environment. Thus, it must be handled properly before discharging into the environment. This chapter discusses the performance evaluation of the sequential batch reactor (SBR) system in managing dyes. It further complied and analyzed the impact of several environmental and operational parameters on the system in dye degradation. The various variables such as cycle time, hydraulic and sludge retention times, aeration, agitation, pH, dissolved oxygen, redox potential, feeding, temperature, hydrodynamic shear force, etc. are described. **Major advances**: Due to their complicated structure and synthetic root source, dyes are often regarded as one of the most difficult parts of textile wastewater to process. Environmental contamination emanating from toxic dye processing by dyeing industries continues to be a challenge. All through dye degradation, a variety of methods have been used, with various levels of efficacy, which may be attributable to differences in dye properties,

A. H. Jagaba (✉) · S. R. M. Kutty · A. Noor
Department of Civil and Environmental Engineering, Universiti Teknologi PETRONAS, Bandar Seri Iskandar, Perak, Malaysia
e-mail: ahmad_19001511@utp.edu.my

A. H. Jagaba · S. Abubakar · I. M. Lawal · I. Umaru · I. Hassan
Department of Civil Engineering, Abubakar Tafawa Balewa University, P.M.B. 0248, Bauchi, Nigeria

M. H. Isa
Civil Engineering Programme, Faculty of Engineering, Universiti Teknologi Brunei, Tungku Highway, 1410 Bandar Seri Begawan, Darussalam, Brunei

A. C. Affam
Civil Engineering Department, School of Engineering and Technology, University College of Technology Sarawak, Persiaran Brooke, 96000 Sibu, Sarawak, Malaysia

N. Aminu
Nigeria Social Insurance Trust Fund, Regional Office, Kaduna, Nigeria

I. M. Lawal
Department of Civil and Environmental Engineering, University of Strathclyde, Glasgow, UK

S. S. Muthu and A. Khadir (eds.), *Dye Biodegradation, Mechanisms and Techniques*, Sustainable Textiles: Production, Processing, Manufacturing & Chemistry,
https://doi.org/10.1007/978-981-16-5932-4_8

discharge conditions, technical potentialities, regulatory obligations, and financial thoughtfulness. Recently, the usage of SBR in dye degradation has continuously received attention from scientists. It is widely preferred in dyes degradation because it is an efficient process. More so, taking advantage of the high SBR versatility, the control parameters can be changed appropriately and the decolorization potential can be recovered. Owing to its operational versatility, shock load resistance, and high biomass retention, this chapter emphasized SBR application and efficiency for dye treatment in the context of environmental conservation. The chapter further stressed the effects of environmental and operational parameters during dyes degradation. In recent decades, dye degradation has been successfully regulated by optimizing a lot of valuable environmental and operational parameters. They assist in the comprehension of bio-flocculation structure, properties, and mechanisms. These optimization strategies for environmental/operational parameters thus pave the route for the production of fewer by-products. The application of RSM to physicochemical dye degradation processes could result in better design and optimization. Researchers focusing on dye degradation, physicochemical processes, and RSM can find the outcome of this chapter extremely beneficial. Looking at the environmental aspect and considering the high content of hazardous intermediate metabolites found in dyes, the SBR system combined with other treatment technologies are more efficient for dye degradation. The content of this chapter is expected to improve readers' fundamental literacy, direct research scholars, and be integrated into upcoming laboratory experiments on SBR systems for dye wastewater treatment.

Keywords Dyes · Degradation · Sequencing Batch Reactor · Environmental parameters · Operational parameters

1 Introduction

1.1 Environmental Pollution by Textile Industry Wastewater

Industrial operations in different mining fields, battery manufacturing, tannery, smelting, electroplating, textile, leather, petroleum processes, etc. are described as the major sources of wastewater [21]. Surface runoff, sewer infiltration, and poor management of urban solid waste also generate wastewater. Organic substances such as dyes present in wastewater pose a great environmental threat for all living organisms [1]. Therefore, reduction in effluent quantity and improving the quality would have major positive effects on land use and human health [6, 32, 58, 84].

1.2 Textile Industry as a Major Dye Source

During the textile production process, the industry is considered the most resource (water and energy) intensive industry, thereby causing intense pollution [18]. It discharges a large quantity of wastewater with complex pollutants composition [77]. Textile wastewater resulting from a finishing, dyeing, and printing industry is a complicated, highly heterogeneous, and dynamic solution emanating from many polluting substances such as salts, degradable organics, enzymes, heavy metals, toxicants, sulfur, refractory organics, nutrients, heat, surfactants, basicity, suspended solids, soaps, solvents, and dyes [40, 46, 58].

Conferring extremely colored water, textile wastewater usually comprises 10–200 mg/L amount of dye, characterized by high temperature, alkaline pH, organic matter (COD) concentration, conductivity values, and color presence [82]. Based on the aforementioned, it can be deduced that textile industrial wastewater could lead to serious environmental challenges because of its carcinogenic, mutagenic, and toxic impacts on aquatic organisms [60]. Therefore, prior to being released into the environment or any potential recycling, it is important to treat the wastewater [7]. However, it can only be reused provided conductivity and organic material are significantly decreased and the color is eliminated [114]. Due to a lack of water and stricter international standards, environmentally sustainable development in the textile industry is highly recommended [43]. This could be achieved through wastewater remediation and recycling to address the enormous environmental effect associated with the industry [12, 34].

1.3 Environmental Effects of Dye

1.3.1 Dye

Dyeing is among the many phases of the textile process where certain compounds like dye, surfactant, salt, formaldehyde, and metals are being used [47]. Dye wastewater discharged by dye processing plants, garment firms, plastic, etc. is raising environmental concerns [67]. The dyeing sector is an extreme user of water that creates high amounts of wastewater [75]. Dyes are culpable for the presence of color in textile industrial effluents, obstructing photosynthetic activity and disturbing habitats in surface and groundwater systems [26].

Dyes are often regarded as perhaps the most troublesome components of textile wastewater to be processed due to their synthetic origins and complex structure [78]. It is possible to classify dyes based on their solubility and chemical characteristics [10]. The common kinds can be grouped as direct, reactive, azo, acidic, and basic dyes [36]. Phthalocyanine derivatives, triphenylmethyl, indigoid, azo anthraquinone, and sulfur are the chemical dyes commonly used on an industrial level [78]. In the textile industry, the fixation of dyes to fabrics partly befalls, thereby producing highly colored wastewater [19, 68]. Dyes wastewater usually containing refractory dyes and organic chemicals, and their metabolite derivatives are mutagenic and toxic that can hang in the air for a long duration [11, 112].

1.3.2 Effect of Dye on Environment

Environmental contamination emanating from toxic dye processing by dyeing industries continues to be a challenge [79, 92]. Dye discharge in aqueous streams not only affects the clarity of natural water and its aesthetic value in terms of color but can also hinder the penetration of sunlight, thus decreasing photosynthetic activity and oxygen bioavailability, restricts the use of water, and can have a toxic impact on marine life [9, 85]. It can also cause hemolytic anemia, hyperbilirubinemia, acute renal failure cancer, and mutation in humans [62]. In general, because of its resistant nature, the considerable quantity of dyes used during the dyeing phase of textile production increases the environmental hazard. To break off the health, aesthetic, and environmental issues emanating from the incessant discharge of contaminated dye wastewater, its treatment has become a complicated environmental problem [49]. Therefore, developing an efficient treatment technique in terms of managing inadequate water supplies and as well the call for environmental conservation becomes paramount [49].

1.3.3 Treatment of Dye

A treatment system capable of COD, color, and conductivity value reduction in dyes wastewater seems to be of great importance in order to recycle the final effluent [23]. But, due to dye wastewater's inhibitory and recalcitrant behavior, it is usually not appropriate to apply traditional biological wastewater treatment as most of the colored dye substances with large molecular mass are immune to biodegradation [2, 72]. Strategies other than the conventional ones employed for dye-containing wastewater degradation for possible reuse categorized into physicochemical, biological, and combined systems have been reviewed in recent times [39]. According to the literature, the numerous physicochemical methods used for decolorization of dyeing effluents include coagulation/flocculation, flotation, precipitation, photocatalytic degradation, ultrafiltration, pyrolyzed petrified sediment, membrane processes, electrochemical processes, ion exchange, adsorption, electrolysis, nanofiltration, reverse osmosis, irradiation, and chemical/photochemical/advanced oxidation are common practices [48, 58, 61, 80]. However, these methods are faced with one or more disadvantages of: (i) operational problems, limited versatility, economical unviability, and environmental unfriendliness, (ii) generation of huge quantity of hazardous sludge, adsorbent regeneration, chemical waste generation, and membrane fouling, (iii) quite inefficient, complicated, expensive, pollutants phase transfer with toxic intermediates, (iv) high operating costs and energy consumption, and (v) interference by other wastewater constituents [7, 19, 29].

In a review by [16], it was concluded that biological treatment methods are effective, reliable, eco-friendly, energy-saving, financially appealing dye degradation techniques with minimum chemicals usage requirements. According to [94], color substances in dyestuffs that contaminated water are a sort of refractory organic material, which can be either solubilized or utilized by microorganisms in traditional biological treatment systems as sources of energy and carbon. A high concentration of influent color can induce partial inhibition of microorganisms that degrade color. Colorant-removal mechanisms in biological systems are either based on degradation, adsorption, or both [92].

In biological treatment systems, a broad variety of microorganisms, particularly bacteria, algae, yeast, and fungi are capable of de-staining and degrading many types of dyes with bacteria and fungi as the most intensively researched [2, 20, 112]. Aerobic systems as part of the various biological processes are commonly utilized for dye degradation. Unfortunately, treatment is normally insufficient. Interestingly, the anaerobic systems can handle large amounts of organic matter, as many dyes are susceptible to reductive change in anaerobic environments. Therefore, the hybrid aerobic and anaerobic process is sufficient for the accomplishment of both color degradation and dye molecule disintegration [23]. In the hybrid environment, dye elimination is usually achieved through bond cleavage; as a result, colorless and dangerous aromatic amines are formed, along with metabolite biodegradation [85].

The typical biological wastewater treatment methods such as activated sludge process, pure culture of decolorizer, oxidation ponds, aerated lagoons, and sequencing batch reactors produce less sludge [13, 111]. Recently, the usage of SBR in dye-containing wastewater treatment has continuously received attention from scientists because it is an efficient process for wastewater treatment. SBR is widely preferred in dyes-containing wastewater treatment [75]. More so, taking advantage of the high SBR versatility, the control parameters can be changed appropriately and the decolorization potential can be recovered [70].

2 Sequential Batch Reactor (SBR) System for Dye Degradation

2.1 *Sequential Batch Reactor (SBR)*

SBRs are a batch-operated form of the activated-sludge process, in which the various conditions are all met at varying periods in a single reactor, basin, or tank. The different treatment stages are performed at fixed and programmable periods, forming a cycle [65]. The process is divided into five (5) steps: filling, reaction, settling, drawing, and idling, with aeration and clarification in between. When aeration is turned off and a drainage system is used to remove the supernatant liquor, sludge settles [31]. The process design of the SBR system is a function of the influent load, biomass mass, and settleability, reactor capacity, aeration system, and the fractions of the overall cycle allotted for the individual treatment steps (Bungay et al. 2007). SBRs are furnished with diffusers, inlet and outlet valves, oxygen supplying kits, and mechanical sludge out take devices to effectively regulate the system [64].

SBR as an alternative to conventional suspended growth wastewater treatment system is a sludge activation mechanism that has been tweaked to integrate both aerobic and anaerobic phases in a single unit for enhanced COD and dye degradation [29, 87]. It solves bulking sludge and low-density bio-sludge problems caused by clarifier large capacity [75].

The traditional SBR is a crucial process for nitrification/denitrification, in which ammonia is initially oxidized to nitrite, followed by the oxidation of nitrite to nitrate and the formation of nitrogen gas [85]. SBR has been widely adopted in landfill leachate, phenolic compounds, nutrients, and various dyes [2]. It has a long history of usage in the treatment of textile waste, particularly for azo dyes elimination because they allow the fungal activity to be maintained over an extended duration and may achieve better results in dye decolorization when compared with batch cultivation [60]. According to [75], SBRs are extremely effective in textile industrial wastewater degradation for organic matter, nutrients, and dying materials [74].

Several kinds of SBR under diverse operational conditions have been largely examined for whole dyes bio-decolorization/mineralization in the literature. These include aerobic, sequential anaerobic–aerobic, sequential anoxic–aerobic bioreactors, and sequential aerobic–anaerobic SBRs [2, 40, 71, 76]. Common aerobic methods utilized in textile wastewater treatment have been discovered not to be successful in degrading dyes and exceeding the adsorption capacity of biomass; this may be due to the dyes' bio-resistant nature [89]. In contrast to that, anaerobic SBR for treating dye degradation has proved to be further efficient than the aerobic SBRs that mostly accommodate organic chemicals and refractory dyes [112].

2.2 *Factors that Affect SBR Systems*

Influent sample characteristics such as dissolved oxygen (DO), hydraulic retention time (HRT), carbon source, pH, redox potential (ORP), sludge retention time (SRT), feed pattern, organic loading rate, anoxic/oxic ratio, temperature, cycle duration, and settleability are all variables that can impact SBR system efficiency [83]. In concurrent N and P removal systems, low temperature is rated a serious problem and is studied since it affects microbial activity in activated sludge negatively [88]. Although the SBR operation is economical, its service needs added proficiency and responsiveness. More so, SBR consistency creates issues like bulking, sludge foaming, and trouble reaching the same starting conditions. It makes biological reaction gauging carried out with standard procedures more difficult. Operational difficulties such as COD fluctuations are also common during system operation [75]. In other to curtail the effect of the aforementioned factors, several approaches have been adopted by researchers to address the weak performance of the aerobic biological elimination of dyes. These include combining SBR with nanofiltration membrane, coupling photo-Fenton with SBR, integrating SBR with photocatalysis, coupled biological SBR-PAC adsorption systems, utilizing GAC to support biofilm attachment and use of novel bacterial consortia [30, 40, 57, 81, 93].

2.3 *Merits of the SBR System*

In terms of textile/dye wastewater decolorization, SBR offers the following advantages: (i) need small footprint and a single basin activity with no demand for a secondary clarifier [31] and operates automatically with better process control abilities; (ii) design, configuration, and process are simple and versatile [2, 19], cheap operation and installation charges [73], great bulking control and high tolerance to different shock loads [54]; (iii) high organic and dye removal efficiencies [92]; (iv) greater reaction rate at the onset of reaction [111]; (v) less accumulation of toxic biodegradable intermediates; (vi) volume required for continuous flow system is almost twice that required for corresponding fill-and-draw system [1]; and (vii) concurrent organics, nitrogen, and phosphorus removal, elevated biomass retention, robustness, toxicity resistance, lower power consumption, and breaking up of SRT from HRT [24, 50, 63]. Most of the aforementioned advantages could also be attained through microorganisms utilization [78] for dyes degradation from industrial discharge. SBR operational flexibility is convenient for dye effluents degradation, considering the fact that reaction time could be linked to the feed load with seldom organic loads. The system does allow settling time to be manipulated but also enables single pulse feeding, leading to an undesirable inventory of transient carbon substrates for filamentous bacterial growth [5].

3 Environmental and Operational Parameters Effects on SBR System

In recent decades, many valuable environmental and operational variables for dye degradation have been successfully implemented [110]. These variables have a strong influence on the effectiveness of the SBR system; as such, there is a direct correlation between the variables and treatment efficiency. These can be determined by looking at how they affect biological nitrification, denitrification, and dephosphatation, as well as substrate storage and utilization, population and structure of microbial communities, biofilm formation, granulation, and toxicity [56]. More so, they aid in understanding floc structure, properties, and bioflocculation mechanisms. Therefore, the implications of these variables should be studied in depth for the purposes of establishing a reliable and effective biological treatment process [99]. Table 1

Table 1 SBR dye removal efficiency

Type of dye	Dye concentration (mg/L)	Focused parameter	Operational capacity of the reactor (L)	Cycle duration (h)	HRT (d)	SRT (d)	Optimal conditions	Percentage removal (%)	References
Reactive Brilliant Red K-2G	500–1100	Salinity: NaCl		12			150 g/L salt, max. decolorization rate 2185.38 mg/(l.d) at 1100 mg/L dye	≈100	[99]
Reactive Brilliant Red KE-3B	200–800	Salinity: NaCl		12			50 g/L salt, max. decolorization rate 392.05 mg/(l.d) at 400 mg/L dye		
Acid Red 18	0–280	Feeding	5.5	24	2.75	12	35 mg/L dyes	44 dyes >85 COD	[29]
Basic Red 46	5–500	HRT, cycle time & salinity	10	8–24	1	20		65 dyes	[7]
Reactive Black 5	10–250	Feeding	1.8	24			467 mg/L COD; 200 mg/L dyes	98.97 dyes 94 COD	[78]
Acid Red 18 (AR18)	50–100	DO	5	6	0.375		4.0 mg/L maximum DO; OLR of 1.1–1.4 kg/m^3	38 dyes >85 COD	[70]
Reactive Blue 19	40	Salinity	7	24	1.83	10	0–30 PACl	71.7 dyes 93 COD	[67]
Reactive Red 159	1000–8000	HRT and SRT	2	24	4–24	7–24	6,500 mg/L of Reactive Red 159; 20d SRT; 8d HRT	97.68 dyes	[96]

(continued)

Table 1 (continued)

Type of dye	Dye concentration (mg/L)	Focused parameter	Operational capacity of the reactor (L)	Cycle duration (h)	HRT (d)	SRT (d)	Optimal conditions	Percentage removal (%)	References
Mixture of MX-8B, MX-5B, and MX-2R	25–100	Cycle time		6–12			Mixture of Procion Red MX-8B, Red MX-5B, and Orange MX-2R. 12 h cycle time,	58 COD	[89]
Remazol Brilliant Violet 5R (RBV 5R)	100	Redox Potential (ORP)	6.5	12		15	Sulfate reducing bacteria (SRB)	83–89 dyes	[20]
Acid Orange 7 (AO7)	50	ORP	8	24			1:40 (substrate/co-substrate)	85 dyes 90 TOC	[13]
Acid Red 18 (AR18)	1000	Feeding	7.7	24	2.2			97 dyes 99 COD	[49]
Acid Red 14 (AR14)	40	Feeding	1.5	6	0.5		20 mg Silva nanoparticles; calcium nitrate to 60 and 120 mg NO_3/L	89 dyes	[25]
Reactive Red 198	20–50	Feeding	10	12–24		15	Max dye removal at 20 mg/L with 16/4 anaerobic/aerobic phases	76–98 dyes 81–94 COD	[45]
Brill Blue KN-R	20–40	Feeding	5.5	24	1.83	10	20 mg/L	57 dyes 97.22 COD	[101]
Methyl Orange (MO)	25–500	HRT	2.5	24	4–14		8 d HRT; 1000 mg/L COD	>75 dyes >85 COD	[111]

(continued)

Table 1 (continued)

Type of dye	Dye concentration (mg/L)	Focused parameter	Operational capacity of the reactor (L)	Cycle duration (h)	HRT (d)	SRT (d)	Optimal conditions	Percentage removal (%)	References
Remazol Brilliant Violet 5R	50–100	DO and glucose	6.5	12		15	20 - 40 mL/min	38 dyes 96 COD	[18]
Acid Orange 7 (AO7)	15–60	Feeding	3	24		15	Anoxic REACT period 16 h	99 AO7 dyes 90–92 COD	[2]
Reactive Red 195 (RR 195)	30–50	C/N/P ratio	5	12		30–50	Alternate anaerobic–aerobic conditions, 800 mg/L influent COD, 50 d SRT, 24 h cycle time, 4 h aerobic phase, inhibition at > 40 mg/L influent color loadings	97 dyes	[23]
Methylene Blue (MB)	4–10	Feeding	4	4	0.333			56 dyes, 93 COD	[62]
Sirius Blue K-CFN (Direct Blue 85:DB)	50–85	Salinity		24			2.5 g salt	60–69 Color, 79 TOC, 80 COD	[85]
Acid Orange 7	125–625	Feeding	2	24			625 mg/L Color 3.5 mg/L DO	98 Color 88 COD	[79]
Reactive Blue 21 (RB21)	50	Aeration	4	24	3	10	Anaerobic–aerobic phases (8 h:13 h)	98 dyes; 98.5 COD	[39]
Direct Red 23	40	Feeding	10	24	3–7.5		0.89 g/L	76 dyes	[95]

(continued)

Table 1 (continued)

Type of dye	Dye concentration (mg/L)	Focused parameter	Operational capacity of the reactor (L)	Cycle duration (h)	HRT (d)	SRT (d)	Optimal conditions	Percentage removal (%)	References
Direct Blue 201	40–160	Feeding	10	24	7.5	28–31	3000 mg/L MLSS, HRT 7.5 d and 40 mg/L of dyes	94–99 dyes 94–97 TKN	[94]
Procion Red H-E7B	250	HRT	2	24	1–10		Best results were obtained at 1 day HRT,	65 dyes 52.4 DOC	[27]
Acid Red 18 (AR18)	500	Aeration	5.5	24	2	18	Alternating anaerobic–aerobic SBR with external feeding	88–98 dyes; 55–91 COD	[9]
Vat Yellow 1	40	Feeding	10	24	3	5–16	0.89 g/L	75.12 dyes; 70.61 COD;	[92]
Acid Orange 7	5 g/L	Redox Potential (ORP)	1	24	1.667	15		90–99 dyes 80 COD	[5]
Orange II dye	100–600	Feeding	5	24			3730 mg/L COD dosage; max. specific decolorization rate at 0.17 g/hr	89 Color	[80]
Blue Bezaktiv S-GLD 150 dye (BB 150)	3–20 g/L	Feeding	8	24		30	Volumetric dye loading rates <15 g dye/m^3d	88–97 dyes; 95–98 COD	[40]
Remazol Brilliant Violet 5R (RBV-5R)	100	Cycle time	6.5	6–48			24 h cycle time	92 dyes > 75 COD	[19]

(continued)

Table 1 (continued)

Type of dye	Dye concentration (mg/L)	Focused parameter	Operational capacity of the reactor (L)	Cycle duration (h)	HRT (d)	SRT (d)	Optimal conditions	Percentage removal (%)	References
Reactive yellow 15 (YD)	10–50	Aeration	3	24			Alternating anaerobic–aerobic and microaerophilic SBR conditions, 51.6 mg/L optimum YD conc. at 80 days SBR operation	89–100 dyes; 79–95 TOC; 92–100 NH_4^+–N	[86]
Tetra-Azo Dye Direct Black 22		Aeration	5	24			Intermittent micro-aeration, upon 58 days, steady-state circumstances for dye and COD removal were developed	81.4 dyes	[68]
Acid Red 14	20–60	SRT	1.5	6		15–25	15 days	90 Color 77 COD	[26]

compiles the efficiency of different dye-treatment reactors. This is to explicitly indicate the role of several operational and environmental variables on dye degradation in an SBR system. The table further states the optimal conditions attained for each parameter during each experimental study. It was clear from the literature that the majority of the treatments were aimed at removing color and COD.

3.1 Cycle Time

Equation (1) is used to describe a loop in SBR, where an entire cycle duration (t_C) is the sum of the fill, react, settle, decant, and idle times.

$$t_C = t_F + t_R + t_S + t_D + t_I \tag{1}$$

where t_F = fill time (h), t_S = settle time (h), t_R = react time (h), t_I = idle time (h), and t_D = decant time (h) [100].

The effect of the cycle duration has been seldom investigated. It can be seen that the continuous reduction in cycle duration led to an improvement in biomass composition as a result of a more plentiful organic portion [38]. Operating an optimized cycle time SBR system could possibly support sludge particles agglomeration with improved settleability and maintain comparable COD and color removal. Excellent phosphorus removal was observed at a short cycle length. This may be attributed to a higher rate of nitrogen eliminated through the nitrite route, which renders biodegradable carbon more available. However, as the cycle length rose, the system's phosphorus percentage removal decreased due to the need for biodegradable organic carbon denitrification continued to increase in conjunction with total nitrification [28].

An aerobic SBR employed as a post-treatment to anaerobically treat a mixture of three different dyes has the potential to further remove pollutants from wastewater. The SBR demonstrated the ability to handle the dye mixture at variable cycle times where the cycle duration of 12 h was observed to be the best at producing excellent effluent while still adhering to SS and COD regulatory limits [89]. In similar research for azo dye removal brilliant violet 5R (RBV-5R) treatment, it has been revealed that reducing anaerobic cycle time did not alter the system's color removal rate. The best SBR efficacy for color and aromatic amine degradation was achieved after a total cycle period of 24 h. The efficiency of COD removal was harmed when the total cycle length was reduced from 48 to 24 h [19]. However, shortening cycle duration from 24 to 8 h did not affect the decolorization rate as reported by [7].

3.2 HRT

HRT could be regarded as a measure of the mean duration whereby wastewater lived in a bioreactor system. The HRT computation for an SBR system is given by:

$$HRT = \frac{(t_C)}{V_F/V_T}\frac{1}{24} \tag{2}$$

where V_F in Eq. (2) represents the amount of wastewater filled and collected effluent for a full cycle, V_T represents the reactor's total working volume, and t_C represents the total cycle length [100]. HRT is an important parameter in biological wastewater and hydrogen production processes [96]. Increment in HRT offers ample time for COD and other system intermediates fractional mineralization [7]. For dyestuff wastewater treatment using the only biological system as SBR, longer HRT is required [108].

The color removal performance of a GAC-SBR system has been reported to be high and stable at an increased HRT [91]. The rise in HRT primarily needed under low temperatures triggers decreased concentration of biomass, endogenous decay rate, specific growth rate, and yield of biomass. Therefore, mindful attention should be given to the choice of the optimal HRT for efficient SBR application in dye wastewater treatment [90]. HRT can be lowered where possible by incorporating membrane modules that enhance system operating flux and permeability. With a decrease in HRT, the specific nitrate reduction rates, specific nitrite and ammonium oxidation rates, sludge volume index, and specific oxygen absorption rates all inflate. Therefore, causing a reduction in the microbial population through the biomass washout, continual HRT lowering could further degrade system efficiency [59].

3.3 SRT

In an activated sludge process, SRT is an important design and operating variable for controlling process parameters, such as effluent water quality, nitrification, sludge volume, oxygen demand, and growth status [107]. It refers to how long an organism settles in a bioreactor on average. According to literature, to keep organisms alive in a batch reactor, the level of net growth must be greater than or equal to the SRT.

SRT can indeed be calculated mathematically using the equation:

$$\text{SRT} = \frac{V \times X_r}{Q \times X_e} \tag{3}$$

where V = useful reactor capacity; Q = effluent quantity per day; X_r and X_e = influent and effluent VSS concentration [88].

SRT obtained from Eq. (3) can be managed by activated sludge wasting regularly. The equation can be utilized to calculate the waste volume:

$$Q_w = \frac{V}{SRT} \tag{4}$$

where Q_w in Eq. (4) = suspended solids wasting rate, L/d; V = working capacity of the reactor [22].

In biological nutrient removal operations, SRT could be utilized to change the microbial population. It takes a much longer duration to achieve huge nitritation rates as SRT is raised. The nitrification start-up period would be shortened because of the comparatively short SRT. Shorter SRTs have a higher nitrite accumulating rate (NAR). Diverse floc morphology may result from using an SBR at varying SRTs. Low SRTs are associated with uneven sludge floc morphology.

Divergent SRTs are likely to have significant differences in effluent SS rates. The potential of the sludge to flocculate varies, and this depends on SRT. A study treating Acid Red 14 showcases the ability of SRT versatility to eliminate color by azo dye depletion in activated granular sludge systems. SRT regulation to 15 days led to a 30% decrease in color removal rate. Nonetheless, increasing SRT to >25 days regressed the impact and aided the total bioconversion of the known aromatic amine through the aerobic reaction period [26]. In another related study treating highly concentrated dye in an AnSBR, the system efficiency maintained firmness for 30 days at short HRT and long SRT [96].

3.4 Feeding

This is the process of loading a reactor with dye wastewater combined with biomass for microbial activity, whether by pumping or by gravity among reactors at low and high water levels. Aerated, static, and mixed fill are all fill strategies that can be managed by a time controller. Continuous vs. pulse feeding does not alter the biomass-dominant bacteria greatly [17]. It is of note that what reactor is fed with (carbon source) and its concentration significantly matters in the SBR system during dye degradation. Because of dye wastewater complexity, external carbon sources sometimes referred to as co-substrates are usually employed for the sustenance of the reactor [95].

Adding Thai rice noodle wastewater increased dye removal efficiency by 30% [94]. While discussing co-substrate and its efficacy, it has been noted that the Reactive Black 5 removal efficiency was enhanced as influent co-substrate concentration increased [78]. Increasing OLR has proved that it increases dye removal efficiency but deteriorates COD [45]. It is a general observation that high initial dye concentrations can cause a reduction in dye percentage removal. This implies that volumetric loading rate can affect dye removal efficiency in SBR [40]. The dye removal efficiency of a bioreactor system is said to decrease with an increase in influent dye concentration [94].

3.5 Shock Loads

An abundance of particulate organic matters might sometimes be nourished to SBRs, which could require extra HRT to process. Therefore, it becomes important to analyze a batch reactor efficiency at shock loading state when dealing with dyes since it is complicated wastewater. Sudden unplanned changes in influent concentration, also known as organic shock loading, can cause treatment efficacy to be disrupted over time. Shock loading causes stress to the bioreactor. This can be standardized by incorporating surplus common co-substrates [37]. The experimental recuperation time is usually longer than the theoretical recuperation time throughout shock loading. This may be because of the toxic compounds available in the wastewater having an inhibitory effect [69].

Hydraulic, organic, toxic, and combined loads are examples of typical shock loads. The toxic shock load occurs by the use of organic solvents to raise pollutant concentrations in a batch reactor blended well above the activated sludge process's threshold level. Various COD concentrations can be applied at different time interludes to produce an organic shock load. By adulterating the reactor wastewater, differences in COD concentrations can be produced. Reducing HRTs of stressed batch reactors generates hydraulic shock loads. This causes a major constraint for the reaction rate. For the combined shock load, two (2) or three (3) of the above state shocks are concurrently applied to the reactor, with varying intensity levels at given cycles, and the reactor's efficacy in wastewater treatment is then assessed [69]. Organic loading of 0.56 and 0.75 kg COD/cum-day rates has exhibited less impact on the treatment of Acid Black 210, as the reactor effectiveness was not restrained at the examined loading rates [72].

3.6 Agitation

The degree of agitation is critical for creating better mixing environments, solubilizing suspended organic compounds, and enhancing mass transfer. These characteristics contribute to a higher rate of substrate consumption, which can shorten the total cycle time. Mechanical stirring, biogas, and liquid recirculation can all be used to provide agitation in an SBR system. The dynamics of volatile acid production and utilization were clearly changed by increasing mass transfer resistance, enabling the device to achieve different observable steady states at a lower rate of agitation [3]. A research reveals that the production of biogas was insufficient to increase the turbulence required to minimize the occurrence of possible fixed zones as well as mass transfer resistance. As a result, an anaerobic SBR was produced, with agitation provided by diaphragm pump recirculation of the effluent. The authors concluded that effluent recirculation can be used as a method of agitation. The maximal recirculation velocity for an AnSBR device utilized during wastewater treatment was tested to assess the effectiveness of recirculation. When the machine was operating at lower speeds, mass transfer was found to be a constraint. Higher velocities, on the other hand, can reduce microbial activity due to excessive shearing, which can harm the flocs in the biomass and trigger granule rupture, resulting in poor solid separation [66].

3.7 Aeration

In aerobic sludge granulation, aeration is a very important parameter (D. [106]. Sluggish aeration levels in SBRs might lower overall nitrite and nitrate levels, lowering the carbon demand of denitrifying bacteria and resulting in far more carbon sources being prepared for denitrification processes. The oxygen-depleted environment can enhance dye biodegradability and lessen refractory substances toxicity, allowing nitrifying and denitrifying bacteria to develop in the SND phase longer. According to the literature, a higher aeration rate helps in rapid aerobic granulation [9]. To maintain the stability of aerobic granules, it is also necessary to render adequate hydraulic sharpening power to prevent filamentous bacteria overgrowth. However, it has some drawbacks, including high costs due to energy demand, inability to remove TN, degradation of anaerobic conditions, and low phosphorus removal, among others. A traditional SBR system's aeration process can be changed, resulting in an intermittent aeration SBR, which is described as a mechanism in which anoxic and aerobic environments are formed by intermittently redoing aeration and non-aeration cycles. The key benefits of using intermittent aeration in SBRs entail improved nitrogen removal and lower operating costs.

Micro-aeration strategies tested for Direct Black 22 tetra-azo dye treatment have enhanced aromatic amine reduction and ecotoxicity. The uninterrupted micro-aeration resulted in four times greater oxygen consumption compared to the intermittent phase. However, intermittent micro-aeration attached to anaerobic digestion became the better choice with far lower oxygen intake compared to the undisturbed micro-aeration process and an extraordinary total acute ecotoxicity removal [68]. Studying the effect of aeration in Reactive Yellow 15 dyes degradation at micro-aerophilic and anaerobic states, the removal of dye was greater than the aerobic conditions, with robust removal of dye for elevated YD concentrations. In the anoxic period, azo dye, ammonium, organic carbon concentrations were significantly reduced treating dye wastewaters in bioreactors [86]. In a related study using Reactive Blue 21 (RB21) dye, the aerobic phase could not significantly eliminate RB21 and the greatest dye degradation happened in the anaerobic stage. Given the phthalocyanine dye's high solubility and molecular structure strength, microorganisms were unable to significantly affect RB21 molecules during the short anaerobic duration of 4 h. Surprisingly, surfactant and COD degradation were significantly improved at the high aerobic phase 17 h [39]. Successful color biodegradation for mono-azo dye Reactive Red 195 (RR 195) has been stated to befall under anaerobic conditions [23].

3.8 Salinity

Salinity is regarded as one of the major elements that can affect the production and composition of EPS. Osmotic pressure usually generated by salinity may trigger the microorganisms to produce exopolysaccharides [104]. It can inhibit cell enlargement and division directly or indirectly. The development, effectiveness, and efficiency of wastewater treatment systems can also be hindered. Increasing salinity in an SBR reactor substantially affects system performance negatively, leading to the collapse of the reactor ecosystem. It releases cellular materials, amounting to soluble COD increase and capable of affecting happily acclimated microorganisms to a steady salinity. The SBR system is delicate about salinity shock, and it is difficult for the main anti-salt microbes to respond well to a higher salt concentration range instantly. However, it is contradictory that some scholars claim that increasing wastewater salinity may boost SVI, while others insist that salinity increase could result in SVI reduction. High salinity is required for a batch reactor to attain a steady short-cut nitrogen elimination as the process rapidly changes to partial nitrification. However, it results in stronger inhibition as it can repress NOB and AOB population.

High salinity also has powerful inhibition on the activated sludge's phosphorus degradation and nitrification capacity [110]. It also contributes to soluble microbial products aggregation and decreases effluent quality. The increase of effluent suspended solid concentration, disintegration of activated sludge flocs, decline of organic matter removal rate, increase of buoyancy force, and inhibition of bacterial metabolism are ways by which salinity negatively affects an SBR treatment system. A study by [105] revealed that although microbial community structure is greatly influenced by salinity, SBRs performance in terms of organic matter extraction is not affected by high salinity. Efficient methods for saline wastewater treatment have been established by the combination of fluidized bed with activated sludge process, cultivation of marine activated sludge utilizing sea mud, and incorporation of high salt-resistant bacteria. Increased aeration rate, reasonably prolonged aeration time, or the inclusion of halotolerant bacteria may also increase the degradation of dye [15].

Hyper-salinity wastewater, as previously mentioned, commonly induced plasmolysis and/or loss of cell function, leading to detrimental consequences on biological treatment techniques. It could be described that azo dye was suitable for the production of a rare presiding species under lower-salt conditions and played a significant part in the arrangement of the microbial community. However, certain halophilic and salt-tolerant species may be enhanced during acclimatization as salinity is elevated. Decolorization of KE-3B was partly impaired under high salt environments, as the decolorization consortium was more susceptible to high salt concentrations. On the contrary, a limited impact was witnessed on K-2G decolorization. More so, with the rise in NaCl concentration, microbial diversity proved to be more prevalent for both dyes [99]. In a related study treating Direct Blue 85 (DB), NaCl exhibited an insignificant impact on the adsorption intensity [85]. Treating Basic Red 46 in an SBR system, COD and color degradation rates are reduced with rising nitrate ion and salt concentration in the wastewater [7]. Adding PACl to an SBR system treating Reactive Blue 19 enhanced effluent TSS and had no notable impact on effluent dyes, COD, and turbidity [67].

3.9 pH

The pH value of the BNR system responds to microorganisms' cell membrane permeability and electrical charge. pH directly impacts microorganism growth and activity in the environment significantly. It is crucial in ensuring the prevailing position of useful bacteria in wastewater treatment methods. As high concentration nitrate-containing wastewater denitrification is controlled by pH, metal toxicity also depends on pH. While studying the connection between pH and the degradation of biological phosphorus, the capacity of anoxic phosphate uptake was observed to be enhanced by pH rise for denitrifying polyphosphate accumulating organisms (DNPAOs) sludge. Low pH can be raised by the alkalinity generation in a treatment process [55]. To improve SBR performance, detail of the microbial population dynamics under different pH is required.

3.9.1 Alkalinity

Alkalinity is strongly associated with nitrogen loss. It exhibits a linear reverse interaction with the concentration of the effluent nitrogen. Another benefit for alkalinity is the simplicity and ease of testing with test kits. Alkalinity <100 mg/L does not portray adequate denitrification, while values greater than 250 mg/L indicate inadequate nitrification. An alkalinity test is usually carried out in an SBR system to observe the system operational situation and subsequently show the end of nitrification/denitrification. Nevertheless, there is currently no real-time alkalinity evaluation. Samples of wastewater had to be obtained from treatment plants and weighed by titration, which resulted in several minutes of time lag. Alkalinity variation amongst influent and effluent ($\Delta \text{Alk}_{\text{inf.-eff.}}$) can be employed for nitrification/denitrification lengths indicator. The influent and effluent alkalinity difference theoretical values can be computed as:

$$\Delta \text{Alk}_{\text{theory}} = 3.57\left([\text{NH}_4^+]_{\text{inf.}} + [\text{NO}_2^-]_{\text{eff.}} + [\text{NO}_3^-]_{\text{eff.}} - [\text{NH}_4^+]_{\text{eff.}}\right)(\text{mg/L}) \quad (5)$$

Since nitrate and nitrite are mostly less than the influent limit, they cannot be incorporated in Eq. (5).

Alkalinity difference resulting from laboratory experiment can thus be computed as:

$$\Delta \text{Alk}_{\text{exper.}} = \text{Alkalinity}_{\text{inf.}} - \text{Alkalinity}_{\text{eff.}}(\text{mg/L}) \quad (6)$$

$\Delta Alk_{exper.}$ in Eq. (6) above was weighed up with ΔAlk_{theory} for ΔAlk accuracy which indicates nitrogen extraction. The existence of alkalinity consumption and formation in a single tank of an SBR system could minimize the difference in alkalinity between the influent and the effluent (B. K. [51].

3.9.2 DO

DO concentration can be measured by microelectrodes to provide important details for SBR optimization. The DO concentration is said to be the most important factor in regulating SND activity SBR. It also affects granular sludge formation and aerobic granular system stability [79]. Thus, altering the DO concentration could alter the aerobic zone and subsequently the bacterial community structure within the reactor. As increased levels of DO ensures total nitrification and eliminate organic carbon, a low DO state is fit for achieving a high degree of SND [102]. Lessened aeration levels are said to be beneficial for conserving energy and denitrification stimulation [109].

In a study where DO concentration was varied during the aerobic stage, the influent solids decreased DO levels in the aerobic period from 2 to 0.8 mg/L and improved phosphorus removal efficiency. However, it consequentially reduced the nitrification levels proving the deficiency of oxygen [28]. For the biodegradation of Acid Red 18 (AR18) dye, aerobic conditions with a 5–6 mg/L DO concentration were not adequate [70]. When the effect of oxygen on the anaerobic biodegradation capacity of blended microbial culture for removal of brilliant violet 5R was investigated, increased oxygen supply had a negative effect on anaerobic color removal efficiency and azo reductase activity of anaerobic microorganisms. In the SBR dyes removal, molecular oxygen was said to have greatly decreased color removal [18]. Similarly, the increase of DO up to 3.5 mg/L did not reveal any effect on color removal by the SBR system treating Acid Orange 7 [79].

3.9.3 Temperature

The sustenance and growth of microorganisms alongside anaerobic degradation of organic substances are highly affected by temperature. As temperature decreases, enzymatic and chemical reactions slow down leading to total growth termination. Temperature has a significant effect on nitrogen elimination community structure in the anammox SBR with NLR, even though it has no impact on granular biomass integrity sustentation and anammox sludge grown on biofilm. Temperature decrease could transform floccular sludge microbial community structure [53]. Great TN and SNDPR performance removal can be accomplished at less temperature [52]. However, the functions of activated sludge microbes could be negatively affected by low temperatures. It has been reported that the SBR system faces numerous hurdles at a temperature ≤10 °C. This results in the declination of denitrification and nitrification levels because of the destabilized microbial activities. Decreasing temperature can result in oversaturated DO, which hinders anoxic denitrification conduct [113].

3.9.4 ORP

A significant determinant that may affect the biodegradability of resolute micropollutants is the redox environment. ORP as a control parameter is essential for treatment process effectiveness optimization. According to the literature, the existing redox conditions for an SBR system are fully aerobic, anoxic/aerobic, and microaerobic [5]. An Orion ORP Probe is mostly used for measuring ORP. In engineering applications, the accuracy of ORP for online monitoring is below expectation which could be due to the low pH sensitivity to nitrogen concentrations, probe fouling, and difficulties in identifying breakpoints in ORP profiles comprising numerous factors [51]. Consumption of oxidized nitrogen forms and residual DO can lead to a decrease in ORP value, thereby changing the protein configuration within the system. Depending on sensor signal change, ORP reduction can be regulated either by raising/lowering the air flow or by switching the aeration on/off [97]. With and without the involvement of co-substrates and nutrients, ORP monitoring was performed in an AnSBR. The extremely active anaerobic biomass found could be due to the bioreactor's low ORP values [80].

Adding variable sulfate quantity to a reactor for Remazol Brilliant R5 treatment changes the ORP values because of sulfate reduction. A larger quantity of sulfate inclusion in bioreactors with lower ORP values has a positive effect on azo dye biodegradation. The authors concluded that the most appropriate ORP value for color removal is anything below −150 mV [20]. In a related study, it was observed that sulfate supplementation into bioreactors fed with acid dye led to sulfate decrease. However, decolorization was not improved [5]. In an SBR treating azo dye AO7, ORP made it easier to regulate the aerobic and anaerobic phases, as the lone control variable [13].

3.9.5 C/N/P Ratios

Microbial breakdown of dyes molecules largely relies on the quantity of carbon, nitrogen, and phosphorus ready for their activity. Even though bacteria cannot produce the necessary enzymes required for carbon utilization with a small amount of nitrogen, a large amount of nitrogen can inhibit bacterial growth (H. Y. [102]. The availability and nature of electron donors conveniently denoted in terms of C/N ratio performs a significant role in biological denitrification [35]. It is a factor in the control of the denitrification efficiency that guides in the design and configuration of the procedure to attain the aimed treatment [103], and could influence microbial population dynamic of aerobic granular sludge during the continuous operation. Thus, applying appropriate proportion is important for the efficient performance of any BNR process [42]. In an SBR treatment system, an influent COD/N ratio that can be calculated by Eq. (7) higher than 10 is recommended for effective nitrogen removal [98].

The influent C/N ratio can be calculated by the equation:

$$\text{C/N ratio} = \text{COD/TN} = c(\text{COD})/c(\text{TN}) \tag{7}$$

where c(COD) and c(TN) stand for influent COD and TN concentrations, respectively [98].

Studies found that elevated COD/N ratios resulted in poor granulation as low COD/N ratios led to microorganism aggregation in reactors. Besides that, a fall in the COD/N ratio can cause an increase in the nitrification rate, rapid carbon deficit, granule destabilization and treatment downturn, major microbial community change, physical structure strength, and settleability [102]. High nitrogen or low carbon due to settling of suspended load at the initial stages can lead to a low C/N ratio [42]. Consequentially, NH_4^+–N elimination performance decreases with a reduction in C/N ratio [35]. In order to restrict the electron supplies for reductive half-reactions, the application of a low C/N ratio has been documented, resulting in the growth of denitrification intermediates (NO_2^-, NO, and N_2O). It has a certain degree of inhibition on denitrifying bacteria [35]. However, in all C/N ratios tested, the utilization of surplus electron donors contributes to electron source wastefulness, increased MLSS, MLVSS, and effluent COD. Decreasing C/N/P led to a consistent improvement in the COD removal efficiency in the SBR to 99.95% [14]. In a related study, reduced COD/N ratios from 20 to 4 boosted bioactivity and maintained settling, while reducing nitrogen removal as a result of carbon deficiency [102].

3.9.6 Mixed Liquor Suspended Solids (MLSS)

MLSS is a highly complex and the most important operational parameter for SBR systems because it significantly influences effluent efficiency. It is the composition of a specified quantity of suspended solids, combined with arriving wastewater. As a result, it should be checked on a regular basis. Sludge bulking occurs when the MLSS value is high, rendering the dye treatment system less efficient. In dye-containing wastewater, the SBR system efficacy improved to 75% by increasing MLSS values up to 2000 mg/L with 90 turbidity and 75% COD removals [75]. Batch experiments showed that there was great potential in applying high MLSS for dye degradation. These oppose the study in which effluent quality significantly drops under high MLSS concentrations [4].

3.9.7 Hydrodynamic Shear Forces

Dye degradation in the SBR system through aerobic granulation can be accomplished whenever fitting shear forces are employed. It has been judged that a specified hydrodynamic shear force value is essential for aerobic granules development, structure, stability, and metabolism. To effectively calculate the shear forces, hydrodynamic shear forces must be considered because of the friction among biomass and liquid exterior [102].

Thus, the shear force (τ) can be calculated by the following equation:

$$\tau = \frac{25\mu F}{D_P S_R}\frac{1-\varepsilon}{\varepsilon^2} + \frac{1.75\rho F^2}{S_R^2}\frac{1}{\varepsilon^2} \tag{8}$$

where S_R = geometric section area (m^2), D_p = equivalent diameter (m) of filling particles, μ = absolute fluid viscosity (kg/m s), ε = bed porosity (m^{3}empties/m^3 bed), F = recirculation flow rate (m^3/s) and ρ = fluid density (kg/m^3).

Equation (8) explains that shear forces are dependent on ε and D_p.

Biomass density and hydrophobicity increase as hydrodynamic shear forces increase. A higher hydraulic shearing force which is mainly affected by DO greatly contributes to the rapid development of aerobic granules [109]. This could be achieved if adequate oxygen is provided by restated and strong shear force to mitigate filament growth for the system to remain stable in the course of long-term operations [102].

3.9.8 Electric Power

A study by [8] reported the negative effects of power interruption as poor TSS and COD removal efficiencies alongside poor sludge settling properties. It also concluded that the longer the interruption time, the greater the recovery time required for the system to attain steady state. In a power-related study, passing electric current density through a functional SBR system successfully removed nitrogen and phosphorus compounds by autotrophic (hydrogenotrophic) denitrification and coagulation processes, respectively [44].

4 Optimization of SBR Operational and Environmental Parameters by Response Surface Methodology (RSM)

RSM applies statistical and mathematical modeling approaches to estimate and optimize essential parameters that influence the behavior of a given response. It is an effective way to construct models, design experiments, consider the relationship between variables, and deduce optimal working variables. It is also used to assess

SBR performance at various operational and environmental conditions, especially in dye-containing wastewater treatment [87]. It is commonly used to improve the treatment ability of synthetic dyes-contaminated wastewater. As a successful technique, RSM also has the potential to model and optimize over four efficient factors, which could lead to more useful and informative performance. Among RSM's different design approaches, such as Central Composite Design (CCD), Box-Behnken Design (BBD), Full Factorial Design, and Doehlert Design (DD), CCD was perhaps the most popular. CCD joined with RSM overwhelms the constraints of several classical approaches. Most researchers have studied ≤4 variables and pH, RT, and C0 were the most common response variables in all procedures, with %R being the most common response.

Currently, the most popular computer program for RSM is Minitab and Design Expert. In recent studies, no particular pattern has been noticed regarding the kind of dyes [36]. The effects of SRT, dye concentration, and HRT were investigated using AnSBR on Reactive Red 159 decolorization employing RSM via a central composite design following non-sterile and anaerobic conditions [96]. The CCD was chosen because of its capacity to support a range of factors and provide a definite prediction. In the study, CCD was fitted applying the second-order polynomial quadratic equation (Eq. (6)) with three independent factors including HRT, dye concentration, and SRT, while the decolorization was the response variable. Model expressions adoption or rejection was carried out on 95% confidence level probability (P) value. The findings were analyzed using variance analysis (ANOVA).

$$Y = \beta_0 + \sum\nolimits_{i=1}^{k} \beta_i X_i + \sum\nolimits_{i=1}^{k} \beta_{ii} X_i^2 + \sum\nolimits_{i \geq j}^{k} \sum\nolimits_{i=1}^{k} \beta_{ij} X_{ij} X_j \quad (9)$$

where i = linear coefficient, j = quadratic coefficient, k = number of factors, X_i = coded experimental levels of the variables, Y = predicted response, β_0 = constant, β_i = linear coefficient, β_{ii} = interactive coefficient (regression coefficients for quadratic effects).

To unveil the fitness of the selected model, the coefficient of determination (R^2) was being utilized. Fisher's F-test was adopted to confirm the statistical significance of the fitted model. Based on the consequences of the independent parameters, contours and 3D plots for the dependent factors were determined. As response variables for mathematical models, dye effluent, decolorization, and rate of decolorization were utilized with dye concentration as a favored AnSBR shift factor. The laboratory experiment conducted with 8 days HRT, 6,500 mg/L dyes, and 20 days SRT achieved 97.68, 142.62 mg/L, and 264.54 mg/L/h for % decolorization, dye effluent, and rate of decolorization, respectively. The results also indicate that there was a high-grade balance between the decolorization efficiencies experimental and predicted values.

Unlike the earlier discussed scenario, all experiments were conducted in SBR mode at 5-day HRT utilizing Box-Behnken design (BBD) as COD, SVI, and decolorization were evaluated for the individual operational state. The impact of process

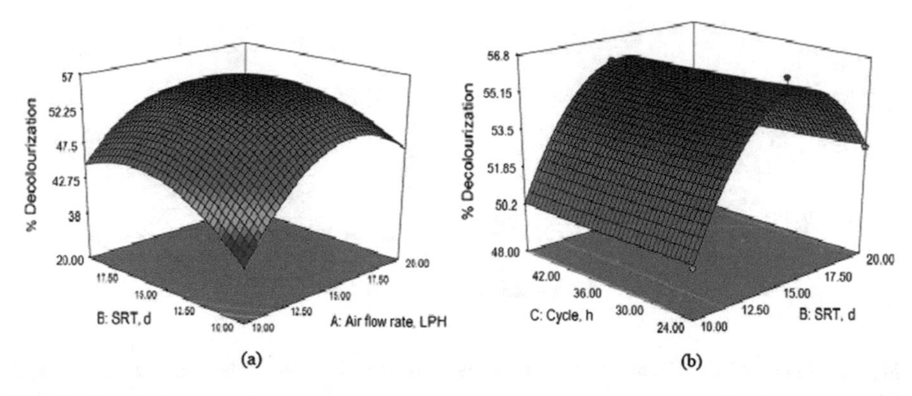

Fig. 1 3D plot of **a** SRT and air flow rate **b** cycle time and SRT effects on dye wastewater decolorization

variables such as cycle time, air flow rate, and SRT was examined by the authors with findings evaluated by ANOVA utilizing the quadratic regression model. For COD, SVI, and decolorization, model *P* values were quite low, reaffirming the significance of the model. The predicted R^2 value agrees with the adjusted R^2 values for the three responses. The SBR performance was investigated at various OLRs under optimal provisions, and the outcomes showed that OLR of 0.165 kg COD/m^{3}d achieved COD reduction and highest decolorization of 79.4% and 71.3%, respectively. The SVI was found to be low indicating the high SBR effectiveness. To further explain the optimization studies, Fig. 1a illustrates decolorization as a function of SRT and air flow rate, confirming that decolorization often rises as the air flow rate rises. However, air flow rate >15.9 LPH decreases decolorization. Figure 1b demonstrated that a rise in SRT value resulted in a rise in decolorization percentage while decolorization was not affected significantly by increasing the cycle duration.

For COD reduction, similar profiles were received as depicted in Fig. 2a, b. As can be observed in Fig. 3a, b rise in SRT led to a reduction in the SVI value. However,

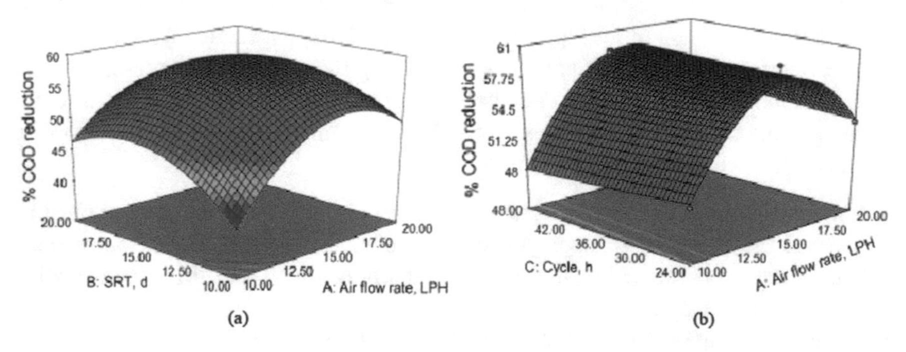

Fig. 2 3D plot of **a** SRT and air flow rate **b** cycle time and air flow rate effects on COD reduction in dye wastewater

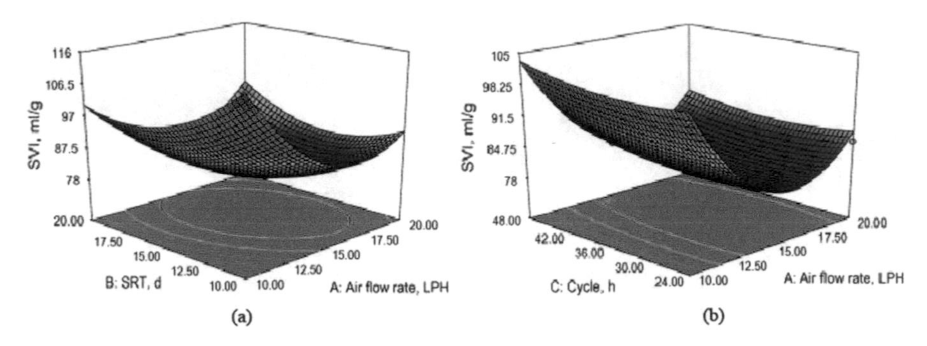

Fig. 3. 3D plot of **a** SRT and air flow rate **b** cycle time and air flow rate effects on SVI in dye wastewater treatment

at elevated SRT, the SVI value rises. This could be due to sludge disintegration at a higher air flow rate. As shown in Fig. 3b, cycle duration had no negative impact on the SVI value. From this study, it can be inferred that RSM could be effectively used for dye wastewater treatment in SBR [87].

5 Conclusion

To break off the health, aesthetic, and environmental issues emanating from the incessant discharge of contaminated dye wastewater into the environment, the treatment of dye-contaminated wastewater has become a big environmental problem. Therefore, developing an efficient wastewater treatment technique in terms of managing limited water supplies and as well the need for environmental conservation becomes paramount. Numerous strategies have been tried for dye degradation, showing different levels of effectiveness and weaknesses as discussed in the chapter. In recent times, the usage of SBR in dye-containing wastewater treatment has continuously received attention from scholars because it is an efficient process for wastewater treatment, taking advantage of the high SBR versatility. This chapter included a thorough evaluation of current research papers, emphasizing the usability and efficacy of the SBR system for dye treatment. The chapter went on to look at the impact of environmental and operational variables on the SBR system and arrived at the following conclusions:

- For effective use of SBR while treating dye wastewater, special attention must be given to the choice of the appropriate environmental and operational parameters. These control parameters can be appropriately changed and the decolorization potential can be recovered.
- These optimization strategies for environmental/operational parameters thus set the stage for SBR to become a successful, long-term, and inexpensive technology with fewer by-products.

- The application of RSM to physicochemical dye removal processes could result in great design and optimization. Researchers focused on dye degradation and physicochemical processes, and RSM would find this article extremely helpful.
- Due to the higher elimination of dangerous intermediate metabolites found in dyes, it is better for the environment to combine sequencing batch reactor system with other treatment technologies that are efficient for dye degradation instead of the conventional SBR process alone.

References

1. Abu-Ghunmi LN, Jamrah AI (2006) Biological treatment of textile wastewater using sequencing batch reactor technology. Environ Model Assess 11(4):333–343. https://doi.org/10.1007/s10666-005-9025-3
2. Al-Amrani WA, Lim PE, Seng CE, Ngah WSW (2014). Factors affecting bio-decolorization of azo dyes and COD removal in anoxic-aerobic REACT operated sequencing batch reactor. J Taiwan Inst Chem Eng 45(2):609–616. https://doi.org/10.1016/j.jtice.2013.06.032
3. Alami AH, Alasad S, Ali M, Alshamsi MJS (2021) Investigating algae for CO_2 capture and accumulation and simultaneous production of biomass for biodiesel production 759:143529
4. Alattabi AW, Harris CB, Alkhaddar RM, Ortoneda-Pedrola M, Alzeyadi AT (2019) An investigation into the effect of MLSS on the effluent quality and sludge settleability in an aerobic-anoxic sequencing batch reactor (AASBR). J Water Process Eng. https://doi.org/10.1016/j.jwpe.2017.08.017
5. Albuquerque MGE, Lopes AT, Serralheiro ML, Novais JM, Pinheiro HM (2005) Biological sulphate reduction and redox mediator effects on azo dye decolourisation in anaerobic-aerobic sequencing batch reactors. Enzyme Microbial Technol 36(5–6):790–799. https://doi.org/10.1016/j.enzmictec.2005.01.005
6. Almahbashi N, Kutty S, Ayoub M, Noor A, Salihi I, Al-Nini A, Ghaleb AJ (2020) Optimization of preparation conditions of sewage sludge based activated carbon
7. Assadi A, Naderi M, Mehrasbi MR (2018) Anaerobic-aerobic sequencing batch reactor treating azo dye containing wastewater: effect of high nitrate ions and salt. J Water Reuse Desalination 8(2):251–261. https://doi.org/10.2166/wrd.2017.132
8. Aygun A, Nas B, Berktay A, Ates H (2014) Application of sequencing batch biofilm reactor for treatment of sewage wastewater treatment: effect of power failure. Desalination Water Treatment 52(37–39):6956–6965. https://doi.org/10.1080/19443994.2013.823354
9. Azizi A, Moghaddam MRA, Maknoon R, Kowsari E (2015) Comparison of three combined sequencing batch reactor followed by enhanced Fenton process for an azo dye degradation: Bio-decolorization kinetics study. J Hazard Mater 299:343–350. https://doi.org/10.1016/j.jhazmat.2015.06.044
10. Baloo L, Isa MH, Sapari NB, Jagaba AH, Wei LJ, Yavari S, Vasu RJ (2021) Adsorptive removal of methylene blue and acid orange 10 dyes from aqueous solutions using oil palm wastes-derived activated carbons 60(6): 5611–5629
11. Bashiri B, Fallah N, Bonakdarpour B, Elyasi S (2018) The development of aerobic granules from slaughterhouse wastewater in treating real dyeing wastewater by Sequencing Batch Reactor (SBR). J Environ Chem Eng 6(4):5536–5543. https://doi.org/10.1016/j.jece.2018.05.020
12. Blanco J, Torrades F, Moron M, Brouta-Agnesa M, Garcia-Montano J (2014) Photo-Fenton and sequencing batch reactor coupled to photo-Fenton processes for textile wastewater reclamation: Feasibility of reuse in dyeing processes. Chem Eng J 240:469–475. https://doi.org/10.1016/j.cej.2013.10.101

13. Buitron G, Martinez KM, Vargas A (2006) Degradation of acid orange 7 by a controlled anaerobic-aerobic sequencing batch reactor. Water Sci Technol 54(2):187–192. https://doi.org/10.2166/wst.2006.503
14. Chao CF, Zhao YX, Jayant K, Ji M, Wang ZJ, Li X (2020) Simultaneous removal of COD, nitrogen and phosphorus and the tridimensional microbial response in a sequencing batch biofilm reactor: with varying C/N/P ratios. Biochem Eng J. https://doi.org/10.1016/j.bej.2019.04.017
15. Chen YJ, He HJ, Liu HY, Li HR, Zeng GM, Xia X, Yang CP (2018) Effect of salinity on removal performance and activated sludge characteristics in sequencing batch reactors. Biores Technol 249:890–899. https://doi.org/10.1016/j.biortech.2017.10.092
16. Chiavola AJWER (2012) Textiles 84(10):1511–1532
17. Ciggin AS, Rossetti S, Majone M, Orhon D (2012) Effect of feeding regime and the sludge age on the fate of acetate and the microbial composition in sequencing batch reactor. J Environ Sci Health Part a-Toxic/Hazardous Substances Environ Eng 47(2):192–203. https://doi.org/10.1080/10934529.2012.640556
18. Cinar O, Demiroz K, Kanat G, Uysal Y, Yaman C (2009) The effect of oxygen on anaerobic color removal of azo dye in a sequencing batch reactor. Clean-Soil Air Water 37(8):657–662. https://doi.org/10.1002/clen.200900095
19. Cinar O, Yasar S, Kertmen M, Demiroz K, Yigit NO, Kitis M (2008) Effect of cycle time on biodegradation of azo dye in sequencing batch reactor. Process Saf Environ Prot 86(B6):455–460. https://doi.org/10.1016/j.psep.2008.03.001
20. Cirik K, Kitis M, Cinar O (2013) The effect of biological sulfate reduction on anaerobic color removal in anaerobic-aerobic sequencing batch reactors. Bioprocess Biosyst Eng 36(5):579–589. https://doi.org/10.1007/s00449-012-0813-2
21. Crini G, Badot P-M (2008) Application of chitosan, a natural aminopolysaccharide, for dye removal from aqueous solutions by adsorption processes using batch studies: a review of recent literature. 33(4):399–447
22. Esparza-Soto M, Nunez-Hernandez S, Fall C (2011) Spectrometric characterization of effluent organic matter of a sequencing batch reactor operated at three sludge retention times. Water Res 45(19):6555–6563. https://doi.org/10.1016/j.watres.2011.09.057
23. Farabegoli G, Chiavola A, Rolle E, Naso M (2010) Decolorization of Reactive Red 195 by a mixed culture in an alternating anaerobic-aerobic Sequencing Batch Reactor. Biochem Eng J 52(2–3):220–226. https://doi.org/10.1016/j.bej.2010.08.014
24. Fongsatitkul P, Elefsiniotis P, Yamasmit A, Yamasmit NJB (2004) Use of sequencing batch reactors and Fenton's reagent to treat a wastewater from a textile industry. 21(3):213–220
25. Franca RDG, Oliveira MC, Pinheiro HM, Lourenco ND (2019) Biodegradation products of a sulfonated azo dye in aerobic granular sludge sequencing batch reactors treating simulated textile wastewater. Acs Sustain Chem Eng 7(17):14697–14706. https://doi.org/10.1021/acssuschemeng.9b02635
26. Franca RDG, Vieira A, Mata AMT, Carvalho GS, Pinheiro HM, Lourenco ND (2015) Effect of an azo dye on the performance of an aerobic granular sludge sequencing batch reactor treating a simulated textile wastewater. Water Res 85:327–336. https://doi.org/10.1016/j.watres.2015.08.043
27. Garcia-Montano J, Torrades F, Garcia-Hortal JA, Domenech X, Peral J (2006) Degradation of Procion Red H-E7B reactive dye by coupling a photo-Fenton system with a sequencing batch reactor. J Hazardous Mater 134(1–3):220–229. https://doi.org/10.1016/j.jhazmat.2005.11.013
28. Ginige MP, Kayaalp AS, Cheng KY, Wylie J, Kaksonen AH (2013) Biological phosphorus and nitrogen removal in sequencing batch reactors: effects of cycle length, dissolved oxygen concentration and influent particulate matter. Water Sci Technol 68(5):982–990. https://doi.org/10.2166/wst.2013.324
29. Hakimelahi M, Moghaddam MRA, Hashemi SH (2012) Biological treatment of wastewater containing an azo dye using mixed culture in alternating anaerobic/aerobic sequencing batch reactors. Biotechnol Bioprocess Eng 17(4):875–880. https://doi.org/10.1007/s12257-011-0673-7

30. Hosseini Koupaie E, Alavi Moghaddam M, Hashemi SJ, Technology (2013) Successful treatment of high azo dye concentration wastewater using combined anaerobic/aerobic granular activated carbon-sequencing batch biofilm reactor (GAC-SBBR): simultaneous adsorption and biodegradation processes. 67(8):1816–1821
31. Jagaba A, Kutty S, Lawal I, Abubakar S, Hassan I, Zubairu I, Ghaleb AJJ (2021a) Sequencing batch reactor technology for landfill leachate treatment: a state-of-the-art review. 282:111946
32. Jagaba AH, Abubakar S, Nasara MA, Jagaba SM, Chamah HM, Lawal IMJ, Chemistry T (2019) Defluoridation of drinking water by activated carbon prepared from tridax procumbens plant (A Case Study of Gashaka Village, Hong LGA, Adamawa State, Nigeria). 7(1):1
33. Jagaba AH, Kutty SRM, Noor A, Birniwa AH, Affam AC, Lawal IM, Kilaco AU (2021b) A systematic literature review of biocarriers: central elements for biofilm formation, organic and nutrients removal in sequencing batch biofilm reactor. 42:102178
34. Jagwani J, Johnson J, Datta M, Lakshmi BJ (2018) Bacterial community dynamics involved in Reactive Orange M2R dye degradation using a real time quantitative PCR and scale up studies using sequence batch reactor. Bioremediat J 22(1–2):43–51. https://doi.org/10.1080/10889868.2018.1476452
35. Jin RF, Liu GF, Li CL, Xu RJ, Li HY, Zhang LX, Zhou JT (2013) Effects of carbon-nitrogen ratio on nitrogen removal in a sequencing batch reactor enhanced with low-intensity ultrasound. Biores Technol 148:128–134. https://doi.org/10.1016/j.biortech.2013.08.141
36. Karimifard S, Moghaddam MRA (2018) Application of response surface methodology in physicochemical removal of dyes from wastewater: a critical review. 640:772–797
37. Khalaf AH, Ibrahim W, Fayed M, Eloffy M (2021) Comparison between the performance of activated sludge and sequence batch reactor systems for dairy wastewater treatment under different operating conditions. Alexandria Eng J 60(1):1433–1445
38. Khalil M, Liu Y (2021) Greywater biodegradability and biological treatment technologies: a critical review. Int Biodeterioration Biodegrad 161:105211
39. Khosravi A, Karimi M, Ebrahimi H, Fallah N (2020) Sequencing batch reactor/nanofiltration hybrid method for water recovery from textile wastewater contained phthalocyanine dye and anionic surfactant. J Environ Chem Eng 8(2). https://doi.org/10.1016/j.jece.2020.103701
40. Khouni I, Marrot B, Amar RB, Technology P (2012) Treatment of reconstituted textile wastewater containing a reactive dye in an aerobic sequencing batch reactor using a novel bacterial consortium. 87:110–119
41. Khouni I, Marrot B, Ben Amar R (2012) Treatment of reconstituted textile wastewater containing a reactive dye in an aerobic sequencing batch reactor using a novel bacterial consortium. Sep Purif Technol 87:110–119. https://doi.org/10.1016/j.seppur.2011.11.030
42. Khursheed A, Gaur RZ, Sharma MK, Tyagi VK, Khan AA, Kazmi AA (2018) Dependence of enhanced biological nitrogen removal on carbon to nitrogen and rbCOD to sbCOD ratios during sewage treatment in sequencing batch reactor. J Cleaner Prod 171:1244–1254. https://doi.org/10.1016/j.jclepro.2017.10.055
43. Klepacz-Smolka A, Pazdzior K, Ledakowicz S, Sojka-Ledakowicz J, Mrozinska Z, Zylla R (2009) Kinetic studies of decolourisation of concentrates from nanofiltration treatment of real textile effluents in anaerobic/aerobic sequencing batch reactors. Environ Protect Eng 35(3):145–155
44. Klodowska I, Rodziewicz J, Janczukowicz W (2018) The influence of electrical current density and type of the external source of carbon on nitrogen and phosphorus efficiency removal in the sequencing batch biofilm reactor. J Ecol Eng 19(5):172–179. https://doi.org/10.12911/22998993/89811
45. Kocyigit H, Ugurlu A (2015) biological decolorization of reactive azo dye by anaerobic/aerobic-sequencing batch reactor system. Global Nest J 17(1):210–219
46. Korenak J, Ploder J, Trcek J, Helix-Nielsen C, Petrinic I (2018) Decolourisations and biodegradations of model azo dye solutions using a sequence batch reactor, followed by ultrafiltration. Int J Environ Sci Technol 15(3):483–492. https://doi.org/10.1007/s13762-017-1406-z
47. Koupaie EH, Moghaddam MRA, Hashemi SH (2012) Investigation of decolorization kinetics and biodegradation of azo dye Acid Red 18 using sequential process of anaerobic sequencing

batch reactor/moving bed sequencing batch biofilm reactor. Int Biodeterior Biodegradation 71:43–49. https://doi.org/10.1016/j.ibiod.2012.04.002
48. Koupaie EH, Moghaddam MRA, Hashemi SH (2013) Evaluation of integrated anaerobic/aerobic fixed-bed sequencing batch biofilm reactor for decolorization and biodegradation of azo dye Acid Red 18: comparison of using two types of packing media. Biores Technol 127:415–421. https://doi.org/10.1016/j.biortech.2012.10.003
49. Koupaie EH, Moghaddam MRA, Hashemi SH (2013) Successful treatment of high azo dye concentration wastewater using combined anaerobic/aerobic granular activated carbon-sequencing batch biofilm reactor (GAC-SBBR): simultaneous adsorption and biodegradation processes. Water Sci Technol 67(8):1816–1821. https://doi.org/10.2166/wst.2013.061
50. Lemus-Gomez LE, Martinez-Trujillo MA, Membrillo-Venegas I, Garcia-Rivero M (2018) Performance analysis of azo dye decolorization by immobilized trametes versicolor in a sequencing batch reactor. Environ Eng Sci 35(12):1322–1328. https://doi.org/10.1089/ees.2018.0033
51. Li BK, Irvin S (2007) The comparison of alkalinity and ORP as indicators for nitrification and denitrification in a sequencing batch reactor (SBR). Biochem Eng J 34(3):248–255. https://doi.org/10.1016/j.bej.2006.12.020
52. Li C, Liu SF, Ma T, Zheng MS, Ni JR (2019) Simultaneous nitrification, denitrification and phosphorus removal in a sequencing batch reactor (SBR) under low temperature. Chemosphere 229:132–141. https://doi.org/10.1016/j.chemosphere.2019.04.185
53. Li Q, Wang SP, Zhang PD, Yu JJ, Qiu CS, Zheng JF (2018) Influence of temperature on an Anammox sequencing batch reactor (SBR) system under lower nitrogen load. Biores Technol 269:50–56. https://doi.org/10.1016/j.biortech.2018.08.057
54. Li SY, Fei XN, Cao LY, Chi YZ (2019) Insights into the effects of carbon source on sequencing batch reactors: performance, quorum sensing and microbial community. Sci Total Environ 691:799–809. https://doi.org/10.1016/j.scitotenv.2019.07.191
55. Li W, Gao M, Zeng F, Liu N, Zhu X, Zhang C (2019) Effect of pH and SRT on denitrifying phosphorus removal in A2N sequencing batch reactor process. Appl Ecol Environ Res 17(3):5737–5751. https://doi.org/10.15666/aeer/1703_57375751
56. Liao BQ, Droppo IG, Leppard GG, Liss SN (2006) Effect of solids retention time on structure and characteristics of sludge flocs in sequencing batch reactors. Water Res 40(13):2583–2591. https://doi.org/10.1016/j.watres.2006.04.043
57. Lim P, Er CJT, Chemistry E (2000) Treatment of dye-containing wastewater by sequencing batch reactor with powdered activated carbon addition. 75(1–2): 75–87
58. Lotito AM, Di Iaconi C, Fratino U, Mancini A, Bergna G (2011) Sequencing batch biofilter granular reactor for textile wastewater treatment. New Biotechnol 29(1):9–16. https://doi.org/10.1016/j.nbt.2011.04.008
59. Lotito AM, Fratino U, Mancini A, Bergna G, Di Iaconi C (2012) Is a sequencing batch biofilter granular reactor suitable for textile wastewater treatment? Water Sci Technol 66(7):1392–1398. https://doi.org/10.2166/wst.2012.312
60. Lourenco ND, Novais JM, Pinheiro HM (2006) Kinetic studies of reactive azo dye decolorization in anaerobic/aerobic sequencing batch reactors. Biotech Lett 28(10):733–739. https://doi.org/10.1007/s10529-006-9051-5
61. Lucas MS, Peres JA, Pigments (2006) Decolorization of the azo dye reactive black 5 by Fenton and photo-Fenton oxidation. 71(3):236–244
62. Ma DY, Wang XH, Song C, Wang SG, Fan MH, Li XM (2011) Aerobic granulation for methylene blue biodegradation in a sequencing batch reactor. Desalination 276(1–3):233–238. https://doi.org/10.1016/j.desal.2011.03.055
63. Mace S, Mata-Alvarez JJI, Research EC (2002) Utilization of SBR technology for wastewater treatment: an overview. 41(23):5539–5553
64. Mahvi AH (2008) Sequencing batch reactor: a promising technology in wastewater treatment. Iran J Environ Health Sci Eng 5(2):79–90
65. Maqbool Z, Shahid M, Azeem F, Shahzad T, Mahmood F, Rehman M, Pollution S (2020) Application of a dye-decolorizing *Pseudomonas aeruginosa* strain ZM130 for remediation

of textile wastewaters in aerobic/anaerobic sequential batch bioreactor and soil columns. 231(8):1–18

66. Maurina GZ, Rosa LM, Beal LL, Baldasso C, Gimenez JR, Torres AP, Sousa MP (2014) Effect of internal recirculation velocity in an anaerobic sequencing batch reactor (ASBR). Braz J Chem Eng 31(4):895–903. https://doi.org/10.1590/0104-6632.20140314s00002895
67. Mehrali S, Moghaddam MRA, Hashemi SH (2010) Removal of reactive blue 19 by adding polyaluminum chloride to sequencing batch reactor system. Iran J Environ Health Sci Eng 7(1):63–70
68. Menezes O, Brito R, Hallwass F, Florencio L, Kato MT, Gauazza S (2019) Coupling intermittent micro-aeration to anaerobic digestion improves tetra-azo dye Direct Black 22 treatment in sequencing batch reactors. Chem Eng Res Des 146:369–378. https://doi.org/10.1016/j.cherd.2019.04.020
69. Mizzouri NS, Shaaban MG (2013) Individual and combined effects of organic, toxic, and hydraulic shocks on sequencing batch reactor in treating petroleum refinery wastewater. J Hazard Mater 250:333–344. https://doi.org/10.1016/j.jhazmat.2013.01.082
70. Moghaddam SS, Moghaddam MRA (2016) Aerobic granular sludge for dye biodegradation in a sequencing batch reactor with anaerobic/aerobic cycles. Clean-Soil Air Water 44(4):438–443. https://doi.org/10.1002/clen.201400855
71. Mohan SV, Rao NC, Sarma PJ (2007) Simulated acid azo dye (acid black 210) wastewater treatment by periodic discontinuous batch mode operation under anoxic–aerobic–anoxic microenvironment conditions. 31(4):242–250
72. Mohan SV, Rao NC, Sarma PN (2009) Simulated acid azo dye wastewater treatment using suspended growth configured sequencing batch reactor (SBR) under anoxic-aerobic-anoxic microenvironment. Appl Ecol Environ Res 7(1):25–34
73. Mojiri A, Ohashi A, Ozaki N, Kindaichi T (2018) Pollutants removal from synthetic wastewater by the combined electrochemical, adsorption and sequencing batch reactor (SBR). Ecotoxicol Environ Saf 161:137–144. https://doi.org/10.1016/j.ecoenv.2018.05.053
74. Ng J, Wong D, Kutty S, Jagaba A (2021) Organic and nutrient removal for domestic wastewater treatment using bench-scale sequencing batch reactor. Paper presented at the AIP Conference Proceedings
75. Ogleni N, Arifoglu YD, Ileri R (2012) Microbiological and performance evaluation of sequencing batch reactor for textile wastewater treatment. Water Environ Res 84(4):346–353. https://doi.org/10.2175/106143011x13233670703323
76. Ong S-A, Toorisaka E, Hirata M, Hano TJ (2005) Treatment of azo dye Orange II in aerobic and anaerobic-SBR systems. 40(8):2907–2914
77. Ong SA, Ho LN, Wong YS, Pakri KAM (2017) Comparative study on the biodegradation of mixed remazol dyes wastewater between integrated anaerobic/aerobic and aerobic sequencing batch reactors. Rendiconti Lincei-Scienze Fisiche E Naturali 28(3):497–501. https://doi.org/10.1007/s12210-017-0622-2
78. Ong SA, Ho LN, Wong YS, Raman K (2012) Performance and kinetic study on bioremediation of diazo dye (reactive black 5) in wastewater using spent GAC-biofilm sequencing batch reactor. Water Air Soil Pollut 223(4):1615–1623. https://doi.org/10.1007/s11270-011-0969-4
79. Ong SA, Toorisaka E, Hirata M, Hano T (2008) Granular activated carbon-biofilm configured sequencing batch reactor treatment of CI Acid Orange 7. Dyes Pigm 76(1):142–146. https://doi.org/10.1016/j.dyepig.2006.08.024
80. Ong SA, Toorisaka E, Hirata M, Hano T (2012) Decolorization of Orange II using an anaerobic sequencing batch reactor with and without co-substrates. J Environ Sci 24(2):291–296. https://doi.org/10.1016/s1001-0742(11)60766-3
81. Rodrigues CS, Madeira LM, Boaventura RA (2014) Synthetic textile dyeing wastewater treatment by integration of advanced oxidation and biological processes–performance analysis with costs reduction. 2(2):1027–1039
82. Rodrigues CSD, Madeira LM, Boaventura RAR (2009) Treatment of textile effluent by chemical (Fenton's Reagent) and biological (sequencing batch reactor) oxidation. J Hazard Mater 172(2–3):1551–1559. https://doi.org/10.1016/j.jhazmat.2009.08.027

83. Rollemberg SLD, Barros ARM, de Lima JPM, Santos AF, Firmino PIM, dos Santos AB (2019) Influence of sequencing batch reactor configuration on aerobic granules growth: engineering and microbiological aspects. J Cleaner Prod. https://doi.org/10.1016/j.jclepro.2019.117906
84. Saeed AAH, Harun NY, Sufian S, Bilad MR, Nufida BA, Ismail NM, Al-Dhawi BN (2021) Modeling and optimization of biochar based adsorbent derived from Kenaf using response surface methodology on adsorption of Cd^{2+}. 13(7):999
85. Santos SCR, Boaventura RAR (2015) Treatment of a simulated textile wastewater in a sequencing batch reactor (SBR) with addition of a low-cost adsorbent. J Hazard Mater 291:74–82. https://doi.org/10.1016/j.jhazmat.2015.02.074
86. Sarvajith M, Reddy GKK, Nancharaiah YV (2018) Textile dye biodecolourization and ammonium removal over nitrite in aerobic granular sludge sequencing batch reactors. J Hazard Mater 342:536–543. https://doi.org/10.1016/j.jhazmat.2017.08.064
87. Sathian S, Rajasimman M, Radha G, Shanmugapriya V, Karthikeyan CJ (2014) Performance of SBR for the treatment of textile dye wastewater: optimization and kinetic studies. 53(2):417–426
88. Sekine M, Akizuki S, Kishi M, Toda T (2018) Stable nitrification under sulfide supply in a sequencing batch reactor with a long fill period. J Water Process Eng 25:190–194. https://doi.org/10.1016/j.jwpe.2018.05.012
89. Singh KS, LeBlanc MM, Bhattacharyya D (2008) Polishing of pretreated dye wastewater using novel sequencing batch reactors. Water Sci Technol 58(2):407–411. https://doi.org/10.2166/wst.2008.416
90. Sirianuntapiboon S, Chairattanawan K (2012) Effects of some operating parameters on the efficiency of a sequencing batch reactor system for treatment of textile wastewater containing acid dyes. Desalinatio Water Treatment 50(1–3):206–219. https://doi.org/10.1080/19443994.2012.719470
91. Sirianuntapiboon S, Chairattanawan K (2016) Comparison of sequencing batch reactor (SBR) and granular activated carbon-SBR (GAC-SBR) systems on treatment textile wastewater containing basic dye. Desalination Water Treatment 57(56):27096–27112. https://doi.org/10.1080/19443994.2016.1167629
92. Sirianuntapiboon S, Chairattanawan K, Jungphungsukpanich S (2006) Some properties of a sequencing batch reactor system for removal of vat dyes. Biores Technol 97(10):1243–1252. https://doi.org/10.1016/j.biortech.2005.02.052
93. Sirianuntapiboon S, Sadahiro O, Salee PJJ (2007) Some properties of a granular activated carbon-sequencing batch reactor (GAC-SBR) system for treatment of textile wastewater containing direct dyes. 85(1):162–170
94. Sirianuntapiboon S, Sansak J (2008) Treatability studies with granular activated carbon (GAC) and sequencing batch reactor (SBR) system for textile wastewater containing direct dyes. J Hazard Mater 159(2–3):404–411. https://doi.org/10.1016/j.jhazmat.2008.02.031
95. Sirlanuntapboon S, Sadahiro O, Salee P (2007) Some properties of a granular activated carbon-sequencing batch reactor (GAC-SBR) system for treatment of textile wastewater containing direct dyes. J Environ Manage 85(1):162–170. https://doi.org/10.1016/j.jenvman.2006.09.001
96. Srisuwun A, Tantiwa N, Kuntiya A, Kawee-Ai A, Manassa A, Techapun C, Seesuriyachan P (2018) Decolorization of Reactive Red 159 by a consortium of photosynthetic bacteria using an anaerobic sequencing batch reactor (AnSBR). Preparative Biochem Biotechnol 48(4):303–311. https://doi.org/10.1080/10826068.2018.1431782
97. Stadler LB, Su LJ, Moline CJ, Ernstoff AS, Aga DS, Love NG (2015) Effect of redox conditions on pharmaceutical loss during biological wastewater treatment using sequencing batch reactors. J Hazard Mater 282:106–115. https://doi.org/10.1016/j.jhazmat.2014.08.002
98. Tan C, Ma F, Qiu S (2013) Impact of carbon to nitrogen ratio on nitrogen removal at a low oxygen concentration in a sequencing batch biofilm reactor. Water Sci Technol 67(3):612–618. https://doi.org/10.2166/wst.2012.554
99. Tan L, Qu YY, Zhou JT, Ma F, Li A (2010) Microbial community shifts in sequencing batch reactors for azo dye treatment. Pure Appl Chem 82(1):299–306. https://doi.org/10.1351/pac-con-09-01-10

100. Thakur C, Mall ID, Srivastava VC (2013) Effect of hydraulic retention time and filling time on simultaneous biodegradation of phenol, resorcinol and catechol in a sequencing batch reactor. Arch Environ Protect 39(2):69–80. https://doi.org/10.2478/v10265-012-0028-2
101. Vaigan AA, Moghaddam MRA, Hashemi H (2009) Effect of dye concentration on sequencing batch reactor performance. Iran J Environ Health Sci Eng 6(1):11–16
102. Wang HY, Song Q, Wang J, Zhang H, He QL, Zhang W, Li H (2018) Simultaneous nitrification, denitrification and phosphorus removal in an aerobic granular sludge sequencing batch reactor with high dissolved oxygen: effects of carbon to nitrogen ratios. Sci Total Environ 642:1145–1152. https://doi.org/10.1016/j.scitotenv.2018.06.081
103. Wang YY, Peng YZ, Stephenson T (2009) Effect of influent nutrient ratios and hydraulic retention time (HRT) on simultaneous phosphorus and nitrogen removal in a two-sludge sequencing batch reactor process. Biores Technol 100(14):3506–3512. https://doi.org/10.1016/j.biortech.2009.02.026
104. Wang ZC, Gao MC, Wang Z, She ZL, Chang QB, Sun CQ, Yang N (2013) Effect of salinity on extracellular polymeric substances of activated sludge from an anoxic-aerobic sequencing batch reactor. Chemosphere 93(11):2789–2795. https://doi.org/10.1016/j.chemosphere.2013.09.038
105. Wu G, Guan Y, Zhan X (2008) Effect of salinity on the activity, settling and microbial community of activated sludge in sequencing batch reactors treating synthetic saline wastewater. Water Sci Technol 58(2):351–358. https://doi.org/10.2166/wst.2008.675
106. Xu D, Liu J, Ma T, Gao Y, Zhang S, Li J (2021) Rapid granulation of aerobic sludge in a continuous-flow reactor with a two-zone sedimentation tank by the addition of dewatered sludge. J Water Process Eng 41:101941
107. Xu SN, Wu DL, Hu ZQ (2014) Impact of hydraulic retention time on organic and nutrient removal in a membrane coupled sequencing batch reactor. Water Res 55:12–20. https://doi.org/10.1016/j.watres.2014.01.046
108. Xu X, Ji FY, Fan ZH, He L, Hu XB, Zhang K (2012) Dyestuff wastewater treatment by combined SDS-CuO/TiO_2 photocatalysis and sequencing batch reactor. J Central South Univ 19(6):1685–1692. https://doi.org/10.1007/s11771-012-1194-z
109. Yan LL, Zhang MY, Liu Y, Liu C, Zhang YD, Liu S, Zhang Y (2019) Enhanced nitrogen removal in an aerobic granular sequencing batch reactor under low DO concentration: role of extracellular polymeric substances and microbial community structure. Biores Technol. https://doi.org/10.1016/j.biortech.2019.121651
110. Ye L, Peng CY, Tang B, Wang SY, Zhao KF, Peng YZ (2009) Determination effect of influent salinity and inhibition time on partial nitrification in a sequencing batch reactor treating saline sewage. Desalination 246(1–3):556–566. https://doi.org/10.1016/j.desal.2009.01.005
111. Yu L, Zhang XY, Wang S, Tang QW, Xie T, Lei NY, Lam MHW (2015) Microbial community structure associated with treatment of azo dye in a start-up anaerobic sequenced batch reactor. J Taiwan Inst Chem Eng 54:118–124. https://doi.org/10.1016/j.jtice.2015.03.012
112. Zhang LL, Sun YH, Guo DL, Wu ZR, Jiang DM (2012) Molecular diversity of bacterial community of dye wastewater in an anaerobic sequencing batch reactor. Afric J Microbiol Res 6(35):6444–6453. https://doi.org/10.5897/ajmr12.718
113. Zhang LQ, Wei CH, Zhang KF, Zhang CS, Fang Q, Li SG (2009) Effects of temperature on simultaneous nitrification and denitrification via nitrite in a sequencing batch biofilm reactor. Bioprocess Biosyst Eng 32(2):175–182. https://doi.org/10.1007/s00449-008-0235-3
114. Zuriaga-Agusti E, Iborra-Clar MI, Mendoza-Roca JA, Tancredi M, Alcaina-Miranda MI, Iborra-Clar A (2010) Sequencing batch reactor technology coupled with nanofiltration for textile wastewater reclamation. Chem Eng J 161(1–2):122–128. https://doi.org/10.1016/j.cej.2010.04.044

Removal of Dyes from Wastewaters in Moving Bed Biofilm Reactors: A Review of Biodegradation Pathways and Treatment Performance

Francine Duarte Castro, Fernanda Ribeiro Lemos, and João Paulo Bassin

Abstract Dye-containing wastewaters are produced in large quantities worldwide. The bright colors they provide, associated with their potential toxicity, carcinogenicity, and mutagenicity, consist of environmental threats. Therefore, remediation of these waste streams is essential. However, the low biodegradability shown by these compounds constitutes an obstacle to the application of effective and low-cost conventional biological wastewater treatment systems. In this context, other advanced biofilm technologies, such as the moving bed biofilm reactor (MBBR), may consist in an economically viable and eco-friendly alternative for achieving high removal of dyes, while also reaching high performance in the concomitant degradation of other pollutants, such as organic matter and nutrients. In this chapter, the MBBR technology is briefly explained, and recent investigations on the treatment of dye-containing wastewaters by this biofilm process are described. Moreover, the most influential operational parameters for color removal improvement are discussed. The use of sequential reactors in multistage processes and the association of MBBR with other technologies are also evaluated. Finally, the kinetics of dye degradation in MBBR is presented. Few studies were found in the literature on the application of MBBR for dye removal, mainly consisting of lab-scale or pilot-scale investigations. A single MBBR as a stand-alone technology does not seem to achieve complete dye mineralization. Nonetheless, the use of a series of MBBR or their combination with other physicochemical or biological processes seems to be a good alternative for dyes remediation. It is clear that more studies on MBBR process optimization are needed, especially for real dye-containing wastewaters, to guarantee the effectiveness and sustainability of large-scale treatment plants.

Keywords Color removal · Biodegradation of dyes · Bioremediation · Biofilm process · Reactor operating conditions

---

F. D. Castro · F. R. Lemos · J. P. Bassin (✉)
Chemical Engineering Program, COPPE, Federal University of Rio de Janeiro, P.O. Box 68502, Rio de Janeiro 21941-972, Brazil
e-mail: jbassin@coppe.ufrj.br

S. S. Muthu and A. Khadir (eds.), *Dye Biodegradation, Mechanisms and Techniques*, Sustainable Textiles: Production, Processing, Manufacturing & Chemistry,
https://doi.org/10.1007/978-981-16-5932-4_9

Abbreviations

AOA	Ammonium-Oxidizing Archaea
AOB	Ammonium Oxidizing Bacteria
An-SBR	Anaerobic Sequencing Batch Reactor
AR18	Acid Red 18
ATP	Adenosine 5'-triphosphate
BOD	Biochemical Oxygen Demand
CAPEX	Capital Expenditures
CAS	Conventional Activated Sludge
COD	Chemical Oxygen Demand
DNB	Denitrifying Bacteria
DO	Dissolved Oxygen
DR75	Direct Red 75
EPS	Extracellular Polymeric Substances
FADH	Flavin Adenine Dinucleotide
GAC	Granular Activated Carbon
HDPE	High-Density Polyethylene
HNO_2	Nitrous Acid
HRT	Hydraulic Retention Time
NaCl	Sodium Chloride
NADH	Nicotinamide Adenine Dinucleotide
NADPH	Nicotinamide Adenosine Dinucleotide Phosphate
NH_3	Unionized ammonia
NH_4^+	Ionized ammonia, or ammonium
NO_2^-	Nitrite
NOB	Nitrite Oxidizing Bacteria
NPs	Nanoparticles
MBBR	Moving Bed Biofilm Reactor
MBR	Membrane Bioreactor
MB-SBR	Aerobic Moving Bed Sequencing Batch Biofilm Reactor
OPEX	Operational Expenditures
pH	Potential of Hydrogen
PLA	Poly (lactic acid)
PRBC	Photo-Rotating Biological Contactor
PU	Polyurethane
PU-AC	Polyurethane-Activated Carbon
PU-DSCM	Polyurethane-Dyeing Sludge Carbonaceous Material
PVC	Polyvinyl Chloride
RB-5	Reactive Black-5
RBC	Rotating Biological Contactors
RR 239	Reactive Red 239 (RR 239)
RO16	Reactive Orange 16
SS	Suspended Solids

TAHNDS	1-2-7-Triamino-8-hydroxy-3-6-naphthalinedisulfate
TDS	Total Dissolved Solids
TSS	Total Suspended Solids
UASB	Upflow Anaerobic Sludge Blanket
UF	Ultrafiltration
WWTP	Wastewater Treatment Plant

1 Introduction

Dyes are natural or synthetic compounds responsible for giving color to a certain material. In their natural form, they have been used since antiquity to dye fabrics, ceramics, and leathers. The production of synthetic dyes, on the other hand, was only boosted in the mid-nineteenth century, after the industrial revolution and the synthesis of the first organic dye by William Perkin, leading to the development of the textile industry [50]. Currently, the annual production of dyes accounts for 8×10^5 t [4], supplying industries of different branches, such as paper and cellulose, plastics, paints, food, and textiles, the latter being the main consumer. The global market of textile dyes accounted for $9.4 billion in 2018 and may reach $15.5 billion in 2026 [61].

Overall, more than 280,000 t of dye-containing wastewaters are produced worldwide on a yearly basis, and generally constitute around 80% of the emissions generated by the textile industry [74]. The high quantities of water used mainly in the dyeing process result in the generation of large volumes of effluents. For example, the dyeing of 1 kg of cotton (most used fiber worldwide) requires 70–150 L of water, in addition to 0.6–0.8 kg of NaCl and 30–60 g of dyes [1, 2]. Thus, the large production and discharge of liquid effluents by this industrial sector threaten the environment [84]. If not properly treated and discharged in water bodies, dye-containing wastewaters may present several risks to the local biota. Even in small concentrations, dyes can provide bright colors, which may block sunlight penetration in aquatic systems, affecting photosynthetic processes and leading to changes in biological cycles [73].

Dyes are highly toxic to all forms of life and are potentially mutagenic and carcinogenic [36, 62]. Besides, they have complex organic aromatic structures, which are not only responsible for the fixation and durability of color but may also act as an obstacle to biodegradation [13]. In a dye, color is given by the electronic transition between several molecular orbitals. These compounds have a chromophore group (–N = N–, –C = C–, –C = O–) which is responsible for the color effect due to electron excitation. Dyes are also composed of an auxochrome (–OH, $-NH_2$, $-NR_2$), which promotes color fixation [25].

Dyes can be classified according to their chemical structure and the nature of the chromophores in "nitroso, nitro, monoazo, diazo, stilbene, diarylmethane, triarylmethane, xanthene, acridine, quinoline, methine, thiazole, indamine, indophenol, azine, oxazine, thiazine, aminoketone, anthraquinone, indigoid, phthalocyanine, and inorganic pigments" [52]. Among textile dyes, the azo type is the most used one, as it provides more intense colors than other classes. However, between 15 and 50% of azo dyes do not remain in the fabric during dyeing, being discarded as wastewaters generally used for irrigation in agriculture in developing countries [60].

Currently, azo dyes represent more than 60% of the dyes used in the textile industry [30]. Approximately 70% of all dyes used in this sector are azo dyes [47]. In view of the large volume of dyes that do not fixate onto the fabrics, color is often the main problem concerning textile effluents. However, the discoloration of these effluents is not a simple task since the color is often not removed by the conventional wastewater treatment processes [45]. Physicochemical methods have shown to be economically disadvantageous for the removal of dyes since they require high energy consumption, and the use of chemicals implies high costs. Furthermore, such methods do not completely remove recalcitrant azo compounds and the generated by-products, leading to the production of chemical sludge [66]. Coagulation/flocculation processes are effective mainly for removing sulfur and dispersant dyes, but they present low removal of acid, direct, reactive, and vat dyes [66]. Conventional aerobic biological processes (e.g., activated sludge), widely used to treat different types of wastewaters due to relatively low cost and effectiveness, are usually inefficient in degrading azo dyes [10, 16, 24]. Moreover, in the presence of dyes, activated sludge deflocculation may be observed [34].

In the activated sludge process, there is a complex composition of filamentous flocs and microorganisms, polymers, and metabolic excreta. Microorganisms can synthesize extracellular polymeric substances (EPS) that lead to the formation of flocs by agglomeration of bacteria. EPS provides a large surface area per volume for microorganism attachment and significantly affects floc settling [35]. Işık and Sponza [34] studied different means of cultivation of activated sludge flocs and observed that as the composition of EPS is influenced by the type of substrate and by the microorganisms in the activated sludge, the sedimentation characteristics also become strongly influenced by the reactor microenvironment, including the dynamics of the microbial population and degradable organic compounds. Flocs grown in wastewater containing easily degradable organics exhibited good sedimentation properties at low sludge volume index values. Flocs grown in wastewater containing organic substrates with greater difficulty in degradation, such as chemicals and dyes, exhibited worse sedimentation properties.

On the other hand, biofilm systems retain bacterial cells in appropriate quantities within the reactor by developing a biofilm adhered to fixed or mobile supports. The biofilm consists of a humid matrix and a variety of soluble and particulate components that include microorganisms and EPS [79]. The moving bed biofilm reactor (MBBR) is an example of a biofilm technology in growing expansion over the world. Recent studies have shown that this biofilm technology may be efficient for the treatment of dye-containing effluents [65, 71, 87]. However, the number of studies addressing the application of the MBBR process for this purpose is limited. Therefore, in this chapter, an overview of the MBBR technology is presented with the main focus on the application of this process on the removal of dyes. The factors affecting the effectiveness of dyes removal, optimum process conditions, and kinetic aspects were addressed in this contribution. The chapter is divided into seven main topics, including a brief description of biofilm reactors,the MBBR technology and its advantages over other biofilm systems; the removal of dyes in MBBR; factors affecting dye removal in MBBR; kinetics of dye removal in MBBR; process combinations; and a comparison between MBBR and other biological wastewater treatment systems for dye removal, considering technological, environmental, and economic aspects.

2 Biofilm Reactors

Biological processes are widely used for wastewater treatment, being divided into two classes: suspended or fixed biomass (biofilms) systems. In the first case, microorganisms cluster in the form of microbial flocs, with a predominance of bacteria. Examples of systems with suspended biomass are activated sludge, membrane bioreactors, stabilization ponds, among others. These processes are very efficient in removing organic matter and nutrients, but some have limitations such as the need for large areas and high sludge production [8].

In the case of biofilm reactors, microorganisms grow attached to a solid surface with a high surface area, leading to the accumulation of a high concentration of biomass in the reactor. A biofilm consists of three-dimensional heterogeneous microbial aggregates containing several microorganisms that compete for the available substrates. They are immobilized in a matrix of EPS, together with cellular products, and grow in a very compact way. As a result, a large amount of biomass can be accumulated in a small reactor volume, leading to high pollutant removal rates. Biofilms are resistant to dehydration and offer protection against predatory organisms, good stability, and resistance to shock loads [28].

Biofilm growth occurs through different processes: free cell transport from the liquid medium to the solid surface and initial fixation; growth, production, and excretion of EPS; fixation of cells to the already formed biofilm; erosion of small particles; and loss of larger aggregates [85]. Cells can detach from the biofilm due to shear forces and also when the environment becomes unfavorable. Due to a balance between the detachment of cells and microbial growth, biofilm thickness varies with time and position. The detached microorganisms start to occupy the bulk phase and, as the suspended microbial community increases, it may also play an important role in pollutants degradation [28].

The transport of the components from the bulk phase to the cells involves several sequential steps: adsorption onto the biofilm surface, diffusion through the stagnant liquid film at the interface between biofilm and liquid phase, and diffusion through the biofilm (Fig. 1). The encapsulated structure that maintains the biofilm cohesion may lead to concentration gradients for all substances. Therefore, ensuring an effective mass transfer in biofilms is an important factor for effective pollutant removal. If the substrate mass transfer is limited, reaction rates will be reduced, compromising the treatment process [46].

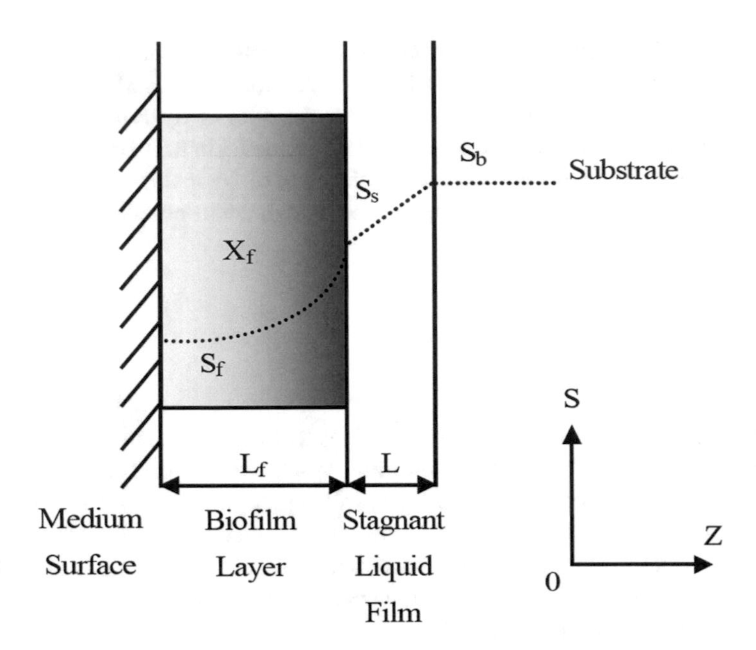

Fig. 1 Substrate diffusion from the bulk phase to the biofilm layer. *Source* Adapted from Lin [46]

Biofilm-based wastewater treatment systems provide great advantages to the treatment process compared to those with suspended growth. Since the microorganisms grow adhered to a surface and are not constantly removed from the reactor along with the liquid effluent, the hydraulic retention time (HRT) is decoupled from the sludge retention time (SRT), allowing the use of a high SRT regardless of the HRT, and without the need of sludge recycle from the secondary clarifier. In addition, by retaining the biomass inside the reactor for longer periods, this type of system facilitates the development of slow-growing organisms. The formation of biofilms also reduces the area required for the installation of the reactors since it simplifies the solid–liquid separation step [8].

Some examples of biofilm processes are trickling filter, submerged aerated filter, rotating biological contactors (RBC), moving bed biofilm reactors (MBBR), among others. The MBBR, in particular, is a relatively recent technology, which stands out over other biofilm systems as it does not present clogging problems and provides lower head loss [9]. Further details on this system are given next.

3 The Moving Bed Biofilm Reactor (MBBR)

The MBBR process was created in Norway between the 1980s and 1990s, being used ever since to treat domestic and industrial wastewaters as an alternative to conventional secondary treatment processes. The MBBR provides high removal rates

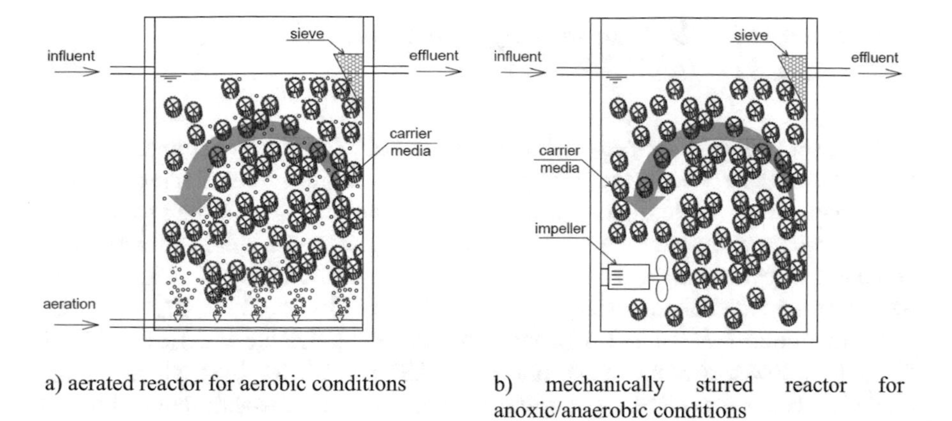

a) aerated reactor for aerobic conditions

b) mechanically stirred reactor for anoxic/anaerobic conditions

Fig. 2 Scheme of aerated MBBR (**a**) and mechanically stirred MBBR (**b**). *Source* Adapted from [63]

of biodegradable organic matter and nitrogen with the advantage of allowing the use of smaller reactor volumes as compared to activated-sludge-based systems [12].

The MBBR technology uses low-density moving carriers, where biofilms are formed. The media is inserted in the reactor and moves freely throughout the reactor volume. It can be applied in both aerobic and anoxic/anaerobic environments, with the mixing and fluidization of the carriers being obtained by diffuse aeration (for aerobic reactors) or mechanical mixing (for anoxic/anaerobic reactors) [40, 42]. Agitation also enables the transport of the substrates to the biofilm and helps to control the biofilm thickness due to the action of shear forces [41]. A scheme of the MBBR process (for both aerobic and anoxic/anaerobic configurations) is displayed in Fig. 2. A sieve is used at the outlet of the reactor to retain the carriers and allow only the treated effluent to pass through.

MBBR has advantages over fixed biomass systems, such as low head loss, absence of clogging, use of moving carrier media with high specific biofilm surface area, and the entire volume of the system available for the biological conversions [8, 39]. Besides, the use of moving media excludes the necessity for sludge recycling and facilitates the subsequent step of solid–liquid separation easier [8]. By introducing carriers to an existing treatment facility, the sludge age can be increased without major changes in the plant, allowing the development of bacteria with low growth rates, such as nitrifiers. It may also favor the production of specific enzymes by the microorganisms, which are necessary for the degradation of certain recalcitrant compounds, such as dyes [40, 49]. Other characteristics of MBBR systems that may favor dye degradation are the presence of anoxic/anaerobic zones inside the biofilm, even for aerated systems, and the resistance to toxic compounds provided by the external protective EPS layer.

3.1 Removal of Dyes in Moving Bed Biofilm Reactors (MBBR)

Although MBBR is already an established technology for wastewater treatment, few studies have been conducted regarding the use of this biofilm process for the treatment of dye-containing wastewaters (Table 1). Among the studies published so far, most of them were conducted on a laboratory or pilot scale, with synthetic wastewater simulating a real scenario used to feed the bioreactors. Process associations have also been investigated, such as ozonation + aerobic MBBR [14, 15, 21, 22, 27, 59], Fenton or photo-Fenton + MBBR [3, 72], anaerobic MBBR + aerobic MBBR [23, 27, 39, 55, 68], upflow anaerobic sludge blanket (UASB) + MBBR [48], photo-rotating biological contactor (PRBC) + aerobic MBBR [80], granular-activated carbon bed (GAC) + MBBR [76].

3.2 Factors that Influence Dye Removal in MBBR

- Dissolved oxygen (DO) concentration

  Biological systems can be kept under anaerobic, anoxic, and aerobic conditions. The redox condition is a crucial factor influencing the biological removal of azo dyes since there are big differences between the physiology of microorganisms that grow in the presence and absence of oxygen, directly influencing the degradation mechanism [57].

  The removal of dyes from wastewaters can occur via adsorption onto the biomass or via biodegradation. The biosorption mechanism, alone, has some limitations since the microbial biomass becomes saturated over time, and there are problems of final disposal of the adsorbent (sludge), which contains the undegraded toxic compounds. However, biosorption is generally the first stage of biodegradation [57].

  There are several hypotheses to explain the mechanisms of biodegradation of dyes. Some of them involve intra- or extracellular enzymes or a non-specific extracellular reduction [18]. In one of the proposed mechanisms, it is suggested that electrons produced during the generation of adenosine 5′-triphosphate (ATP), in catabolic reactions, are transferred to the dye (by means of enzymes and coenzymes), which acts as a final electron acceptor, inducing the chromophore breakage (e.g., azo bond). The electron transfer can occur by enzymatic pathway directly (Fig. 3a) or indirectly by means of redox mediators (Fig. 3b), which are produced during the cellular metabolism of certain substrates or added to the system. These mechanisms will be further described in the sequence. Another hypothesis relates the chromophore breakage to the reducing action of end products of the cellular catabolism, such as inorganic compounds, leading to color removal (Fig. 3c) [54, 57]. For instance, when H_2S is present in the reaction

Table 1 Summary of previous work on dye removal in moving bed biofilm reactor (MBBR): operational details and main results

Main wastewater characteristics	Inoculum	Process description	Main results	Reference
Synthetic wastewater. 100 mg/L of Chemistar Turq Blue dye, COD = 780 mg/L, BOD = 198 mg/L	*Microbacterium marinilacus*, obtained from textile sludge	Fluidized bed Fenton + aerobic MBBR. Carriers: made of PVC; working volume (MBBR) = 7.875 L	Fenton: BOD/COD increase from 0.25 to 0.52. MBBR: maximum removals of 87.22% (COD) and 80% (BOD), at pH = 7.33, filling ratio of 67.07%, and HRT = 2.5 days	[3]
Synthetic wastewater. Reactive Orange 16 (RO16) (25 mg/L), glucose COD = 400 mgO_2/L	Sludge from municipal WWTP	Process 1: Dye ozonation + aerobic MBBR; Process 2: aerobic MBBR. For both: HRT = 6 h; working volume = 200 mL; carrier: K1; filling ratio: 40%	Process 1: >97% color removal in 5 min. Process 2: color removal not achieved; COD removal: >90 ± 1%; Ammonium removal: >97 ± 2%	[14]
Synthetic wastewater. RO16, glucose COD of 0–800 mgO_2/L	Sludge from municipal WWTP	Process 1: Anaerobic MBBR (5–25 mg/L RO16, HRT = 6–12 h); Process 2: Ozonization (100–500 mg/L RO16) + aerobic MBBR (HRT = 6 h). For both cases: working volume = 200 mL; carrier: K1; filling ratio: 40%	Process 1: up to 61 ± 18% color removal and 93 ± 2% COD removal, for HRT = 12 h, inlet glucose COD of 800 mgO_2/L and [RO16] = 5 mg/L. Process 2: removals of 98 ± 1% (ammonium) and 46% (ozonation products) for 500 mg/L of RO16	[15]

(continued)

Table 1 (continued)

Main wastewater characteristics	Inoculum	Process description	Main results	Reference
Synthetic wastewater. 50 mg/L of Reactive Red 239 (RR 239), 375 mg/L of glucose	Sludge from municipal WWTP	Process 1: aerobic MBBR; Process 2: Ozonation + aerobic MBBR. HRT = 6 h, working volume = 300 mL, carriers: K1, filling ratio = 40%	Process 1: No color removal. Process 2: Inhibition of nitratation. Heterotrophic bacteria could adapt to the ozonized dye while nitrifiers could not. Ammonium removal: 41%	[22]
Synthetic wastewater. 50 mg/L of Reactive Red 239 (RR 239), glucose COD of 400 mg/L	Sludge from municipal WWTP	Ozonation + two aerobic MBBRs in series (total HRT = 6 h, working volume of 300 mL each) containing K1 carriers (filling ratio: 40%)	COD removals: up to 91% ($MBBR_1$) and 94% ($MBBR_2$). Low ammonium removal. Biomass detachment in the $MBBR_2$ increased, by increasing ozonation time	[21]
Synthetic wastewater. Reactive brilliant red X-3B, COD, and color: 500 mg/L and 400 (Pt–Co unit)	Sludge from municipal WWTP	Anaerobic MBBR + aerobic MBBR (HRT_1 = 11 h and HRT_2 = 5 h, effective volume of 60 L, filling ratio of 35 vol.%, carrier made of polyethylene) + hollow-fiber membrane microfiltration	Average removals (overall): 90% (color), 85% (COD) and 95% (SS). COD removals: 20–35% ($MBBR_1$) and 60–70% ($MBBR_2$)	[23]
Real textile wastewater. COD: 730–965 mg/L, BOD_5: 118–143 mg/L, color: 142–189°, COD:P = 100:1	Sludge from a WWTP treating textile wastewater	Anaerobic MBBR (2 L) + aerobic MBBR (2 L) + ozonation + aerobic MBBR (4 L). Carrier: cylindrical, made of polyethylene. Filling ratio: 60% (v/v). Sludge retention time (MBBRs): 10 days	Optimum conditions: 14 h HRT for both $MBBR_1$ and $MBBR_2$, 14 min ozonation and HRT of 10 h for $MBBR_3$. Total removals: 94.3% (COD), 97.8% (SS), 85.3% (ammonia), and 96.3% (color)	[27]

(continued)

Table 1 (continued)

Main wastewater characteristics	Inoculum	Process description	Main results	Reference
Bemacid Rot (200 mg/L), glucose as carbon source	*Cerioporus squamosus* White-Rot-Fungi	Aerobic MBBR (12 L). Carriers: 3 L of a mix of 88% high-density polyethylene, 5% talcum, and 7% cellulose, plus 20 PUFs carriers	Interactions between –OH and $-NH_2$ with the nitrogen from the azo bond were identified, resulting in amino groups. Partial degradation of marginal methyl groups was also observed. The biodegradation of the dye was confirmed	
Synthetic wastewater. Glucose (1.5 g/L), lactose (1.5 g/L), urea (116.5 mg/L), and the azo dye Acid Red 18 (AR18) (100, 500, or 1000 mg/L). COD/N/P = 100/2/0.3	Granulated sludge (UASB) and activated sludge (municipal WWTP)	Anaerobic sequencing batch reactor (An-SBR) + aerobic moving bed sequencing batch biofilm reactor (MB-SBBR). Carrier: 2H-BCN017KL. Filling ratio: 50%. Cycle duration: 24 h. Volume of wastewater: 2 L. Agitation speed: 100 rpm. T = 35 ± 0.2 °C (anaerobic) and T = 22 ± 2 °C (aerobic)	Up to 98% discoloration and more than 80% COD removal. More than 80% of the amine 1-naphthylamine-4-sulfonate was removed aerobically. 65–72% of the dye degradation products were mineralized	[39]
Real textile effluent (diluted twofold), containing mainly azo dyes and indigo dye. COD 650 ± 80 mg/L	No details provided	Aerobic MBBR (2 L). Carrier: K1. HRT: 24 h. DO: 2.0 ± 0.4 mg/L. Initial SRT: 15 days. Initial MLSS: 12,500 mg/L. T = 30 to 55 °C	Maximum removals: 70.1% (COD) at 50 °C, 39.1% (ammonium) at 35 °C. Total EPS increased up to 45 °C (1,748 mg/L) and then decreased. Thermophilic communities changed with temperature	[43]

(continued)

Table 1 (continued)

Main wastewater characteristics	Inoculum	Process description	Main results	Reference
Real wastewater. COD = 800–2500 mg/L, BOD_5 = 500–2000 mg/L, color [times] = 300–800	Municipal flocculent sludge	Anaerobic microorganism carrier-UASB (1.72 m^3) + Porous bio-gel (PU sponge cube) MBBR (2.82 m^3) with sludge recycling. Specific surface area > 4000 m^2/m^3. Filling ratio: 15% (UASB) and 30% (MBBR). HRT: 9 h each. Influent pH: 6.5–7.5	COD removals: 37.0 ± 7.5% (UASB) and 53 ± 12.7% (MBBR). Color removal: 84.0% (overall), 50.3% (UASB), and 13.7% (MBBR). Ammonium concentration: 20 mg/L (inlet), 45 mg/L (after UASB), 3.49 ± 0.54 mg/L (after MBBR)	[48]
Real textile wastewater. COD = 608 mg/L, color = 553 Pt–Co unit	Activated sludge from a dyeing WWTP	Anaerobic MBBR + anaerobic MBBR + aerobic MBBR (15 L each). Filling ratio = 20% (v/v). Total HRT of 44 h. Carriers: polyurethane-activated carbon (PU-AC) foam	Average COD removals: 56% ($MBBR_1$), 22% ($MBBR_2$) and 8% ($MBBR_3$), 85% (overall). Color removal: 50% (overall), 37% ($MBBR_1$), 13% ($MBBR_2$) and 1% ($MBBR_3$)	[55]
Real textile wastewater. COD = 643 mg/L, color = 950 Pt–Co unit	White-rot fungus, *Phanerochaete chrysosporium* (ATCC #24,725)	Aerobic MBBR (15 L, HRT = 24 h) + aerobic MBBR (15 L, HRT = 24 h) + chemical coagulation. Carrier: Polyurethane-Dyeing Sludge Carbonaceous Material (PU-DSCM) foam. Filling ratio: 20%	Removals in the MBBRs: 79% of COD and 54% of color at MLSS concentration of 2,900 mg/L and HRT = 48 h (total). Coagulation with alum (1.55 mg alum/mg COD): 95.7% removal of COD and 73.4% removal of color (overall)	[56]

(continued)

Table 1 (continued)

Main wastewater characteristics	Inoculum	Process description	Main results	Reference
Synthetic wastewater. 100 mg/L of Reactive Black-5 (RB-5), starch (COD = 1,000 mg/L)	Sludge from a Textile WWTP	Process 1: Ozonation + aerobic MBBR (5 L). Process 2: aerobic MBBR (5 L). Carrier: K1, filling ratio: 20%, DO > 4 mg/L, HRT: 6 to 24 h	Process 1: Maximum color (96.9%) and COD (89.13%) removals achieved in 24 h. Process 2: maximum COD (83.6%) and color (81.2%) removals achieved in 24 h	[59]
Synthetic wastewater. 100 mg/L of Direct Red 75 (DR75), 2 g/L of yeast extract. COD = 2347 mg/L	Rice husks are used as a source of microorganisms	Anaerobic MBBR (Kaldnes K1 carriers). T = 30 ± 0.5 °C or 21 ± 2 °C; filling ratio: 50%; working volume: 2L; pH = 6; stirring speed = 100 rpm; HRT = 24, 48 or 12 h	85% decolorization at HRT = 48 h, 30 ± 0.5 °C, and 45% at 21 ± 2 °C. Maximum COD removal (19.55%) achieved at 30 ± 0.5 °C and HRT = 48 h	[65]
Real textile effluent. COD = 900 mg/L, color = 3200 Pt–Co unit	Activated sludge from a WWTP	Anaerobic MBBR + aerobic MBBR + aerobic MBBR + chemical coagulation. Working volume: 15 L (each), filling fraction: 20% (v/v), total HRT; 44 h. Carriers: polyurethane-activated carbon (PU-AC) foam	Average COD removal = 94.9% (29.1% in the $MBBR_1$, 50.9% in the $MBBR_2$, 5.5% in the $MBBR_3$ and 9.2% by coagulation). Average color removal: 97.4% (54.9% in the $MBBR_1$, 8.6% in the $MBBR_2$, 5.9% in the $MBBR_3$ and 27.9% by coagulation)	[68]
Real textile wastewater. DBO_5 = 298 mg/L, COD = 667, spectral area = 47%	Activated sludge from an industrial WWTP	Aerobic MBBR. Working volume: 1 L; carrier: specific area of 750 m^2/m^3; filling ratio: 10%; organic loading rates: 0.3–9.0 g COD/L.d; HRT = 1.6–61.7 h	Maximum removal efficiencies obtained at HRT = 61.7 h and organic load of 0.3 gCOD/L.d: 65%(COD), 94% (BOD_5), 82% (TSS), 58% (spectral area) and 87% (toxicity)	[69]

(continued)

Table 1 (continued)

Main wastewater characteristics	Inoculum	Process description	Main results	Reference
Synthetic wastewater. Congo red, beef extract (2.0 mg/L), peptone (1.0 mg/L), and 0.2% (w/v) glucose	*Bacillus* sp. MH587030.1	Aerobic MBBR. Carriers: polyurethane foam-polypropylene; working volume = 2.0 L; flow rate = 25 mL/h; HRT = 132 h	Optimum conditions (batch MBBR): pH = 7.0, initial dye concentration = 50 mg/L, and filling ratio = 45%. Maximum decolorization: 95.7% (continuous operation)	[71]
Real textile wastewater. Ramazol Red F 3B, Olive R and Ramazol Turquoise blue CG. 210 mg/L of BOD, 980 mg/L of COD	*Azoarcus* bacteria from textile sludge	Solar photo-Fenton (pH 3, 30 min., $[Fe^{2+}]$ = 4 mg/L, and 20 mM H_2O_2) + aerobic MBBR (11.25 L, agitation speed of 30 rpm, PVC carriers with surface area of 350 m^2/m^3) with and without a magnetic field (4–14 mT)	Pretreatment: BOD:COD increase from 0.21 to 0.54. Optimal conditions (MBBR): pH = 7, filling ratio = 62%, HRT = 2.4 d. Maximum removals: 68.9% of BOD, and 80% of COD. Under the magnetic field of 12 mT (attractive), for 12 h exposure, the removals increased to 87.4% (COD) and 87% (BOD)	[72]
Real wastewater. color = 1,813 Pt–Co units, COD = 1,682 mg/L, BOD = 399 mg/L	Sludge from sewage treatment plants and dairy animals excrement	Granular activated Carbon bed (GAC) + MBBR. HRT_1 = 12 h. Inlet flowrate: 0.09 m^3/s. Carriers: made of polypropylene. Filling ratio: 67%. HRT_2 = 4, 6, 8, or 10 h	Maximum removals: 90% (COD) and 95% (BOD), for HRT = 4 h. Color, COD, BOD, and TSS removals: mostly on the GAC tank. TDS removal: primarily at the MBBR. Color levels reduced to ~500 Pt–Co units, TDS to ~4,000 mg/L, and TSS to ~100 mg/L	[76]

(continued)

Table 1 (continued)

Main wastewater characteristics	Inoculum	Process description	Main results	Reference
Acid Red 1, Reactive Black 5, yeast extract (100 mg/L), sucrose, and/or sodium acetate	*R. palustris* W1 isolated from sludge samples	Photo-rotating biological contactor (PRBC) + aerobic MBBR (working volumes: 1 L and 2 L, HRT: 5 h and 10 h, respectively). pH = 7.0. T = 35°C. System I: 1000 mg/L sucrose + 1000 mg/L sodium acetate as co-substrate. System II: 500 mg/L of sucrose	Maximum color removal: 91.0% (mixed systems with RB5 and AR1, with abundant co-substrate). Discoloration took place mostly at the MBBR	[80]
Real pretreated textile wastewater. COD = 1000–2000 mg/L, BOD_5/COD = 0.26–0.3	Activated sludge mixed liquor suspended solids (MLSS)	Lab-scale MBBRs (working volume: 5 L). T = 30–35°C, DO = 4 ± 1 mg/L. Stirring: 550 rpm. Biocarrier: made of HDPE, zinc nanoparticles (Zn NPs) and poly (lactic acid). Contents of Zn NPs: 0, 10, 15, 20 and 30 wt%. Specific surface area: 620 ± 20 m^2/m^3. Filling ratio: 40%	Optimum performance: 20 wt% Zn NPs. It showed higher attached solids concentration (2460 mg/L) and biofilm dehydrogenase activity (~350 μg TPF/g), higher ammonia nitrogen and COD removal efficiencies, and higher biodiversity	[83]
Real textile wastewater. COD = 2000 mg/L, color = 700 Pt–co/L, BOD = 400 mg/L	Aerobic sludge from the textile industry WWTP	Process 1: Conventional Activated Sludge (CAS) (4 L, HRT = 2 d). Process 2: aerobic MBR (hollow fiber UF membrane, 20 L). Process 3: aerobic MBBR (filling ratio: 30%, BIOFILL C-2 carriers with a specific surface of 590 m^2/m^3). T = 25 °C. OLR = 1 kg COD/m^3.d in all cases	Best removals: MBR (COD: 91%, TSS: 99.4%, and color: 80%, at HRT = 1.3 d). MBBR and CAS: COD removals of 82% versus 83%, for HRT of 1 d versus 2 d. TSS removal: 73% MBBR versus 66% CAS. MBBR: the most economical and most feasible at full-scale, with the lowest environmental impacts	[87]

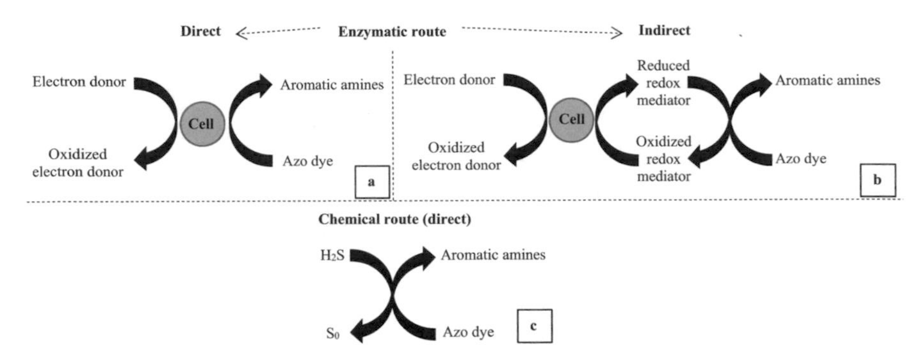

Fig. 3 Possible mechanisms related to the biological degradation of azo dyes: direct enzymatic route (**a**), indirect enzymatic route (**b**), and chemical route (**c**). *Source* Adapted from [54]

medium, it reacts with monoazo dyes in a molar proportion of 2:1, leading to the formation of 2 mol of sulfur (S_0) and 2 mol of aromatic amines [88].

When the mechanism is intracellular, the removal of the dye depends on its diffusion across the cell membrane. This transport can be impaired when the dye has a high molar mass, and there are sulfonated groups in the molecule [18]. In such cases, it is likely that the mechanism is extracellular. Hence, for the azo bond to be broken, the cell must establish a link between the intracellular respiratory chain and the extracellular dye molecule. To this end, electron carriers must be located outside the cell, enabling contact with the dye [57].
The enzymes involved in reducing azo dyes are known as azo reductases. These enzymes catalyze the azo bond reduction only in the presence of reduction equivalents (coenzymes FADH, NADH, and NADPH) [70] (Fig. 4). When electrons are released during ATP production, the reduction equivalents promote the electron transfer to the azo reductases, which then catalyze azo dye reduction directly or

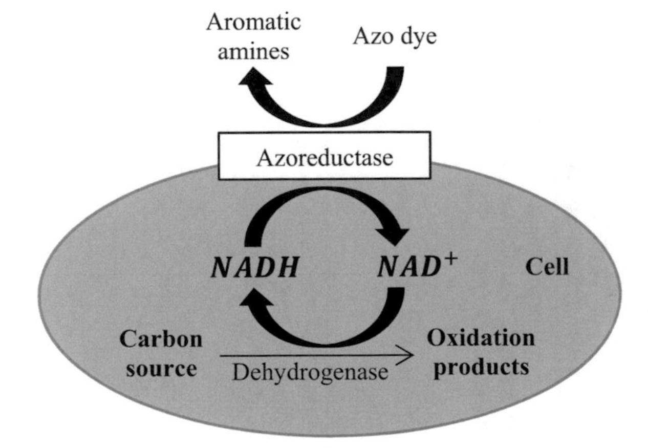

Fig. 4 Extracellular direct removal of azo dyes. *Source* Adapted from [57]

by means of redox mediators. Azo reductases can be synthesized both in the presence and absence of oxygen [18]. However, for these enzymes to be produced in aerobic systems, a long adaptation period in the presence of a simple azo substance is required. After that period, a compound-specific azo reductase is produced, and the azo dye can then be removed in the presence of oxygen.

On the other hand, in anaerobic systems, the process is not specific to an azo compound. In that case, any added azo dye can be removed, with greater or lesser efficiency, depending on the dye molecular structure and reactor operating conditions used. Therefore, anaerobic processes are more used than aerobic for the purpose of biological dye removal [57].

Anthraquinones represent the second most important group of textile dyes (after azo dyes) [19, 90]. They are used for dyeing cellulosic fabrics, wool, and polyamide fibers. Another class of dyes, triphenylmethane, is used to dye nylon, nylon modified with polyacrylonitrile, wool, and silk. These dyes are usually resistant to light, temperature, and biodegradation, so they accumulate in the environment [64, 90]. In addition to azoreductases, laccases and peroxidases are the most important enzymes capable of degrading dyes.

Laccases are part of the family of multicopper oxidase enzymes that catalyze the oxidation of countless substrates in water, through a reaction mechanism involving radical formation [11]. These enzymes are mostly of fungal and vegetal origin, although some have been identified in bacteria and insects [6, 26]. The most useful and the most researched laccases in biotechnology applications are of fungal origin. Physiologically, laccases have several functions, such as lignolysis, pigment formation, detoxification, and pathogenesis, which result from the ability of enzymes to oxidize a wide variety of aromatic substrates (e.g., polyphenols and diamines) and inorganic compounds [26, 32].

Peroxidases are a group of multiple versatile and stable heme-containing enzymes that use hydrogen peroxide or electron acceptor of organic hydroperoxides (R-OOH) to catalyze the oxidation of various substances. Peroxidases show great potential to be environmental biocatalysts, being one of the most researched groups of enzymes. They can successfully degrade synthetic dyes with a high redox potential, such as anthraquinone and azo dyes [32].

According to Table 1, anaerobic conditions were used in most of the studies, in which biological color removal was achieved [15, 23, 27, 39, 48, 55, 65, 68, 81]. Castro et al. [14] and Park et al. [55] evaluated the performance of aerobic MBBRs in dye degradation and reported that no significant color removal was observed in the presence of oxygen. Therefore, the redox condition is a key factor to achieve the biological removal of dyes. For removing color, a favorable environment, usually in the absence of oxygen, is required to reduce the chromophores. Oxygen has a high redox potential and can replace the dye by acting as the final electron acceptor, inhibiting discoloration [57].

However, in some cases, the discoloration can be achieved in aerobic conditions, for instance, by using selected microorganisms, such as white-rot-fungi strains (*Cerioporus squamosus* and *Phanerochaete chrysosporium*) [56], *Bacillus* sp.

[71], and the photosynthetic bacterium *Rhodopseudomonas palustris* [80]. Inoculating an MBBR with sludge from a WWTP treating dye-containing wastewaters is also a strategy to achieve aerobic color removal, possibly because of the existence of a microbial community adapted to dyes and, therefore, able to synthesize specific enzymes and use the dye as a carbon source [69, 87].

- Agitation
  Agitation is an important factor to be considered, especially in biofilm and anaerobic systems. The agitation speed can influence both biofilm thickness and mass transfer, affecting the biodegradation of dyes.
  The intrinsic nature of MBBRs requires agitation to achieve bed fluidization. Agitation is also important to provide shear forces that are crucial to guarantee a good balance between biofilm attachment and detachment. If the biofilm becomes too thick, substrate transport to the inner biofilm layers can be affected. However, an excessively high stirring speed can enhance microbial detachment to such a level that the concentration of attached solids is substantially reduced. The detached biomass contributes to an increase in the concentration of suspended solids which, in the absence of sludge recycling, can be washed out (if the cell growth rate is slower than the rate of liquid effluent discharge from the reactor) [5].
  Moreover, high agitation speed results in high mass transfer rates between the reaction medium and the cells, and also between the surrounding air and the medium [70]. At the same time that it may benefit the substrate transport within the biofilm, high agitation intensity may lead to increases in DO concentration, negatively affecting the mechanism of anaerobic dye removal [37]. Stirring speeds ranging from 30 to 500 rpm have been reported (Table 1). However, very few studies regarding the influence of this parameter on dye removal in MBBRs have been conducted. Considering the high molecular weight and size of many dyes, which can lead to low diffusion rates [29], the optimum coordination between agitation and biofilm thickness has to be defined to improve color removal.

- Type of carrier media, material modifications, and filling ratio
  The type of media used in the MBBR affects its performance. Two of the most influential carrier characteristics are its specific surface area and the material that it is made. Carriers showing a high specific surface area may provide higher concentrations of attached solids within the reactor and allow more microorganisms to be placed at the biofilm-bulk phase interface. In fact, the specific surface area available in the carriers for biofilm development has a greater impact on the performance of attached growth processes than the reactor volume itself [53]. The composition of such carriers can also offer more or less affinity toward microorganism attachment and can even act as a substrate source [83].
  The carriers used in dye biodegradation in MBBRs are described in Table 1, being mostly made of plastic (e.g., polyethylene and polypropylene), which satisfy the low-density requirements, but usually present low hydrophilicity and low biological affinity, leading to low growth rates and cell detachment [17]. The

most-reported biomedia used in MBBRs for dyes removal are the Kaldnes® K1 and polyurethane (PU) foams. Kaldnes® K1 media has a high specific surface area of 690 m^2/m^3 (total), or 500 m^2/m^3 (effective) [53], while PU cubic sponges may present specific surface areas of more than 4000 m^2/m^3 [48].

- Wang et al. [83] reported new support made of high-density polyethylene (HDPE), Zn nanoparticles (NPs), and poly (lactic acid) (PLA). The authors observed that the presence of Zn NPs in concentrations of up to 20 wt% helped improving organic matter and ammonium nitrogen removals in an aerobic MBBR treating real pretreated textile wastewater. Dehydrogenase activity was also improved and the biodiversity increased. Zn ion was found to stimulate *Planctomycetes*, which possibly helped improving nitrogen removal. According to the authors, Zn is one of the essential micronutrients used for cofactors and enzymes production and could be beneficial to stimulate the growth of dye-degrading bacteria. At concentrations below the optimal, microbial activity can decrease, while toxic effects can be observed at high doses. Since it is difficult to provide optimum Zn concentrations to all biofilm layers, its incorporation into the biomedia material and its controlled release from it may be an alternative to improve the bioreactor performance.
  Adopting an adequate media filling ratio is another important factor when operating MBBRs. The filling ratio is a parameter that indicates the volume of the reactor that is occupied by the carriers. A low filling ratio usually leads to low concentrations of attached solids within the reactor. However, it should also not exceed 70% (v/v) in order to avoid hydrodynamic problems. An excessive amount of carriers may hinder the effective homogenization of the reaction medium [53]. Sonwani et al. [71] studied the effect of this parameter on the bioremediation of Congo red dye, finding the optimum value of 45% (v/v). Percentages ranging from 30 to 67% (v/v) were found in this review.
- pH
  The azo dye color removal process is strongly dependent on the pH of the medium. Efficient discoloration usually occurs for pH between 6 and 10 [38], decreasing rapidly in strongly acidic or basic environments [89].
  The pH of the medium may be linked to the transport of dye through the cell membrane (rate-limiting step for color removal) [38]. Most enzymes have an optimum pH at which their activity is maximum. The reaction rate decreases as the pH value moves away from the optimum value. These enzymes may possess ionic groups on their active sites, which are subjected to ionization (Eq. 1). By changing the pH, the equilibrium state is disturbed, causing a shift toward the right or left side of Eq. 1, according to Le Chatelier's principle [5]. Hence, the enzymes may be found either in the acidic or basic state depending on the pH, which also happens with the dyes. The linkage between enzyme and substrate and, therefore, dye biodegradation, depends upon the formation of compatible ionic forms [71]. As the reduction of the azo bond tends to generate amines with a lower pH than the original dye, buffer solutions are normally used in the reaction medium [57].

$$HA \leftrightarrow H^{+} + A^{-} \tag{1}$$

The pH also affects the chemical equilibrium of other species in the solution, such as unionized and ionized ammonia (NH_3/NH_4^+), and nitrite and nitrous acid (NO_2^-/HNO_2). At high pH (above 9.4), NH_3 formation is favored, while at low pH, the NH_4^+ compound is predominant. Considering that NH_3 is a volatile molecule, nitrogen loss via stripping can occur at high pH, and the gaseous emissions of the process can increase. HNO_2 formation, on the other hand, is favored in acidic conditions. Both NH_3 and HNO_2 can inhibit the activity of nitrifiers, hindering the performance of aerobic processes and leading to lower nitrification rates. When nitrification is envisaged, a pH around 7.0–8.0 is usually recommended [31].

Biofilm formation is also affected by the pH of the reaction medium. Variations in the pH value decrease EPS excretion and influence its structure and properties, ultimately leading to cell lysis and death [20].

- Temperature

  Temperature is a factor commonly known to influence diffusion, solubility, reaction rates, and cell and enzyme activities. Thus, it plays a crucial role in the microbial decolorization of dyes. As previously mentioned, an efficient substrate diffusion is essential to guarantee a satisfactory performance of MBBRs. By increasing temperature, the resistance to mass transfer decreases and diffusion rates increase. However, when dealing with biological systems, temperature control must be done very carefully. Temperatures above 45 °C commonly lead to enzyme denaturation and loss of cell viability. According to Pearce et al. [57], the optimum temperature for color removal is between 35 and 45 °C. Santos-Pereira et al. [65] reported an increase in the color removal of Direct Red 75 (DR75), in an anaerobic MBBR, from 45% at 21 $\pm$ 2 °C to 85% at 30 $\pm$ 0.5 °C (Table 1).

  Carbon and nitrogen removals are affected by temperature variations as well. Li et al. [43] investigated how temperature affects the effectiveness of an aerobic MBBR treating real textile wastewater from a dyeing company. The following average COD removals were obtained: 60.7% (at 30 °C), 63.2% (at 35 °C), 69.8% (at 40 °C), 54.2% (at 45 °C), 70.1% (at 50 °C), and 41.5% (at 55 °C). Very low COD removal levels were reported for temperatures over 55 °C. Thermotolerant microorganisms were eliminated by increasing temperature up to 40 °C, but a new community of thermotolerant microbes was formed when the temperature was increased to 50 °C. Average NH_3-N removal efficiencies were 33.3% (at 30 °C), 39.1% (at 35 °C), 38.5% (at 40 °C), 28.0% (at 45 °C), 17.5% (at 50 °C), and 11.5% (at 55 °C). Hence, NH_3-N removal was more affected by temperature than COD removal. The same authors observed that the total amount of soluble EPS increased up to 45 °C when it reached maximum levels of 1,748 mg/L and then decreased. This decrease was associated with an inhibitory effect of high temperatures on EPS production and cell damage. Thermophilic communities changed with the increasing temperature, mainly including genera *Caldilinea* (from 35 °C to 45 °C) and *Rubellimicrobium* and *Pseudoxanthomonas* (>50 °C). The thermophilic species *Geobacillus thermoglucosidasius, Pseudoxanthomonas taiwanensis, Geobacillus thermo denitrificans,* and *Rubellimicrobium thermophilum* were identified over 50 °C and contributed to organic matter removal at 50 °C

and 55 °C. The ammonium oxidizing bacteria (AOB) *Nitrosomonas eutropha* was only observed at 35 °C and not at higher temperatures. *Zunongwangia profunda*, which can produce high amounts of EPS, was identified at 45–50 °C.
In aerobic processes, temperature control is also important to guarantee adequate concentrations of dissolved oxygen (DO) and avoid ammonia nitrogen volatilization [43, 75]. Since biological degradation of pollutants is usually carried out at mild temperatures, the effect of temperature on DO concentrations is not commonly a factor of concern.
The sensitiveness of a biological process toward temperature variations can be measured by the factor Q_{10} (Eq. 2) [43].

$$Q_{10} = \left(\frac{R_1}{R_2}\right)^{\frac{10}{(T_2 - T_1)}} \tag{2}$$

where Q_{10} = temperature coefficient (measures the rate of change in a biological process, due to temperature variation); R_1 = rate of pollutant removal at the temperature T_1; R_2 = rate of pollutant removal at the temperature T_2.

- Dye concentration
High concentrations of dye negatively affect color removal due to the toxic effects they may cause on bacteria and the blocking of active enzyme sites by the dye molecule. According to its concentration, there may also be an inappropriate proportion of cell biomass and dye. In addition, aromatic amines formed by azo bond breakage in anaerobic processes can have toxic and inhibitory effects on microorganisms. Therefore, by increasing the dye concentration, the formation of aromatic amines can increase, enhancing the toxic effects [58].
As a result of the toxicity of the dyes and their degradation products, the production of new cells remains low at high dye concentrations [38]. Koupaie et al. [39] observed a reduction in the biofilm mass of an anaerobic system with the increase in Acid Red dye 18 concentration. Wang et al. [81] also reported a decreased growth rate of *R. palustris* W1 by increasing the initial Reactive Black 5 (RB5) concentration. The same authors reported a decrease in the decolorization kinetic constant (K) value from 0.114 to 0.064 h^{-1} by increasing the dye concentration from 50 to 1,000 mg/L.
Sonwani et al. [71] modeled the effect of azo dye concentration, pH, and carrier filling ratio on color removal by *Bacillus* sp. in an aerobic MBBR. The dye concentration was found to be the most influential factor, having a strong negative coefficient on the modeled equation, which described dye removal efficiency. Castro et al. [15] reported that a low azo dye inlet concentration (5 mg/L of RO16) and an HRT of 12 h were necessary to achieve high color removals on an anaerobic MBBR. However, according to Pearce et al. [57], very low concentrations of the dye also influence the process, as the substrate identification by the specific microbial enzymes is hindered under this condition.
- Dye structure

The efficiency of dye degradation is directly linked to the structure of the molecule. Simpler molecules with lower molar mass tend to be metabolized more easily. In this way, monoazo dyes, for example, are removed faster than those that have more than one azo bond (e.g., diazo and triazo dyes) [38]. Azo dyes also tend to release nitrogen after azo bond breakage. Therefore, after discoloration, an increase in nitrogen concentrations in the bulk is often observed [27, 48], while this may not happen for other types of dyes, having different chromophores.

Dyes with electronegative substituents, such as $-SO_3H$ groups and $-SO_2NH_2$ in the ortho and para positions (with respect to the azo bond), usually present higher removal rates, as they lead to a more effective resonance effect and make the azo bond more susceptible to reduction. In contrast, when the substituent is an electron donor, such as –NH-triazine, or when electronegative substituents are in the target position, removal becomes slower [70]. Azo compounds having hydroxyl or amino groups are more easily broken down than those containing methyl, methoxy, sulfo, or nitro groups [58].

Sulphonated groups can hinder dye removal if the mechanism is intracellular, as they can block the molecule's passage through the cell membrane. Therefore, in this case, the higher the number of sulfonated groups, the lower the dye removal efficiency [57]. The steric effect must also be taken into account since the presence of voluminous substituents in the vicinity of the chromophore may hinder access to it [70].

In addition, the toxicity of dyes depends on their chemical nature and the characteristics of their degradation products. Such toxicity can affect not only dye removal but also other biochemical processes within the MBBR, such as nitrogen removal. While Castro et al. [15] reported that no long-term inhibition of nitrifiers was observed after feeding an aerobic MBBR with ozonated RO16 solutions in concentrations of up to 500 mg/L, Dias et al. [22] observed a completely opposite behavior for RR239: its ozonation products disturbed the activity of nitrifying bacteria, completely inhibiting nitratation, at much lower dye concentrations (50 mg/L). Furthermore, nitrifiers did not seem to adapt to the presence of RR 239 ozonation products, since ammonium removal remained low (41%), even after 90 days of reactor operation. In a subsequent study, Dias et al. [21] reported that low ammonium removal was associated with the reduced enzymatic activity of nitrifiers in the presence of chlorine-containing triazine compounds, resulting from RR 239 ozonation.

Dye toxicity can cause higher biofilm detachment rates, leading to small biofilm thickness and low contents of attached solids, therefore influencing removal patterns [14, 15, 21, 39]. The contents of polysaccharides and proteins, which are directly related to the biofilm characteristics, are also influenced by the properties of the dye and its degradation products. Shin et al. [68] observed lower EPS and protein contents in the first reactor of a series of three MBBRs (anaerobic + aerobic + aerobic), while pollutant removal was also lower.

Conversely, dye degradation products can also act as mediators, improving color removal (Fig. 3b). Wang et al. [80] identified the compound 1–2-7-triamino-8-hydroxy-3–6-naphthalinedisulfate (TAHNDS) in the effluent of an anaerobic photo-rotating biological contactor (PRBC) treating a mixed-dye system, containing RB5 and AR1. The effluent of the PRBC was subsequently fed to an aerobic MBBR. According to the authors, the TAHNDS can act as a mediator to improve azo bond breakage. 1-Amino-2-naphthol produced during the dye metabolization was further degraded into quinone, a more effective redox mediator. As a result, discoloration took place mostly in the MBBR. Quinones may also accelerate denitrification [44]. Therefore, denitrification may also take place even in aerobic MBBRs in the presence of such compounds, and due to the presence of anoxic zones within the biofilm [15].

- Organic substrate (electron donor)
  The carbon source in biological systems is essential for survival and microbial growth. The primary substrate also acts as a donor of electrons for the azo bond breakage. These electrons, derived from the substrate oxidation, are transferred to azo dyes, resulting in their reduction and discoloration [70]. Some widely used organic substrates are acetate, ethanol, yeast extract, and glucose [58].
  The type of electron donor used can influence color removal. According to Khan et al. [38], the addition of glucose or acetate ions can induce azo bond breakage. Wang et al. [81] tested the effect of different carbon sources (formate, acetate, sodium lactate, propionate, butyrate, oxalate, and glutamine) on the removal of color from RB5-containing wastewater by the autotrophic bacterium *Rhodopseudomonas palustris* W1, isolated from an anaerobic MBBR. The highest discoloration rates were observed for lactate and glutamine.
  The amount of substrate used is also important. There must be enough carbon to meet cellular needs and act as a donor of electrons for discoloration. However, excessive quantities may cause cells to consume the primary substrate rather than the dye [70]. According to van der Zee and Villaverde [78], two pairs of electrons are needed to reduce the azo bond, which results in a theoretical requirement of 32 mg of COD per mmol of azo dye.
  Castro et al. [15] observed a decrease in the kinetic constant for Reactive Orange 16 (RO16) discoloration upon increasing initial glucose concentration in a batch trial. However, under continuous operation, color removal from RO16-containing wastewater in an anaerobic MBBR increased by increasing glucose concentration. The authors attributed this behavior to the increase in solids concentration within the reactor. Wang et al. [80] also observed an increase in color removal when the co-substrate (sucrose and sodium acetate) concentration was increased in a continuous system composed by a photo-rotating biological contactor (PRBC) followed by an aerobic MBBR (Table 1).
- HRT
  In general, the color removal rates in anaerobic systems are low [77]. As a result, higher HRT values are needed to achieve high color removal efficiency. This behavior may be related to the greater activity of enzymes responsible for the reduction of chromophores at higher HRT values. However, very high HRT can

lead to toxic effects caused by the dye and its degradation products, thereby decreasing the discoloration efficiency. Moreover, it can affect the COD removal, hydrolysis efficiency, and biodegradability [27]. In the literature, a very wide range of HRT values is found for MBBRs treating dye-containing wastewaters (from 5 to 132 h) (Table 1). HRT lower than 11 h was mostly used in aerobic MBBRs.

The HRT also directly influences the organic load, therefore affecting microbial growth and color removal. According to Sonwani et al. [71], by reducing the HRT, the effective attached growth of microorganisms on the carrier media surface is hindered. Castro et al. [15] concluded that increasing the HRT from 6 to 12 h was essential to achieve a higher color removal of RO16 since it allowed higher biomass growth within the reactor.

Overall, MBBR seems to achieve better performances in terms of color removal than conventional activated sludge (CAS), for lower HRT, while COD and TSS removals also remain high. This allows the construction of more compact units and lowers energy consumption, reducing operational costs and environmental impacts [87].

- Composition of the microbiota

Pure and mixed microbial communities have been applied in MBBRs for dye removal. Mixed cultures were mostly developed from inocula collected at municipal or industrial WWTP (Table 1). In particular, full-scale reactors treating dye-containing wastewaters are prone to naturally select microbial strains, which are resistant to the potentially toxic effect of dyes. Nonetheless, other sources of microorganisms may also be used. Santos-Pereira et al. [65] developed the microbiota of a lab-scale anaerobic MBBR by cultivating microbes from rice husks. Vaidhegi et al. [76] inoculated an MBBR with sludge from sewage treatment plants and dairy animals excrement. The use of a mixed culture in biofilm systems allows the coexistence of both heterotrophic and autotrophic communities, which are specialized in the removal of different pollutants. For instance, Liu et al. [48] treated real wastewater from a dyeing factory in an UASB-aerobic MBBR system using sludge from a municipal WWTP. The authors attributed the removal of dyes to *Shewanella* spp., both in the anaerobic and aerobic reactors. Nitrogen removal was associated with the presence of the ammonia-oxidizing archaea *Nitrososphaera* spp., the nitrite-oxidizing bacteria *Arcobacter* spp., and the denitrifying *Hydrogenophaga* spp. The presence of denitrifiers on the aerobic MBBR was also reported. *Protocatella* spp., *Acetoanaerobium* spp., and *Proteiniclasticum* spp. promoted the conversion of organic matter into acetic acid, which was further degraded into methane by the methanogenic archaea *Methanothrix* spp. and *Methanosarcina* spp.

By adopting pure cultures, on the other hand, the removal of a target pollutant can be optimized. White-rot fungi [56] and *Bacillus* sp. [71] are known for achieving high color removal efficiencies from wastewaters under aerobic conditions. Wang et al. [81] and [80] have also shown the effectiveness of the photosynthetic bacterium *Rhodopseudomonas palustris* in the discolorization of the azo dye Reactive Black 5.

- Other factors

  Additional factors that may influence biological color removal from dye-containing matrices include salt and micronutrient concentrations, as well as the presence or absence of a magnetic field.

  Textile wastewaters are known to have high concentrations of salts. High salinity affects the osmotic pressure, causing cell dehydration, plasmolysis, and death. Moreover, by increasing salt concentrations, carrier media coverage by biofilms may decrease, as well as the charge of EPS, leading to biofilm compaction and porosity reduction. Shifts in bacterial communities may also occur [82]. High salt concentrations can also inhibit and denature enzymes (e.g., azoreductases) and reduce substrate transfer rates [81]. In addition, COD and ammonium removal may be affected [86].

  High concentrations of micronutrients (e.g., Zn, Cu, Fe, Mn, Ni, Co, etc.) are also detrimental to dye biodegradation. However, at optimum quantities, their presence is essential since the cofactors act as an enzyme activator and therefore having a decisive role in enzymatic processes and metabolic pathways [5, 83]. In biofilm systems, since resistance to mass transfer may be observed, enzymatic processes can be slowed down due to the low availability of micronutrients. Wang et al. [83] demonstrated how color removal improved by incorporating Zn nanoparticles into the support material and ensuring its controlled release (Table 1).

  Biological processes can also be affected by magnetic fields, which influence bacterial movement and physicochemical properties of the wastewater, inducing colloidal particles to agglomerate [72].

3.3 *Kinetics of Dye Removal in MBBR*

The MBBR is, essentially, a continuous reactor. The mass balance for the limiting substrate in a continuous bioreactor is given by Eq. (3) [67].

$$\begin{aligned}\{\text{Rate of variation in the mass of substrate within the reactor}\} &= \{\text{Incoming mass flow of substrate}\} \\ &- \{\text{Outgoing mass flow of substrate}\} \\ &+ \{\text{Rate of substrate consumption by cells}\}\end{aligned} \quad (3)$$

Equation (2) can be mathematically translated by Eq. (4).

$$\frac{dS}{dt} = \frac{Q}{V}S_0 - \frac{Q}{V}S + (-r_s) \quad (4)$$

where $\frac{dS}{dt}$ is the rate of variation in the mass of substrate within the reactor; S is the limiting substrate concentration in the effluent of the reactor; t is time; Q is the volumetric flow rate; V is reactor volume; S_0 is the inlet concentration of the limiting substrate; $(-r_s)$ is the rate of substrate consumption; X is the cell concentration.

Correspondingly, the mass balance for the microorganisms is expressed by Eqs. (5) and (6).

$$\{\text{Rate of variation in the mass of cells within the reactor}\} = \{\text{Incoming mass flow of cells}\} - \{\text{Outgoing mass flow of cells}\} + \{\text{Rate of cells growth}\} \quad (5)$$

$$\frac{dX}{dt} = \frac{Q}{V}X_0 - \frac{Q}{V}X + r_x \quad (6)$$

where $\frac{dX}{dt}$ is the rate of variation in the mass of cells within the reactor; X is the concentration of the cells in the effluent of the reactor; X_0 is inlet concentration of cells; r_X is the rate of cell growth.

If operated in a steady-state mode, which is usually the case for MBBRs running for a long time, and considering that no bacterial cells are fed to the reactor ($X_0 = 0$) (the first reactor of a series, no recycle), Eqs. (4) and (6) become Eqs. (7) and (8):

$$r_s = \frac{Q(S_0 - S)}{V} \quad (7)$$

$$\frac{QX}{V} = r_x \quad (8)$$

Equations (7) and (8), when combined, result in Eq. (9).

$$r_s = \frac{r_X(S_0 - S)}{X} \quad (9)$$

The substrate consumption rate is described by different models in the literature, such as Monod model (Eq. 10)

$$r_S = \frac{\mu_{max} SX}{K_S + S} \quad (10)$$

where μ_{max} is the maximum specific substrate removal rate; K_s is the half-velocity coefficient for the substrate.

This set of equations (Eqs. 7 to 10) allows designing an appropriate reactor for a given process. Equation (8) evidences the importance of defining r_X, through which the reactor volume can be calculated for a given flow rate. The rates r_X and r_S are related by Eq. (9), while r_S depends on two reaction constants, μ_{max} and K_S, which can be estimated through batch tests by measuring the substrate and cell concentrations over time.

For a batch reactor, there are no influent or effluent flows. Therefore, Eq. (4) can be simplified and transformed into Eq. (11).

$$\frac{dS}{dt} = (-r_s) = -\frac{\mu_{max} SX}{K_S + S} \tag{11}$$

For a small period of time, the concentration of cells can be considered constant, and Eq. (11) can be simplified as Eq. (12).

$$\frac{dS}{dt} = -\frac{R_{max} S}{K_S + S} \tag{12}$$

with $R_{max} = \mu_{max} X$, representing the maximum substrate removal rate.

For low substrate concentrations (S < < Ks), Eq. (12) can be simplified and integrated, resulting in a first-order reaction rate (Eq. 13). For low substrate concentrations (S > > Ks), Eq. (12) becomes a zero-order reaction (Eq. 14) [33].

$$S_t = S_0 e^{-\frac{R_{max}}{K_S} t} = S_0 e^{-k_1 t} \tag{13}$$

$$S_t = S_0 - R_{max} t = S_0 - k_0 t \tag{14}$$

where: S_t= substrate concentration at reaction time t; S_0= initial substrate concentration; k_1 = first-order kinetic constant ($\frac{R_{max}}{K_S}$); k_0 = zero-order kinetic constant (R_{max});

Alternatively, Eqs. (7) and (10) can be put together (Eq. 15) and rearranged into a linear form (Eq. 16) (Modified Stover–Kincannon model), allowing the determination of kinetic parameters in a continuous mode operation [71].

$$\frac{Q(S_0 - S)}{V} = \frac{\mu_{max} SX}{K_S + S} = \frac{R_{max} S}{K_S + S} \tag{15}$$

$$\frac{V}{Q(S_0 - S)} = \frac{K_S}{R_{max} S} + \frac{1}{R_{max}} \tag{16}$$

Table 2 displays the studies in which the kinetics of dye degradation in MBBR was investigated.

3.4 Combination of Processes

In order to take advantage of the best characteristics of each type of process, many authors have used the combination of different methods for treating textile effluents (Table 1). Advanced oxidation processes (AOPs), for example, can achieve high levels of color removal in a short time [21] and may improve dye biodegradability [72]. However, the high cost associated with AOPs to reach the effluent discharge standards can restrict their application. AOPs have been applied both before the

Table 2 Kinetics of dye removal in MBBR systems

Type of dye	Operation mode	Results	Reference
Reactive Orange 16	Batch	Anaerobic biodegradation of the dye (5.3 or 6 mg/L) and glucose presented second-order kinetics and the degradation constants increased by decreasing the initial COD. For initial COD of 400 mgO_2/L, $k_{dye} = 0.0095$ L/mg.h and $k_{COD} = 0.0003$ L/mg.h. For 200 mgO_2/L, $k_{dye} = 0.0177$ L/mg.h and $k_{COD} = 0.0009$ L/mg.h (Reactor volume = 0.2 L)	[15]
Real textile wastewater	Continuous	Kinetic tests showed a kinetic constant of 0.0048 $m^{-2}.h^{-1}$ and maximum specific substrate consumption of $r_s/X = 0.027d^{-1}$ (Reactor volume = 2 L). Best removal efficiencies were achieved at lower organic loading rates	[69]
Congo red	Continuous	The kinetics of Congo red biodegradation resulted in (Ks.Q/V) = 0.253 g/L·day and $R_{max} = 0.263$ g/L·day (Reactor volume = 2 L)	[71]
Reactive Black 5 and Acid Red 1	Batch	RB5 decolorization (120 mg/L) in both single and mixed dye media (RB5 + AR1) followed first order kinetic (k_1 of 0.123 and 0.118/h, respectively). AR1 (110 mg/L) decolorization followed zero-order kinetic (k_0 = 4.3 and 5.5 mg/L.h, for single and mixed dye systems, respectively) (Reactor volume = 2 L)	[80]

MBBR, for color removal and improvement of the wastewater biodegradability, and after it, for polishing (Table 1). When used as a pre-treatment, its operational conditions can affect the nature of by-products and dye toxicity. Dias et al. [21] reported an increase in biomass detachment from an aerobic MBBR when RR239 ozonation time was increased from 12 to 20 min, which contributed to reducing biofilm thickness. Therefore, an in-depth study of the best set of operational conditions is needed for each process.

Biological processes have a relatively low cost, but usually do not meet the disposal standards in one single step [10]. Thus, the combination of anaerobic–aerobic bioreactors has been widely found in the literature. During the anaerobic metabolism of azo dyes, potentially toxic aromatic amines are produced, which are usually not removed in the absence of oxygen. Therefore, a subsequent aerobic process is often used to mineralize azo compounds [78]. During the aerobic post-treatment, less aromatic and more polar substances are formed [39]. Moreover, the adoption of a series of MBBRs also leads to higher removals of organic matter and nitrogen. At high inlet organic matter concentrations, there is usually a predominance of heterotrophic bacteria over autotrophic bacteria, leading to higher COD removals. Meanwhile, at low organic matter concentrations, autotrophic communities such as nitrifiers develop well. Therefore, by adopting a series of two aerobic MBBRs, COD removal usually takes place in the first reactor, while ammonium removal rates are higher in the second one [7].

Gong [27] implemented a series of anaerobic MBBR–aerobic MBBR–ozonation–aerobic MBBR, and concluded that while the first MBBR was important for improving the biodegradability of real textile wastewater, the second and third MBBRs contributed to increasing COD and ammonium removals. Shin et al. [68] adopted a series of anaerobic MBBR–aerobic MBBR–aerobic MBBR–coagulation. Also, in this case, COD removal took place mainly in the aerobic MBBRs (56.4%), while the anaerobic MBBR and the coagulation processes were responsible for the removal of 82.8% of the color.

3.5 Comparison Between MBBR and Other Biological Wastewater Treatment Systems

The type of reactor used for dye remediation has a great influence on color removal efficiency. According to van der Zee and Villaverde [78], biological reactors with greater biomass retention capacity may perform better in the removal of azo dyes than those with less cell retention capacity. The growth of microorganisms in suspension or attached to a carrier can also influence the process. When the biomass grows adhered to a media, the microorganisms grow partially protected from external predators, load shocks, and temperature and pH variations [8, 57]. Thus, the performance of fixed biomass systems can be higher than those in which microorganisms grow in suspension.

Mohan et al. [51] conducted a comparative study on the use of these two types of processes in the treatment of wastewater containing the acid black azo dye 10B. The authors employed two sequencing batch reactors, operating separately: one with suspended biomass and the other with biofilm, with alternating cycles of anoxic–aerobic–anoxic conditions. It was observed that the biofilm system achieved greater color removal than that with suspended biomass. According to Mohan et al. [51], biofilms induce the formation of aerobic and anoxic zones along with the biofilm thickness, leading to a spatial distribution of microorganisms. In this type of system, besides the longer biomass residence times, a high level of metabolic activities is maintained in the reactor.

Yang et al. [87] recommended the use of MBBR over a conventional activated sludge process and MBR for the treatment of textile wastewaters due to satisfactory performance, relatively low costs, and environmental friendliness. The MBBR was the most economical technology at the industrial scale, saving 68.4% of the capital expenditures (CAPEX) and having the same operating expenditures (OPEX) as MBR. It also showed the lowest environmental impacts (15 out of 18 midpoint categories: ozone depletion, human toxicity, photochemical oxidant formation, particulate matter formation, ionizing radiation, terrestrial acidification, freshwater eutrophication, terrestrial ecotoxicity, marine ecotoxicity, agricultural land occupation, urban land occupation, natural land transformation, water depletion, metal depletion, fossil depletion,and 3 out of 3 endpoint categories: human health, ecosystems, and resources) and was selected as the most feasible technology for industrial scale. The treated effluent was reused for textile dyeing.

4 Conclusions and Future Perspectives

The intense global commercialization of textile dyes caused an increased emission of polluting wastewaters containing these harmful compounds into the receiving waters. Therefore, there is a major concern regarding the quality of such wastewaters that reaches the aquatic environment since most dyes are toxic and hardly biodegradable. Conventional biological methods, such as activated sludge, are usually not capable of completely removing dyes and their associated colors from the wastewaters and traditional aerobic biological processes are inefficient to promote dye degradation. Moreover, operating problems such as deflocculation may occur in suspended biomass-based reactors in the presence of these compounds, making solid–liquid separation more difficult.

In contrast, biofilm systems are more resilient to toxic loads and may enable a better degradation of dye molecules. The moving bed biofilm reactor (MBBR) stands out as a good alternative for the bioremediation of toxic wastewaters, being robust, effective, economical, and environmentally friendly. The key characteristics of MBBR, including the high concentration of specialized biomass within the reactor, high sludge age, the protected environment provided by the biofilm, and the presence of different redox zones (anoxic/anaerobic and aerobic) within the biofilms, even for aerated processes, may enhance the biodegradation of dyes. Color removal in MBBR can be achieved in both aerobic and anaerobic conditions, and depends on microbial adaptation period, microbial strains, hydraulic retention time (HRT), co-substrate concentration, and dye concentration and molecular structure. For mixed microbial cultures, color removal is higher in MBBRs subjected to anaerobic conditions. Other factors that influence dye biodegradation in these reactors are agitation, type of carrier, pH, temperature, HRT, solids concentrations, micronutrients, and salinity levels.

Due to the diversity of dyes and the wide concentration ranges they are found in real dye-containing wastewaters, difficulties are faced in the operation and process optimization. Therefore, in-depth studies on the kinetic and mechanism of dye biodegradation are needed, which may help define the best combination of process-specific parameters. In addition, identifying microbial strains capable of degrading specific dyes, and describing the effect of widely used dyes and their degradation products on microbial behavior is crucial for improving dye removal. Investigations on best process combinations are also recommended since a single MBBR as a stand-alone technology is not capable of mineralizing dyes. Future studies should also address technical–economic comparisons between different reactor associations to better understand their benefits and drawbacks in a holistic approach.

References

1. Al-Ghouti MA, Khraisheh MAM, Allen SJ, Ahmad MN (2003) The removal of dyes from textile wastewater: a study of the physical characteristics and adsorption mechanisms of diatomaceous earth. J Environ Manag 69:229–238. https://doi.org/10.1016/j.jenvman.2003.09.005
2. Allégre C, Moulin P, Maisseu M, Charbit F (2006) Treatment and reuse of reactive dyeing effluents. J Membr Sci 269:15–34. https://doi.org/10.1016/j.memsci.2005.06.014
3. Anju F, Sosamony KJ (2016) Treatment of pre-treated textile wastewater using moving bed bio-film reactor. Proc Technol 24:248–255. https://doi.org/10.1016/j.protcy.2016.05.033
4. Ayadi I, Souissi Y, Jlassi I, Peixoto F, Mnif W (2016) Chemical synonyms, molecular structure and toxicological risk assessment of synthetic textile dyes: a critical review. J Dev Drugs 5:1–4. https://doi.org/10.4172/2329-6631.1000151
5. Bailey JE, Ollis DF (1986) Biochemical engineering fundamentals, 2nd edn. McGraw-Hill, New York
6. Baldrian P (2006) Fungal laccases—occurrence and properties. FEMS Microbiol Rev 30(2):215–242. https://doi.org/10.1111/j.1574-4976.2005.00010.x
7. Bassin JP, Dezotti M, Sant'Anna GL Jr (2011) Nitrification of industrial and domestic saline wastewaters in moving bed biofilm reactor and sequencing batch reactor. J Hazard Mater 185:242–248. https://doi.org/10.1016/j.jhazmat.2010.09.024
8. Bassin JP, Dezotti M (2018) Moving Bed Biofilm Reactor (MBBR). Advanced biological processes for wastewater treatment. In: Dezotti M, Lippel G, Bassin JP (eds) Advanced biological processes for wastewater treatment: emerging, consolidated technologies and introduction to molecular techniques. Springer International Publishing, pp 37–74. https://doi.org/10.1007/978-3-319-58835-3_3.
9. Bassin JP, Kleerebezem R, Rosado AS, Van Loosdrecht MCM, Dezotti M (2012) Effect of different operational conditions on biofilm development, nitrification, and nitrifying microbial population in moving-bed biofilm reactors. Environ Sci Technol 46:1546−1555. https://doi.org/10.1021/es203356z
10. Beyene HD (2014) The potential of dyes removal from textile wastewater by using different treatment technology, a review. Int J Environ Monit Anal 2(6):347–353. https://doi.org/10.11648/J.IJEMA.20140206.18.
11. Bourbonnais R, Paice MG, Freiermuth B, Bodie E, Borneman S (1997) Reactivities of various mediators and laccases with Kraft pulp and lignin model compounds. Appl Environ Microbiol 63(12):4627–4632. https://doi.org/10.1128/AEM.63.12.4627-4632.1997

12. Calderón K, Martín-Pascual J, Poyatos JM, Rodelas B, González-Martínez A, González-López J (2012) Comparative analysis of the bacterial diversity in a lab-scale moving bed biofilm reactor (MBBR) applied to treat urban wastewater under different operational conditions. Biores Technol 121:119–126. https://doi.org/10.1016/j.biortech.2012.06.078
13. Calvete T, Lima EC, Cardoso NF, Vaghetti JCP, Dias SLP, Pavan FA (2010) Application of carbon adsorbents prepared from Brazilian-pine fruit shell for the removal of reactive orange 16 from aqueous solution: kinetic, equilibrium, and thermodynamic studies. J Environ Manag 91:1695–1706. https://doi.org/10.1016/j.jenvman.2010.03.013
14. Castro FD, Bassin JP, Dezotti M (2017) Treatment of a simulated textile wastewater containing the Reactive Orange 16 azo dye by a combination of ozonation and moving-bed biofilm reactor: evaluating the performance, toxicity, and oxidation by-products. Environ Sci Pollut Res 24(7):6307–6316. https://doi.org/10.1007/s11356-016-7119-x
15. Castro FD, Bassin JP, Alves TLM, Sant'Anna GL, Dezotti M (2020) Reactive Orange 16 dye degradation in anaerobic and aerobic MBBR coupled with ozonation: addressing pathways and performance. Int J Environ Sci Technol. https://doi.org/10.1007/s13762-020-02983-8
16. Chan YJ, Chong MF, Law CL, Hassell DG (2009) A review on anaerobic–aerobic treatment of industrial and municipal wastewater. Chem Eng J 155:1–18. https://doi.org/10.1016/j.cej.2009.06.041
17. Chen S, Cheng X, Zhang X, Sun D (2012) Influence of surface modification of polyethylene biocarriers on biofilm properties and wastewater treatment efficiency in moving-bed biofilm reactors. Water Sci Technol 65(6):1021–1026. https://doi.org/10.2166/wst.2012.915
18. Chengalroyen MD, Dabbs ER (2013) The microbial degradation of azo dyes: minireview. World J Microbiol Biotechnol 29:389–399. https://doi.org/10.1007/s11274-012-1198-8
19. Christie RM (2007) Environmental aspects of textile dyeing. Woodhead Publishing
20. di Biase A, Kowalski MS, Devlin TR, Oleszkiewicz JA (2020) Physicochemical methods for biofilm removal allow for control of biofilm retention time in a high rate. MBBR Environ Technol. https://doi.org/10.1080/09593330.2020.1843078
21. Dias NC, Alves TLM, Azevedo DA, Bassin JP, Dezotti M (2020) Metabolization of by-products formed by ozonation of the azo dye Reactive Red 239 in moving-bed biofilm reactors in series. Braz J Chem Eng 37:495–504. https://doi.org/10.1007/s43153-020-00046-6
22. Dias NC, Bassin JP, Sant'Anna GL, Dezotti M (2019) Ozonation of the dye Reactive Red 239 and biodegradation of ozonation products in a moving-bed biofilm reactor: revealing reaction products and degradation pathways. Int Biodeterior Biodegrad 144:104742. https://doi.org/10.1016/j.ibiod.2019.104742
23. Dong B, Chen H, Yang Y, He Q, Dai X (2013) Treatment of printing and dyeing wastewater using MBBR followed by membrane separation process. Desalin Water Treat 52(22–24):4562–4567. https://doi.org/10.1080/19443994.2013.803780
24. Forgacs E, Cserháti T, Oros G (2004) Removal of synthetic dyes from wastewaters: a review. Environ Int 30:953–971. https://doi.org/10.1016/j.envint.2004.02.001
25. Ghaly AE, Ananthashankar R, Alhattab M, Ramakrishnan VV (2014) Production, characterization and treatment of textile effluents: a critical review. J Chem Eng Process Technol 5(1). https://doi.org/10.4172/2157-7048.1000182
26. Giardina P, Faraco V, Pezzella C, Piscitelli A, Vanhulle S, Sannia G (2010) Laccases: a never-ending story. Cell Mol Life Sci 67(3):369–385. https://doi.org/10.1007/s00018-009-0169-1
27. Gong X-B (2016) Advanced treatment of textile dyeing wastewater through the combination of moving bed biofilm reactors and ozonation. Sep Sci Technol 51(9):1589–1597. https://doi.org/10.1080/01496395.2016.1165703
28. Goode C (2010) Understanding biosolids dynamics in a moving bed biofilm reactor. PhD thesis, Department of Chemical Engineering and Applied Chemistry, University of Toronto, Toronto, Ontario, Canadá
29. Guo CJ, de Kee D (1991) Effect of molecular size and free volume on diffusion in liquids. Chem Eng Sci 46(8) 2133–2141. https://doi.org/10.1016/0009-2509(91)80171-T
30. Gürses AM, Günes AK, Gürses MS (2016) Classification of dye and pigments. In: Gürses AM, Açikyildiz M, Günes AK, Gürses MS (eds) Dyes and pigments. Springer, Cham, pp 31–45. https://doi.org/10.1007/978-3-319-33892-7_3

31. Hoang V (2013). MBBR ammonia removal: an investigation of nitrification kinetics, biofilm and biomass response, and bacterial population shifts during long-term cold temperature exposure. Master Thesis. Department of Civil Engineering, University of Ottawa. https://ruor.uottawa.ca/bitstream/10393/24041/3/Hoang_Valerie_2013_thesis.pdf.
32. Hofrichter M et al (2010) New and classic families of secreted fungal heme peroxidases. Appl Microbiol Biotechnol 87(3) 871–897. https://doi.org/10.1007/s00253-010-2633-0
33. Işik M, Sponza DT (2005) A batch study for assessing the inhibition effect of Direct Yellow 12 in a mixed methanogenic culture. Process Biochem 40:1053–1062. https://doi.org/10.1016/j.procbio.2004.03.011
34. Işik M, Sponza DT (2004) Decolorization of azo dyes under batch anaerobic and sequential anaerobic/aerobic conditions. J Environ Sci Health Part A 39(4):1107–1127. https://doi.org/10.1081/ese-120028417
35. Jorand F, Boue-Bigne F, Block JC, Urbain V (1998) Hydrophobicity/hydrophilic properties of activated sludge exopolymeric substances. Water Sci Technol 37(4–5) 307–315. https://doi.org/10.1016/S0273-1223(98)00123-1
36. Kabra AN, Khandare RV, Govindwar SP (2013) Development of a bioreactor for remediation of textile effluents and dye mixture: a plant-bacterial synergistic strategy. Water Res 47:1035–1048. https://doi.org/10.1016/j.watres.2012.11.007
37. Kalme S, Ghoda EG, Gov Ndwar S (2007) Red HE7B degradation using desulfonation by Pseudomonas desmolyticum NC M 2112. Int Biodeterior Biodegrad 60:327–333. https://doi.org/10.1016/j.ibiod.2007.05.006
38. Khan R, Bhawana P, Fulekar MH (2013) Microbial decolorization and degradation of synthetic dyes: a review. Rev Environ Sci Biotechnol 12:75–97. https://doi.org/10.1007/s11157-012-9287-6
39. Koupaie EH, Moghaddam MRA, Hashemi SH (2011) Post-treatment of anaerobically degraded azo dye Acid Red 18 using aerobic moving bed biofilm process: enhanced removal of aromatic amines. J Hazard Mater 195:147–154. https://doi.org/10.1016/j.jhazmat.2011.08.017
40. Leyva-Díaz JC, Calderón K, Rodríguez FA, González-López J, Hontoria E, Poyatos JM (2013) Comparative kinetic study between moving bed biofilm reactor-membrane bioreactor and membrane bioreactor systems and their influence on organic matter and nutrients removal. Biochem Eng J 77:28–40. https://doi.org/10.1016/j.bej.2013.04.023
41. Leyva-Díaz JC, Martín-Pascual J, González-López J, Hontoria E, Poyatos JM (2013) Effects of scale-up on a hybrid moving bed biofilm reactor – membrane bioreactor for treating urban wastewater. Chem Eng Sci 104:808–816. https://doi.org/10.1016/j.ces.2013.10.004
42. Li S, Cheng W, Wang M, Chen C (2011) The flow patterns of bubble plume in an MBBR. J Hydrodyn 23(4):510–515. https://doi.org/10.1016/S1001-6058(10)60143-6
43. Li C, Zhang Z, Li Y, Cao J (2015) Study on dyeing wastewater treatment at high temperature by MBBR and the thermotolerant mechanism based on its microbial analysis. Process Biochem 50(11):1934–1941. https://doi.org/10.1016/j.procbio.2015.08.007
44. Li H, Guo J, Lian J, Xi Z, Zhao L, Liu X, Zhang C, Yang J (2014) Study the biocatalyzing effect and mechanism of cellulose acetate immobilized redox mediators technology (CE-RM) on nitrite denitrification. Biodegradation 25(3) 395-404. https://doi.org/10.1007/s10532-013-9668-8
45. Lin SH, Lin CM (1993) Treatment of textile waste effluents by ozonation and chemical coagulation. Water Res 27(12):1743–1748. https://doi.org/10.1016/0043-1354(93)90112-U
46. Lin YH (2008) Kinetics of nitrogen and carbon removal in a moving-fixed bed biofilm reactor. Appl Math Model 32:2360–2377. https://doi.org/10.1016/j.apm.2007.09.009
47. Lipskikh OI, Korotkova EI, Khristunova YEP, Barek J, Kratochvil B (2018) Sensors for voltammetric determination of food azo dyes—a critical review. Electrochim Acta 260:974–985. https://doi.org/10.1016/j.electacta.2017.12.027
48. Liu Y, Wang N, Wei Y, Dang K, Li M, Li Y, Li Q, Mu R (2020) Pilot study on the upgrading configuration of UASB-MBBR with two carriers: Treatment effect, sludge reduction and functional microbial identification. Process Biochem. https://doi.org/10.1016/j.procbio.2020.09.007

49. Mannina G, di Trapani D, Viviani G, Ødegaard H (2011) Modelling and dynamic simulation of hybrid moving bed biofilm reactors: model concepts and application to a pilot plant. Biochem Eng J 56:23–36. https://doi.org/10.1016/j.bej.2011.04.013
50. Menda M (2011). Corantes e pigmentos. http://www.crq4.org.br/quimicaviva_corantespigmentos.
51. Mohan SV, Reddy CN, Kumar AN, Modestra JA (2013) Relative performance of biofilm configuration over suspended growth operation on azo dye based wastewater treatment in periodic discontinuous batch mode operation. Biores Technol 147:424–433. https://doi.org/10.1016/j.biortech.2013.07.126
52. Nikfar S, Jaberidoost M (2014) Dyes and Colorants. Encycl Toxicol 252–261. https://doi.org/10.1016/b978-0-12-386454-3.00602-3
53. Ødegaard H, Gisvold B, Strickland J (2000) The influence of carrier size and shape in the moving bed biofilm process. Water Sci Technol 41(4–5):383–391. https://doi.org/10.2166/wst.2000.0470
54. Pandey A, Singh P, Iyengar L (2007) Bacterial decolorization and degradation of azo dyes. Int Biodeterior Biodegrad 59:73–84. https://doi.org/10.1016/j.ibiod.2006.08.006
55. Park HO, Oh S, Bade R, Shin WS (2010) Application of A2O moving-bed biofilm reactors for textile dyeing wastewater treatment. Korean J Chem Eng 27(3):893–899. https://doi.org/10.1007/s11814-010-0143-5
56. Park HO, Oh S, Bade R, Shin WS (2011) Application of fungal moving-bed biofilm reactors (MBBRs) and chemical coagulation for dyeing wastewater treatment. KSCE J Civ Eng 15(3):453–461. https://doi.org/10.1007/s12205-011-0997-z
57. Pearce CI, Lloyd JR, Guthrie JT (2003) The removal of colour from textile wastewater using whole bacterial cells: a review. Dyes Pigm 58:179–196. https://doi.org/10.1016/S0143-7208(03)00064-0
58. Popli S, Patel UD (2015) Destruction of azo dyes by anaerobic–aerobic sequential biological treatment: a review. Int J Environ Sci Technol 12:405–420. https://doi.org/10.1007/s13762-014-0499-x
59. Pratiwi R, Notodarmojo S, Helmy Q (2018) Decolourization of remazol black-5 textile dyes using moving bed bio-film reactor. IOP Conf Ser: Earth Environ Sci 106:012089. https://doi.org/10.1088/1755-1315/106/1/012089
60. Rehman K, Shahzad T, Sahar A, Hussain S, Mahmood F, Siddique MH et al (2018) Effect of Reactive Black 5 azo dye on soil processes related to C and N cycling. PeerJ 6:e4802. https://doi.org/10.7717/peerj.4802
61. Reports and Data (2019). Textile Dyes Market To Reach USD 10.13 Billion By 2026. https://www.globenewswire.com/news-release/2019/09/12/1914626/0/en/Textile-Dyes-Market-To-Reach-USD-10-13-Billion-By-2026-Reports-And-Data.html
62. Robinson T, McMullan G, Marchant R, Nigam P (2001) Remediation of dyes in textile efluent: a critical review on current treatment technologies with a proposed alternative. Biores Technol 77:247–255. https://doi.org/10.1016/S0960-8524(00)00080-8
63. Rusten B, Kolkinn O, Ødegaard H (1997) Moving bed biofilm reactors and chemical precipitation for high efficiency treatment of wastewater from small communities. Water Sci Technol 35(6). https://doi.org/10.1016/s0273-1223(97)00097-8
64. Rys P, Zollinger H (1972) Fundamentals of the chemistry and application of dyes. Wiley-Interscience, New York, p 196
65. Santos-Pereira GC, Corso CR, Forss J (2019) Evaluation of two different carriers in the biodegradation process of an azo dye. J Environ Health Sci Eng 17(2):633–643. https://doi.org/10.1007/s40201-019-00377-8
66. Saratale RG, Saratale GD, Chang JS, Govindwar SP (2011) Bacterial decolorization and degradation of azo dyes: a review. J Taiwan Inst Chem Eng 42:138–157. https://doi.org/10.1016/j.jtice.2010.06.006
67. Schmidell W, Lima UA, Aquarone E, Borzani W (2001). Biotecnologia industrial, vol 2. Editora Edgard Blücher Ltda, São Paulo

68. Shin DH, Shin WS, Kim Y-H, Ho Han M, Choi SJ (2006) Application of a combined process of moving-bed biofilm reactor (MBBR) and chemical coagulation for dyeing wastewater treatment. Water Sci Technol 54(9):181–189. https://doi.org/10.2166/wst.2006.863
69. Soler CR, Xavier CR (2015) Tratamento de efluente de indústria têxtil por reator biológico com leito móvel. RBCIAMB 38:21–30. https://doi.org/10.5327/Z2176-947820155714
70. Solís M, Solís A, Perez HI, Manjarrez N, Flores M (2012) Microbial decolouration of azo dyes: a review. Process Biochem 47:1723–1748. https://doi.org/10.1016/j.procbio.2012.08.014
71. Sonwani RK, Swain G, Giri BS, Singh RS, Rai BN (2020) Biodegradation of Congo red dye in a moving bed biofilm reactor: performance evaluation and kinetic modeling. Biores Technol 122811. https://doi.org/10.1016/j.biortech.2020.122811
72. Sosamony KJ, Soloman PA (2018) Treatment of pretreated textile wastewater using modified Mbbr. Int J Eng Technol 7(3.8):106. https://doi.org/10.14419/ijet.v7i3.8.16843.
73. Sponza DT, Işik M (2005) Toxicity and intermediates of C.I. Direct Red 28 dye through sequential anaerobic/aerobic treatment. Process Biochem 40:2735–2744. https://doi.org/10.1016/j.procbio.2004.12.016
74. Talha MA, Goswami M, Giri BS, Sharma A, Rai BN, Singh RS (2018) Bioremediation of Congo red dye in immobilized batch and continuous packed bed bioreactor by Brevibacillus parabrevis using coconut shell bio-char. Biores Technol 252:37–43. https://doi.org/10.1016/j.biortech.2017.12.081
75. Tromans D (1998) Temperature and pressure dependent solubility of oxygen in water: a thermodynamic analysis. Hydrometallurgy 48:327–342. https://doi.org/10.1016/S0304-386X(98)00007-3
76. Vaidhegi K, Selvam SA, Kumar AM (2018). Treatment of dye waste water using Moving Bed Biofilm Reactor & Granular Activated Carbon [MBBR-GAC]. J Adv Res Dyn Control Syst 10:08-Special Issue
77. van der Zee FB (2001) Anaerobic azo dye reduction. PhD thesis, University of Wageningen, Holland. https://edepot.wur.nl/121282
78. van der Zee FP, Villaverde S (2005) Combined anaerobic–aerobic treatment of azo dyes—a short review of bioreactor studies. Water Res 39:1425–1440. https://doi.org/10.1016/j.watres.2005.03.007
79. WEF/ASCE/EWRI (2009) Design of municipal wastewater treatment plants, 5ª ed, Manual of Practice No. 8, Water Environment Federation, Alexandria, Virginia
80. Wang X, Cheng X, Sun D (2011) Interaction in anaerobic biodecolorization of mixed azo dyes of Acid Red 1 and Reactive Black 5 under batch and continuous conditions. Colloids Surf, A 379(1–3):127–135. https://doi.org/10.1016/j.colsurfa.2010.11.065
81. Wang X, Cheng X, Sun D, Qi H (2008) Biodecolorization and partial mineralization of Reactive Black 5 by a strain of *Rhodopseudomonas palustris*. J Environ Sci 20(10):1218–1225. https://doi.org/10.1016/s1001-0742(08)62212-3
82. Wang J, Liu Q, Wu B, Hu H, Dong D, Yin J, Ren H (2020) Effect of salinity on mature wastewater treatment biofilm microbial community assembly and metabolite characteristics. Sci Total Environ 711:134437. https://doi.org/10.1016/j.scitotenv.2019.134437
83. Wang F, Zhou L, Zhao J (2018) The performance of biocarrier containing zinc nanoparticles in biofilm reactor for treating textile wastewater. Process Biochem 74:125–131. https://doi.org/10.1016/j.procbio.2018.08.022
84. Wang DM (2016) Environmental protection in clothing industry. In: Zhu L (ed) Sustainable development: proceedings of the 2015 International Conference on sustainable development (ICSD2015), pp 729–735. World Scientific Publishing Co Pte Ltd., Singapore. https://doi.org/10.1142/9789814749916_0076
85. Xavier JB, Picioreanu C, Almeida JS, van Loosdrecht MCM (2003). Monitorização e modelação da estrutura de biofilmes, Biomatemática—Modelação da estrutura de biofilmes. Boletim de biotecnologia 76:2–13. http://biofilms.bt.tudelft.nl/pdf/2002_jxavier_biofilmes.pdf
86. Xu M, Zhou W, Chen X, Zhou Y, He B, Tan S (2020) Analysis of the biodegradation performance and biofouling in a halophilic MBBR-MBR to improve the treatment of disinfected saline wastewater. Chemosphere 22:128716. https://doi.org/10.1016/j.chemosphere.2020.128716

87. Yang X, López-Grimau V, Vilaseca M, Crespi M (2020) Treatment of Textile Wastewater by CAS, MBR, and MBBR: a comparative study from technical, economic, and environmental perspectives. Water 12(5):1306. https://doi.org/10.3390/w12051306
88. Yoo ES, Libra J, Adrian L (2001) Mechanism of decolorization of azo dyes in anaerobic mixed culture. J Environ Eng 127(9):844–849. https://doi.org/10.1061/(asce)0733-9372(2001)127:9(844)
89. Zhao M, Sun PF, Du LN, Wang G, Jia XM, Zhao YH (2014) Biodegradation of methyl red by Bacillus sp. strain UN2: decolorization capacity, metabolites characterization, and enzyme analysis. Environ Sci Pollut Res 21:6136–6145. https://doi.org/10.1007/s11356-014-2579-3
90. Zollinger H (2003) Color chemistry: syntheses, properties, and applications of organic dyes and pigments. 3rd rev. ed. Verlag Helvetica Chimica Acta; Wiley-VCH. Zurich, Weinheim, p 637. https://doi.org/10.1002/anie.200385122

Hybrid Bioreactors for Dye Biodegradation

Swathi Desireddy and Sabumon Pothanamkandathil Chacko

Abstract Pollution due to dye effluents is a raising concern worldwide. The release of partially treated or untreated dye effluents into the environment is a major source of water pollution leading to eutrophication, death of aquatic species, and aesthetic problems. Dyes and their degradation components are recalcitrant in nature, and subsequent bio-accumulated toxic compounds are carcinogenic, mutagenic, and teratogenic in nature. Several physicochemical and biological treatment methods were employed to treat dye containing wastewaters. However, none of the techniques were able to completely mineralize recalcitrant dyes. Application of any single technique to the wide range of dye effluents seems to be impractical, and therefore it is vital to develop hybrid systems by combining biological systems with other suitable and promising treatment methods to achieve efficient dye degradation.

This chapter gives a review of hybrid bioreactors (i) integrating physicochemical and biological processes, (ii) integrating advanced oxidation processes and biological processes, (iii) integrating various biological processes, and (iv) novel bioreactors. These hybrid bioreactors are beneficial in terms of technical and economic feasibility to achieve efficient dye degradation and expected to make more footprints in future.

Keywords Dye effluents · Hybrid bioreactors · Decolourization · Degradation · Mineralization · Aerobic · Anaerobic · Membrane bioreactor · Recalcitrant · Azo dyes · Advanced oxidation process

1 Introduction

In recent years, a strong growth in industrial sector was witnessed which is a boon to the national economies of several countries. Specially, in developing countries

---

S. Desireddy · S. Pothanamkandathil Chacko (✉)
School of Civil Engineering, Vellore Institute of Technology (VIT), Chennai, India
e-mail: pcsabumon@vit.ac.in

S. Desireddy
e-mail: desireddy.swathi@vit.ac.in

S. S. Muthu and A. Khadir (eds.), *Dye Biodegradation, Mechanisms and Techniques*, Sustainable Textiles: Production, Processing, Manufacturing & Chemistry,
https://doi.org/10.1007/978-981-16-5932-4_10

it provides ample employment opportunities. However, the associated exponential increase in water pollution has raised the concern worldwide and crucial importance is laid on research related to treatment of wastewaters containing organic, toxic, and hazardous substances [7].

Dye effluents from industrial sectors have become a threat in recent years. Reports projected that around 10,000 different kinds of dyes are being utilized in various industries that would require a yearly production of 7 lakh tons [82]. Textile industry is the major consumer of dyes and about 15% of dyestuff is lost from the dye baths as the currently available techniques do not support their complete exhaustion. Also, varied dye groups namely anionic (direct dyes, acid dyes, and reactive dyes), cationic (basic dyes) and nonionic dyes (disperse dyes) are extensively being used in several other industries, including cosmetics, paper printing, leathers, detergents, pharmaceuticals, and food sectors. The effluents from these industries are composed of alkalinity, chemical oxygen demand (COD), biochemical oxygen demand (BOD), total suspended solids (TSS), along with dyestuff [69].

The release of untreated dye effluents into the aqueous environment leads to eutrophication, death of aquatic species, and aesthetic problems. Dyes and their degradation components are recalcitrant in nature due to their fused aromatic structures and subsequent bioaccumulation of these toxic components affect human health if entered into food chain due to their carcinogenic, mutagenic, and teratogenic nature [58]. Hence, several countries have enforced strict discharge standards and protocols concerned to release of dye effluents.

In line with this, though few industries with ecological concern are employing certain strategies such as recovery and reuse, the extremely recalcitrant nature of dyes makes it difficult to meet the discharge standards. Also, the influent characteristics of wastewaters emerging from textile industries varies on daily basis based on the products manufactured as many different dyes are used [28]. Therefore, the wastewater treatment plants with conventional treatment methods can not completely degrade the dyes. Also, the wastewater treatment plants mostly employ either physicochemical or biological methods.

Physicochemical techniques include adsorption, membrane filtration, coagulation, ion exchange, oxidation, flocculation-flotation, electro-flocculation, electrokinetic ultra-filtration, reverse osmosis, neutralization, electrochemical destruction, irradiation, precipitation, ozonation and katalox treatment [69]. These techniques have certain drawbacks like (i) high operational costs, (ii) high energy and chemical requirements, (iii) excess sludge generation, (iv) requires regeneration of adsorbents, (v) cannot handle large volumes of influents, (vi) not applicable to wide range of dyes, (vii) uses hazardous chemicals as reagents, and (viii) generates toxic secondary metabolites [77]. On the other hand, biological dye degradation methods are cost efficient compared to physicochemical methods and holds the following advantages: (i) easy operation, (ii) ecofriendly, and (iii) less sludge generation [25]. However, most of the biological methods are inefficient in achieving complete dye degradation.

In developing countries traditional aerobic activated sludge process is most widely used owing to its cost effectiveness. However, it is ineffective in degradation of dyes. On the other hand, anaerobic processes degrades dyes more swiftly but generates toxic aromatic amines which are resistant to further degradation and accumulates in the system. Complete mineralization of these intermediates can be achieved in aerobic process [84]. Therefore, a combination of anaerobic and aerobic processes can be employed to achieve effective degradation and mineralization of dyes.

Recent research focused on Advanced Oxidation Processes (AOPs), but they are not suitable for influents with high organic loading rates and high concentrations of dye stuff and hence need a preliminary treatment step [65]. Industrial wastewaters generally contain high organic compounds, acids, volatile organics, hydrocarbons, and alkalis along with dyes and dyestuff [66]. A few advanced systems combining the AOPs with traditional physicochemical and biological methods were also developed. These combination systems improve the degradation efficacy and also meets the discharge standards. Nevertheless, these combined systems still operate multiple processes sequentially. But the recently developed hybrid dye degradation systems integrate these multiple processes to work in a synergy. These hybrid bioreactors can treat dye effluents in less time with huge cost and space savings [77]. Figure 1 depicts the possible combinations in treatment of dye effluents by hybrid bioreactors.

This chapter gives a brief review of hybrid bioreactors reported in the literature to treat dye effluents. Emphasis is laid on hybrid systems integrating biodegradation with physicochemical methods, AOPs, and combinations of biological processes within themselves. Finally, novel hybrid bioreactors employing the latest techniques are also briefly described.

2 Hybrid Bioreactors Integrating Physicochemical and Biological Processes

Current wastewater treatment systems generally use physicochemical and biological techniques. As described earlier each of these processes holds their own pros and cons. To overcome the underlying issues several hybrid reactor systems integrating the principles of various physicochemical and biological processes have been developed. The important criteria is to select a suitable combination of methods to completely eliminate the curbs of single processes. Few of the noteworthy bioreactors integrating physicochemical processes are detailed below.

2.1 Biodegradation-Adsorption

Adsorption is widely used in treating dye effluents due to its rapid removal and economic benefits. Several adsorbents, namely activated carbon, steel plant slag, china clay, silica, bentonite, coal, fly ash, and few natural low cost adsorbents like wheat straw, maize cob, water hyacinth, peat, chitosan, microbial biomass, corn cob shreds, woodchips, have been used [69]. Though adsorption is a rational method, it

Fig. 1 Broad spectrum of combinations in treatment of dye effluents by hybrid bioreactors

does not totally remove toxic dyes and specially the inexpensive adsorbents have low adsorption capabilities and huge amounts are required. Also, the adsorbents alone cannot be used in practical situations when huge volumes of effluents are being treated.

Conversely, combining adsorption and biodegradation in bioreactors promises reliable, cost-effective and environmentally friendly approach to remove dyes [6]. Synergy between microbes and adsorbents can be advantageous to degrade dye effluents with high COD. These adsorbents swiftly adsorbs the dyes and provides neutral dye conditions which is otherwise toxic to microbes. In standalone physical adsorption systems separate regeneration process has to be carried out. But in hybrid bioreactors amended with adsorbents the extracellular biodegradation of adsorbed dye components occur internally which is mediated by enzymes [28]. Hence internal regeneration of adsorbents and dye removal occurs simultaneously. Generally, other than dyestuff certain bacterial cells, enzymes, organic components, and oxygen are also known to adsorb on adsorbents which aids in regeneration.

Few adsorbents are also known to act as catalysts and accelerate the anaerobic biodegradation of dyes by involving in shuttling reducing equivalents from electron donating co-substrates to azo linkages. For example the activated carbon contains surface carbonyl or quinone groups which acts as redox mediator [47]. Other redox mediators can also be immobilized on suitable adsorbents which when integrated into the bioreactor gives added advantage of enhancing the anaerobic biodegradation of dye at less hydraulic retention time along with adsorption of dyestuff [48].

In yet another process called as biosorption, live or inactivated microbial cells are used as adsorbents in bioreactors. Adsorption of dye components on dead cells involve adsorption, deposition, and ion-exchange. Whereas, decolourization by living cells involve several complex mechanisms such as surface adsorption, ion-exchange, complexation, chelation and micro-precipitation [11]. Microbial cell membranes contain polysaccharide, protein and lipid components with discrete functional groups which interact with dyestuff [69]. It allows the removal of COD and colour from textile wastewater in a single step with no additional physicochemical treatment.

There are also reports of immobilizing adsorbents such as fly ash on microbial cells by techniques like lattice-binding, adsorption, and membrane-entrapment which improves the dye degradation at lesser hydraulic retention time by increasing the surface contact of dyestuff and microbes [23]. Roy et al. [57] used *Pseudomonas* attached to fly ash as support material and fluidized in dye effluents to treat Congo red dye which helps in high biomass retention and attained high dye removal. Details of hybrid reactors integrating adsorption and biodegradation reported in the literature are given in Table 1.

2.2 Biodegradation/Coagulation-Flocculation

Coagulation and flocculation is being used mostly for treatment of dye effluents owing to its ease of applicability, fast reaction, and low cost. It is used solely as a main stream process or as a pre/post treatment step [64]. Textile wastewaters generally contain suspended organic compounds. Coagulants with opposite charge aids in neutralizing the suspended matter to form flocs which can then easily be removed from the system. Different kinds of coagulants namely pre hydrolyzed metal salts, hydrolyzing metal salts, and synthetic cationic polymers were stated [38]. To obtain the efficient colour removal pH, coagulant dose, reaction time, and temperature are to be considered as the most important factors affecting the process.

Apart from the associated benefits, the generation of excess sludge which contains unutilized coagulants and coagulated dye components is a main concern of this method [76]. Presently very little emphasis is given to auxiliary treatment of secondary sludge and is discharged into waterbodies thereby leading to pollution. To address these concerns coagulation can be combined with biodegradation to achieve

Table 1 Comprehensive review of hybrid bioreactors integrating adsorption and biodegradation

Reactor	Adsorbent	Dye	Microbial species	Advantages	References
Packed column system	Granular activated carbon biofilm	Acid orange 7	Mixed microbial culture	Effective dye degradation Highly tolerant to dye toxicity	[51]
Fixed bed reactor	Fungal mycelium	Reactive Black 5 (RB5)	*Trametes pubescens* immobilized on stainless steel sponges	Feasible in real scale applications Efficient and relatively simple bioprocess	[19]
Fixed bed bioreactors	F400 Bone char Bamboo activated carbon	Reactive Black 5	*Aeromonas sp.* (AM262155)	Hydraulic retention time of less than 15 min Internal bio-regeneration of adsorbents	[34]
Anaerobic/aerobic granular activated carbon-sequencing batch biofilm reactor (GAC-SBBR)	Granular activated carbon	Acid Red 18 (AR18)	Anaerobic granular sludge and activated sludge	High concentrations of dye can be treated	[33]
Batch experiments	Mesoporous high surface area activated carbon	Acid Orange 10 (AO10)	*Pseudomonas putida* (CICC 21,906)	Easy immobilization Can treat high concentrations of dye	[86]
Immobilized batch and continuous packed bed bioreactor	Coconut shell bio-char	Congo red	*Brevibacillus parabrevis*	High efficiency Low operational costs Practically viable	[1]
Batch reactor	Fe Core Shell Bio-nanocomposites (Fe_3O_4@MIL-100)	Acid Orange 10 (AO10)	*Pseudomonas putida* (surface engineered)	Efficient dye degradation Easy solid liquid separation of adsorbents due to magnetic properties	[21]

(continued)

Table 1 (continued)

Reactor	Adsorbent	Dye	Microbial species	Advantages	References
Lab scale suspended and immobilized bioreactors	Periphyton biofilms (epiphyton, metaphyton and epilithon)	Crystal violet	*Cyanobacteria, Proteobacteria, Bacteroides, Actinobacteria, Verrucomicrobia*	Mineralization of dye and decreased toxicity Effluents reuse for irrigation	[63]
Fluidized bed bioreactor	Fly ash	Congo red	*Pseudomona sp.*	Low retention time High biomass concentration No bed clogging	[57]
Packed bed bioreactor	Casuarina seed biochar	Methylene blue	Immobilized *Alcaligenes faecalis*	Adsorbent can be reused in several cycles	[7]
Hybrid bioreactor	Coconut shell biochar	Methylene blue	Immobilized *Alcaligenes sp.*	Excellent COD and dye removal	[24]
Batch and continuous packed bed reactors	Corncob biochar	Brilliant green	*Brevibacillus parabrevis*	High efficiency Low operational cost Feasible for large scale application	[26]

complete removal of dyes. Applying coagulants before biodegradation is beneficial to effluents with high alkalinity as few coagulants cannot be used after the biodegradation due to near neutral pH. Conversely, implementing coagulation after the biodegradation step helps in lessening the coagulant dose and subsequent sludge generation [28].

Recently, non-toxic biopolymers such as chitosan are employed in coagulation and flocculation. They are effective in removing dyestuff and are also not toxic to microbial species involved in the biodegradation as compared to synthetic coagulants [55]. Identifying novel and less toxic coagulants to be employed in synergy with biodegradation opens new possibilities in textile wastewater treatment applications. Comprehensive review of hybrid reactors combining biodegradation and coagulation reported in the literature is given in Table 2.

2.3 Biodegradation/Membrane Filtration

Membrane separation technique uses a thin permeable or semipermeable polymeric membrane that acts as a barrier retaining certain particles and allows the flow of permeate. The filtration is directed by partial pressure variance on permeate and feed sides [78]. Based upon the pore sizes of membranes and associated degree of filtration, they are classified as microfiltration, ultrafiltration, nanofiltration, and reverse osmosis.

To reuse textile industry effluents membrane separation methods like nanofiltration and low pressure reverse osmosis are constantly being used. These filtration processes removes majority of colour and produces permeate which allows the reuse of water and unutilized salt escaped from dye baths [36].

However, the inorganic membrane fouling is a concern and regenerating it adds up to the operational costs. Also the safe disposal of reject stream is a laborious and tedious task. Hence implementation of biològical process prior to membrane separation improves the membrane life and also generates less sludge. Many combinations, including sand filtration, and multimedia filtration were employed between biodegradation and nanofiltration step [28].

However, these independent and additive units are not economic, and hence hybrid bioreactors called as membrane bioreactors (MBRs) were developed which integrates membrane separation and aerobic/anaerobic biological process in a single unit. MBRs are regarded as technical revolution and viable technique for end-of-pipe treatment of dye effluents owing to its advantages like less footprint, exceptional effluent quality, retention of enzymes, less sludge generation, robust and flexible to extension, and attractive for degrading recalcitrant dyestuff [15, 37].

In MBRs, membranes play a fundamental role in solid liquid separation by retaining suspended particles and adsorbing soluble particles. To further boost the reactor efficacies, numerous variations in MBRs such as (i) Anoxic-Aerobic MBR, (ii) Bioaugmented MBR, (iii) Biofilm MBR, (iv) Thermophilic MBR, and (v) Enzymatic MBR have been developed [77]. In few cases where extremely concentrated dye

Table 2 Comprehensive review of hybrid bioreactors integrating coagulation and biodegradation

Process/reactor	Coagulant	Dye	Microbial species	Advantages	References
Coagulation—anaerobic and Aerobic bioprocess (Up-flow anaerobic fixed film reactor followed by aerobic reactor)	Ferric chloride Lime	Common textile effluent treatment plant wastewater	*Pseudomonas aeroginosa*	Decreased aeration costs Less excess sludge production Achieved 89% decolourization and 94% COD removal	[49]
Coagulation—flocculation followed by biodegradation (Sequential Batch Reactor)	Lime Alum Magnesium chloride	Disperse and reactive dye effluent	Mixed aerobic culture	Complete removal of COD, BOD, and colour	[18]
Coagulation and solid state fermentation (Solid state fermentation composting bioreactor)	$ZnCl_2$	Reactive Red 120 (RR120) Real textile wastewater	Mixed microbial consortium	99% colour removal Dye sludge decolorization by microbial consortia Effective removal of TOC, COD, BOD, TSS and TDS Ecofriendly and low cost	[38]
Coagulation-biodegradation (Bench scale stirred tank reactor)	*Moringa oleifera* seed powder	Real textile wastewater	Novel consortium-BBA (B. cereus PAB1, B. thuringiensis PAB2, and mycelium of *Aspergillus tamarii* PAB)	98% colour removal Less toxic products after coagulation and biodegradation	[5]
Biodegradation—coagulation (Batch reactors)	Tannin	Effluent from dyeing and laundry industry	Mixed microbial culture	96% colour removal 79% decrease in turbidity	[10]

(continued)

Table 2 (continued)

Process/reactor	Coagulant	Dye	Microbial species	Advantages	References
Biodegradation—Coagulation/flocculation (Pilot scale textile wastewater treatment plant)	$FeCl_3$	Industrial textile wastewater	*Sphingomonas paucimobilis, Bacillus sp.*, and Filamentous bacteria	82% decolorization 76% COD removal	[4]

effluents were being treated, MBRs were amended with suitable adsorbents to control the shock loading rates and subsequently decrease toxicity to microbial species [29]. Recently, several other novel hybrid MBRs implementing AOPs, adsorbents, and white rot fungi in various combinations were reported to increase the efficacy and decreases the biofouling [61]. Comprehensive review on hybrid MBRs employed in treatment of textile wastewaters is given in Table 3.

Table 3 Comprehensive review of hybrid membrane bioreactors treating textile wastewaters

Process (reactor)	Dye	Microbial species	Advantages	References
Biodegradation-MBR with nano filtration membrane and RO membrane filtration	Reactive blue 19 Reactive blue 49 Reactive black 5	*Trametes versicolor* KCTC 16,781	Effective mixed dye decolorization	[41]
Biosorption by powdered activated carbon—adsorption, cake layer filtration, and biodegradation (Membrane coupled fungal reactor)	Poly S119 Acid Orange II	White-rot fungus *C. versicolor* NBRC 9791 Mixed microbial community	Stable and efficient decolouration of both dyes	[31]
Anaerobic MBR with suspended immobilized biocells (PVA-Calcium alginate pellets)	Reactive Black 5 (RB5)	*Bacillus cereus*	Efficient performance Low membrane biofouling	[83]
Anoxic/aerobic type sequencing batch MBR system	Reactive Blue B-2GLN	Aerobic floc sludge	Promising aeration Simultaneous COD (98%), nitrogen (97%) and colour (99%) removal	[81]
Bioaugmented aerobic MBR containing a Granular activated carbon-packed anaerobic zone	Acid orange II Poly S 119 Direct brilliant yellow Reactive orange 16	White-rot fungi *Coriolus versicolor* NBRC 9791	Can handle high dye loading rates Stable decolouration and significant TOC removal Good long term performance	[30]
Hybrid MBR by internal micro-electrolysis (Fe ions)	Anthraquinone dye effluent Reactive brilliant blue X-BR	Mixed microbial culture	Alleviation of membrane fouling	[54]

(continued)

Table 3 (continued)

Process (reactor)	Dye	Microbial species	Advantages	References
MBR amended with pellets and biofilm	Acid orange 7	Attached biofilm and spherical pellets of *Coriolus versicolour* (NBRC 9791)	Resistance to bacterial inhibition 95% decolouration at hydraulic retention time of 1 d A bacterial post-treatment is suggested to improve TOC removal	[32]
Aerobic degradation and detoxification in submerged MBR	Acid Red B	Activated sludge bioaugmented by yeast Fungi—bacteria mixed consortia *Candida tropicalis* TL-F1	96% colour and 97% TOC removal Tolerates high shock loading Less eco toxicity of effluents Long term performance	[43]
Fungal membrane bioreactor integrated into membrane photocatalytic (TiO_2 and ZnO) reactor	Textile wastewater from reactive washing unit	*Phanerochaete chrysosporium*	Efficient colour (93%) and COD removal (99%)	[17]
Lab/pilot scale MBR	Reactive Blue	Fungal biomass (*Rhizopus arrhizus* and *Aspergillus versicolor*)	91% colour and 90% COD removal by mixed fungal consortia	[2]
UV/ozonation/photocatalysis/MBR (Oxidized membrane bioreactor system)	Real textile wastewater	*Proteobacteria* *Bacteroidetes* *Firmicutes*	Mitigates fouling Competent COD and TSS removal	[61]

3 Hybrid Bioreactors Integrating AOPs and Biological Processes

AOPs are deliberated as prominent techniques to treat textile industry effluents. Recently, various AOPs namely ozone-UV, ozone-TiO_2-H_2O_2, H_2O_2-UV, photocatalysis, photo(solar)-Fenton, wet air oxidation, super critical water oxidation and electro-chemical oxidation were reported [73]. AOPs are being employed in treatment of various wastewater applications as AOPs can oxidize, decolorize, mineralize, and degrade organic contaminants [8]. AOPs comprise a chemical oxidation route that generates reactive oxygen species which act as strong oxidants. AOPs can ultimately mineralize the dyestuff by converting non-biodegradable organic matter to simpler non-toxic inorganic molecules like carbon, hydrogen, nitrogen, and sulphur [52].

Although AOPs can efficiently mineralize the toxic dye components, it has certain drawbacks like requirement of energy, costly reagents, and generation of chemical sludge [60]. To overcome these problems and to achieve complete mineralization of recalcitrant dyes researchers have developed hybrid systems combining AOPs targeting partial degradation which improves the efficiency of biological unit. By doing so the metabolites can easily be degraded in the bioreactor and concurrently saves the high cost associated with complete mineralization by AOPs [50].

Several AOPs integrated with biological reactors have been reported. Ozonation or partial oxidation prior to biodegradation improves the removal of COD, toxicity, and thus increases the biodegradable value [53] compared to biodegradation followed by AOPs [72]. However, based on the characteristics of influent wastewaters, optimum selection and time of pre-oxidation step has to be determined. Extended oxidation will not increase any further mineralization, and thus it is important to implement least possible pre-oxidation time and proceed to biodegradation [28]. Nevertheless, in some cases an extended oxidation period may be required making the biodegradation redundant. Characteristics of textile industry effluents varies a lot, and if large amounts of biodegradable components are present in wastewaters, pre-oxidation is not necessary as it will just lead to consumption of chemicals. Thus biodegradation followed by AOP should be implemented. However, hybrid systems with internal recycling between pre-oxidation and biodegradation stages might be a solution to decrease chemical doses in these kind of situations [44].

Further improvements like immobilizing TiO_2 in photo-catalysis to improve biocompatibility, new reactor configurations like biofilms, sequencing batch reactors (SBRs), and MBRs combined with AOPs have been reported [61]. Details of a few of the hybrid bioreactors combining AOPs and biodegradation are given in Table 4. A thorough evaluation of influent characteristics and each of the AOPs could help in selecting a suitable pre-oxidation step to further ease the biodegradation step.

4 Hybrid Bioreactors Integrating Various Biological Processes

Few researchers integrated different aerobic and anaerobic biological processes to gain the benefits of both the processes. These integrated bioreactors are beneficial as simultaneous decolorization and degradation occurs without the requirement of much nutrients since mixed microbial culture can obtain carbon, nitrogen and oxygen from degradation of dyes. Also metabolites produced by few strains of microbes are utilized by other strains thereby leading to effective degradation of dyes [71].

Table 4 Comprehensive review of hybrid bioreactors integrating advanced oxidation processes and biodegradation

Process	Reactor	Dye	Microbial species	Advantages	References
Fentons oxidation—aerobic biodegradation	Sequencing batch reactor	Reactive Black 5 (RB5) Reactive Blue 13 (RB13) Acid Orange 7 (AO7)	Cow dung slurry used as seed biomass	Partial but significant mineralization of dyes	[72]
Fenton's reaction—aerobic biological process	Batch experiments	Reactive black 5 (RB5)	*Candida oleophila*	91% of colour removal at high dye concentration (500 mg/L)	[46]
Activated sludge system—ozonation	Ozonated activated sludge reactor	Orange II	Mixed microbial culture	Decreased foot print Effective colour and COD removal Less ozone required	[75]
Sequential ozonation and biological treatment	Aerobic batch reactors	Remazol black	Mixed microbial culture immobilized on activated carbon	Efficient and economic Eliminates toxicity	[16]
O_3/H_2O_2-biodegradation	Batch incubations	Reactive Red 120	River sediments as seed biomass	Complete mineralization	[20]
Electrochemical oxidation and aerobic biodegradation (bioelectro oxidation process)	Sequential biological oxidation reactor	Procion Scarlet	*Pseudomonas sp.* *Micrococcus*	Reusable effluents Non-biocompatible organic effluent to biocompatible organic compounds Less energy consumption	[62]

(continued)

Table 4 (continued)

Process	Reactor	Dye	Microbial species	Advantages	References
Biodegradation by yeast—photocatalysis by UV/TiO_2	Batch reactors	Reactive Black 5 (RB5)	*Candida tropicalis* JKS2	Efficient removal of aromatic rings produced through biological treatment Cost effective Effective COD removal	[35]
Fentons reaction—biological oxidation	Sequencing batch jacketed reactor	Procion Deep Red H-EXL Procion Yellow H-EXL Astrazon Blue FGGL Dianix Orange K3G Dianix Blue KFBL	Activated sludge	Effluent meets the discharge limits Reduced operational costs Decreased chemical consumption	[56]
Fenton's oxidation with biological oxidation	Biological aerated filter system	Azaleine Methyl Red	Seed sludge from aeration tank	Effective substitute for treatment of non-biodegradable textile effluents Economical and effective method	[87]
H_2O_2/sunlight—biological degradation	Static batch reactor	Reactive Red 180 (RR 180) Reactive Black 5 (RB 5) Remazol Red (RR)	*Aeromonas hydrophila* SK16	100% decolorization Easily adaptable in large scale	[74]
Biodegradation-ozonation-adsorption	Packed bed bioreactor	Congo Red	*Providencia stuartii* immobilized on Arjuna seed biochar	Complete decolorization of Congo red Cost effective	[27]

A hybrid system comprising of an anaerobic bio-filter, an anoxic unit and an aerobic MBR to degrade reactive orange dye was employed to achieve efficient COD and nitrogen removal along with mineralization of aromatic amines. Also methane gas was produced from the anaerobic reactor. However, sulfonated aromatic amines were found to be recalcitrant to biodegradation [67]. An integrated system with continuous flow containing aerobic continuous stirred tank reactor and fixed film bioreactor for the treatment of indigo dye was employed by [39]. They attained high COD (96%) and colour (97%) with influent COD concentration of 1180 mg/L. Their results were superior compared to other aerated biological systems.

A hybrid system to treat print and dye effluents by integrating (i) hybrid anaerobic baffled microbial reactor (hydrolysis and acidification) to increase the biodegradation capacity and (ii) a high performance cross flow aerobic sludge reactor was employed by [9]. They tried to optimize the operation conditions by response surface methodology and achieved good COD and colour removals at 14 h hydraulic retention time that meet the discharge standards. [65] worked on a hybrid anaerobic and aerobic sequential batch reactor system to treat synthetic and real textile effluents. Effective removals of COD (99%), nitrogen (99%) and dye (78%) were obtained in the hybrid reactor which is significant compared to independent applications. The final effluent can be reused in agricultural and industrial activities. Also, the anaerobic process produced viable energy source of biogas.

A sequential anaerobic and aerobic system for biodegradation of methylene blue was studied. An anaerobic hybrid unit containing upflow anaerobic sludge blanket reactor and an anaerobic filter was integrated with a submerged aerobic fixed film reactor. In the presence of acetic acid as co-substrate the hybrid system was able to attain 90% decolourization and complete removal of intermediates catechol, quinone, amino pyrine, and 1,4 diamino benzene which were generated in anaerobic reactor thereby achieving complete mineralization [22].

A complete decolourization and mineralization of azo dye Acid Yellow-36 was attained by using a hybrid upflow anaerobic sludge filter bed and aerobic activated sludge reactor with mixed microbial consortia. At anaerobic conditions, in the presence of sodium acetate dye was degraded (98%) and the intermediate aromatic amine 4-aminobenzenesulfonic acid generated was subsequently mineralized in aerobic conditions [3]. Similarly, [68] achieved simultaneous COD removal, biogas production, and decolourization of reactive black 5 in anaerobic aerobic sequential process. [59] achieved a removal of 85% of COD and 90% of colour in a hybrid column upflow anaerobic fixed bed reactor by immobilizing the microbes on an inert support media.

A modified internal circulation anaerobic reactor attached with an external hydraulic circulation unit to treat highly concentrated dye effluents was fabricated [80]. This hybrid system provides high sludge retention and efficient mixing of biomass, substrate and dye effluents. The anaerobic sludge flocs developed could remove 85% of COD without build-up of volatile fatty acids. Biogas generation of 1.7 L/L d gives value addition to this hybrid bioreactor.

Analysis of available literature shows that hybrid biological dye degradation systems are efficient and cost effective. However, the biodegradation of recalcitrant dyes may require long reaction time compared to that of other hybrid treatment systems [70].

5 Novel Hybrid Bioreactors

Few researchers tried to develop novel hybrid bioreactors combining various biological physicochemical and advanced oxidation processes. Damodar et al. [14] used a hybrid membrane photocatalytic reactor with a submerged flat polytetrafluoroethylene membrane unit immersed into a TiO_2 containing photocatalytic reactor to treat reactive dye. Optimum removals of 100% of colour, 93% of total organic carbon, and 85% of COD were attained at 4 h hydraulic retention time. These reactor conditions maintain TiO_2 in suspension and thus avoids the membrane fouling as the catalyst do not get deposited. Also, catalyst can be reused multiple times as catalyst does not get degraded as in the case of individual systems.

Liu et al. [45] used a hybrid electrolysis and anaerobic process using zero valent iron bed. The zero valent iron provides electrons to the reactor and helps in degrading recalcitrant dye components. High voltage in hybrid reactor was shown to aid in the effective removals of COD and colour. Also the integration of electric field and zero valent iron enhanced the microbial diversity in the hybrid system. Zhang et al. [85] used Fe graphite plate electrode which enhanced the colour (84%) and COD (85%) removal besides protecting the iron bed from rusting as in the case of scrap iron usage. Enhanced growth of dominant microbial species was witnessed which were able to degrade the dye even when electrodes were not supplied with electricity.

The performance of hybrid acidogenic bioreactor integrated with biocatalyzed electrolysis unit to decolorize acid red G was evaluated by [79]. The presence of external voltage enhances the anoxic dye degradation and COD removal. Stable removals of COD, dye, and nitrogen were attained when domestic wastewater was used. Results indicate that this hybrid bioreactor can be employed to treat real wastewaters. 13. Cui et al. [13] used domestic wastewaters as electron donors to decolourize azo dye acid orange 7. This system also simultaneously treats domestic wastewater besides eliminating the need of supply of organic components. Later, [12] successfully coupled bio-electro-catalyzed unit with traditional anaerobic reactor to achieve easy scale up and reduce operational costs. This hybrid anaerobic reactor with electrode biofilm (*Geobacter* and *Syntrophus*) units was proven to be effective in treatment of azo dye effluents and achieved 10% increase in dye degradation and 20% increase in COD removal compared to anaerobic reactor alone. A major drawback of using bio-electro-chemical systems for treating recalcitrant dyes is lack of proper electron donors.

With the advancement in nanotechnology, [42] demonstrated the application of a simple one step 3D self-assembled nanoflowers (graphene oxide/carbon nanotubes/copper phosphate) with immobilized laccase for treatment of crystal violet and neutral red dyes. These nanoflowers help in stable and effective immobilization of enzymes and also increases the laccase activity. The bio-electrode performance of immobilized nanoflowers showed efficient removal of organic dyes. A hybrid SBR-Nanofiltration system to decolorize textile wastewater containing reactive blue 2, and sodium dodecyl sulphate as surfactant was used by [40]. Surfactant enhanced the dye and COD removal. Nanofiltration accounted to 96% of colour and 97% of COD removal. Synthesis of novel composites opens up new possibilities in textile dye effluent treatment.

6 Conclusion

The textile industry effluents generally contain redundant and hazardous materials which ultimately lead to inefficient performance of wastewater treatment applications. To combat with the stringent discharge standards and to overcome the disadvantages associated with individual treatment systems, a myriad of hybrid bioreactors were developed in recent past. Based on the available literature it is evident that hybrid bioreactors integrating various physicochemical and AOPs with biodegradation are most potential and viable options. Several advantages like efficient decolourization along with mineralization, reuse of salts and water, less sludge generation, and low cost are offered by hybrid bioreactors. Thus, these hybrid bioreactors have immense role to play in real wastewater applications. Selection of proper combination of technologies based on case specific circumstances is the key to achieve better treatment efficiency. On proper selection, hybrid bioreactors can be considered as an effective end-of-the-pipe treatment approach for dye effluents.

Acknowledgements We gratefully acknowledge Department of Science and Technology (DST), Government of India, for supporting this work through the research grant DST/TM/WTI/WIC/2K17/82/(G).

References

1. Abu Talha M, Goswami M, Giri BS, Sharma A, Rai BN, Singh RS (2018) Bioremediation of Congo red dye in immobilized batch and continuous packed bed bioreactor by Brevibacillus parabrevis using coconut shell bio-char. Biores Technol 252:37–43. https://doi.org/10.1016/j.biortech.2017.12.081
2. Acikgoz C, Gül ÜD, Özan K, Borazan AA (2016) Degradation of Reactive Blue by the mixed culture of Aspergillus versicolor and Rhizopus arrhizus in membrane bioreactor (MBR) system. Desalin Water Treat 57:3750–3756. https://doi.org/10.1080/19443994.2014.987173
3. Ahmad R, Mondal PK, Usmani SQ (2010) Hybrid UASFB-aerobic bioreactor for biodegradation of acid yellow-36 in wastewater. Biores Technol 101:3787–3790. https://doi.org/10.1016/j.biortech.2009.12.116

4. Ayed L, Ksibi IE, Charef A, Mzoughi RE (2020) Hybrid coagulation-flocculation and anaerobic-aerobic biological treatment for industrial textile wastewater: pilot case study. J Text Inst 1–7. https://doi.org/10.1080/00405000.2020.1731273
5. Bedekar PA, Bhalkar BN, Patil SM, Govindwar SP (2016) Moringa oleifera-mediated coagulation of textile wastewater and its biodegradation using novel consortium-BBA grown on agricultural waste substratum. Environ Sci Pollut Res 23:20963–20976. https://doi.org/10.1007/s11356-016-7279-8
6. Bedekar PA, Kshirsagar SD, Gholave AR, Govindwar SP (2015) Degradation and detoxification of methylene blue dye adsorbed on water hyacinth in semi continuous anaerobic–aerobic bioreactors by novel microbial consortium-SB. RSC Adv 5:99228–99239
7. Bharti V, Vikrant K, Goswami M, Tiwari H, Sonwani RK, Lee J, Tsang DCW, Kim K-H, Saeed M, Kumar S, Rai BN, Giri BS, Singh RS (2019) Biodegradation of methylene blue dye in a batch and continuous mode using biochar as packing media. Environ Res 171:356–364. https://doi.org/10.1016/j.envres.2019.01.051
8. Castro E, Avellaneda A, Marco P (2014) Combination of advanced oxidation processes and biological treatment for the removal of benzidine-derived dyes. Environ Prog Sustain Energy 33:873–885. https://doi.org/10.1002/ep.11865
9. Chen Z-B, Cui M-H, Ren N-Q, Chen Z-Q, Wang H-C, Nie S-K (2011) Improving the simultaneous removal efficiency of COD and color in a combined HABMR–CFASR system based MPDW. Part 1: optimization of operational parameters for HABMR by using response surface methodology. Biores Technol 102:8839–8847. https://doi.org/10.1016/j.biortech.2011.06.089
10. Costa AFS, Albuquerque CDC, Salgueiro AA, Sarubbo LA (2018) Color removal from industrial dyeing and laundry effluent by microbial consortium and coagulant agents. Process Saf Environ Prot 118:203–210. https://doi.org/10.1016/j.psep.2018.03.001
11. Crini G (2006) Non-conventional low-cost adsorbents for dye removal: A review. Biores Technol 97:1061–1085. https://doi.org/10.1016/j.biortech.2005.05.001
12. Cui D, Cui M-H, Lee H-S, Liang B, Wang H-C, Cai W-W, Cheng H-Y, Zhuang X-L, Wang A-J (2017) Comprehensive study on hybrid anaerobic reactor built-in with sleeve type bioelectrocatalyzed modules. Chem Eng J 330:1306–1315. https://doi.org/10.1016/j.cej.2017.07.167
13. Cui M-H, Cui D, Gao L, Wang A-J, Cheng H-Y (2016) Azo dye decolorization in an up-flow bioelectrochemical reactor with domestic wastewater as a cost-effective yet highly efficient electron donor source. Water Res 105:520–526. https://doi.org/10.1016/j.watres.2016.09.027
14. Damodar RA, You S-J, Ou S-H (2010) Coupling of membrane separation with photocatalytic slurry reactor for advanced dye wastewater treatment. Sep Purif Technol 76:64–71. https://doi.org/10.1016/j.seppur.2010.09.021
15. Dasgupta J, Sikder J, Chakraborty S, Curcio S, Drioli E (2015) Remediation of textile effluents by membrane based treatment techniques: A state of the art review. J Environ Manage 147:55–72. https://doi.org/10.1016/j.jenvman.2014.08.008
16. de Souza, S.M.d.A.G.U., Bonilla, K.A.S., de Souza, A.A.U. (2010) Removal of COD and color from hydrolyzed textile azo dye by combined ozonation and biological treatment. J Hazard Mater 179:35–42. https://doi.org/10.1016/j.jhazmat.2010.02.053
17. Deveci EÜ, Dizge N, Yatmaz HC, Aytepe Y (2016) Integrated process of fungal membrane bioreactor and photocatalytic membrane reactor for the treatment of industrial textile wastewater. Biochem Eng J 105:420–427. https://doi.org/10.1016/j.bej.2015.10.016
18. El-Gohary F, Tawfik A (2009) Decolorization and COD reduction of disperse and reactive dyes wastewater using chemical-coagulation followed by sequential batch reactor (SBR) process. Desalination 249:1159–1164. https://doi.org/10.1016/j.desal.2009.05.010
19. Enayatzamir K, Alikhani HA, Rodríguez Couto S (2009) Simultaneous production of laccase and decolouration of the diazo dye Reactive Black 5 in a fixed-bed bioreactor. J Hazard Mater 164:296–300. https://doi.org/10.1016/j.jhazmat.2008.08.032
20. Fahmi MR, Abidin CZA, Rahmat NR (2011) Characteristic of colour and COD removal of azo dye by advanced oxidation process and biological treatment. In: International conference on biotechnology and environment management, pp 13–18

21. Fan J, Chen D, Li N, Xu Q, Li H, He J, Lu J (2018) Adsorption and biodegradation of dye in wastewater with Fe3O4@MIL-100 (Fe) core–shell bio-nanocomposites. Chemosphere 191:315–323. https://doi.org/10.1016/j.chemosphere.2017.10.042
22. Farooqi IH, Basheer F, Tiwari P (2017) Biodegradation of methylene blue dye by sequential treatment using anaerobic hybrid reactor and submerged aerobic fixed film bioreactor. J Inst Eng (India): Ser A 98:397–403. https://doi.org/10.1007/s40030-017-0251-x
23. Fetyan NA, Azeiz AA, Ismail I, Salem T (2017) Biodegradation of Cibacron Redazo dye and industrial textile effluent by pseudomonas Aeruginosa immobilized on chitosan-Fe_2O_3 composite. J Adv Biol Biotechnol 1–15. https://doi.org/10.9734/JABB/2017/31332
24. Geed SR, Samal K, Tagade A (2019) Development of adsorption-biodegradation hybrid process for removal of methylene blue from wastewater. J Environ Chem Eng 7. https://doi.org/10.1016/j.jece.2019.103439
25. Ghosh A, Dastidar MG, Sreekrishnan TR (2017) Bioremediation of chromium complex dyes and treatment of sludge generated during the process. Int Biodeterior Biodegradation 119:448–460. https://doi.org/10.1016/j.ibiod.2016.08.013
26. Giri BS, Gun S, Pandey S, Trivedi A, Kapoor RT, Singh RP, Abdeldayem OM, Rene ER, Yadav S, Chaturvedi P, Sharma N, Singh RS (2020) Reusability of brilliant green dye contaminated wastewater using corncob biochar and Brevibacillus parabrevis: hybrid treatment and kinetic studies. Bioengineered 11:743–758. https://doi.org/10.1080/21655979.2020.1788353
27. Goswami M, Chaturvedi P, Kumar Sonwani R, Dutta Gupta A, Rani Singhania R, Shekher Giri B, Nath Rai B, Singh H, Yadav S, Sharan Singh R (2020) Application of Arjuna (Terminalia arjuna) seed biochar in hybrid treatment system for the bioremediation of Congo red dye. Biores Technol 307. https://doi.org/10.1016/j.biortech.2020.123203
28. Hai FI, Yamamoto K, Fukushi K (2007) Hybrid treatment systems for dye wastewater. Crit Rev Environ Sci Technol 37:315–377. https://doi.org/10.1080/10643380601174723
29. Hai FI, Yamamoto K, Nakajima F, Fukushi K (2012) Application of a GAC-coated hollow fiber module to couple enzymatic degradation of dye on membrane to whole cell biodegradation within a membrane bioreactor. J Membr Sci 389:67–75. https://doi.org/10.1016/j.memsci.2011.10.016
30. Hai FI, Yamamoto K, Nakajima F, Fukushi K (2011) Bioaugmented membrane bioreactor (MBR) with a GAC-packed zone for high rate textile wastewater treatment. Water Res 45:2199–2206. https://doi.org/10.1016/j.watres.2011.01.013
31. Hai FI, Yamamoto K, Nakajima F, Fukushi K (2008) Removal of structurally different dyes in submerged membrane fungi reactor—Biosorption/PAC-adsorption, membrane retention and biodegradation. J Membr Sci 325:395–403. https://doi.org/10.1016/j.memsci.2008.08.006
32. Hai FI, Yamamoto K, Nakajima F, Fukushi K, Nghiem LD, Price WE, Jin B (2013) Degradation of azo dye acid orange 7 in a membrane bioreactor by pellets and attached growth of Coriolus versicolour. Biores Technol 141:29–34. https://doi.org/10.1016/j.biortech.2013.02.020
33. Hosseini Koupaie E, Alavi Moghaddam M, Hashemi S (2013) Successful treatment of high azo dye concentration wastewater using combined anaerobic/aerobic granular activated carbon-sequencing batch biofilm reactor (GAC-SBBR): simultaneous adsorption and biodegradation processes. Water Sci Technol 67:1816–1821. https://doi.org/10.2166/wst.2013.061
34. Ip AW, Barford JP, McKay G (2010) Biodegradation of Reactive Black 5 and bioregeneration in upflow fixed bed bioreactors packed with different adsorbents. J Chem Technol Biotechnol 85:658–667. https://doi.org/10.1002/jctb.2349
35. Jafari N, Kasra-Kermanshahi R, Soudi MR, Mahvi AH, Gharavi S (2012) Degradation of a textile reactive azo dye by a combined biological-photocatalytic process: Candida tropicalis Jks2-Tio 2/Uv. Iran J Environ Health Sci Eng 9:33. https://doi.org/10.1186/1735-2746-9-33
36. Jegatheesan V, Pramanik BK, Chen J, Navaratna D, Chang C-Y, Shu L (2016) Treatment of textile wastewater with membrane bioreactor: A critical review. Biores Technol 204:202–212. https://doi.org/10.1016/j.biortech.2016.01.006
37. Judd S (2008) The status of membrane bioreactor technology. Trends Biotechnol 26:109–116. https://doi.org/10.1016/j.tibtech.2007.11.005

38. Kadam AA, Lade HS, Lee DS, Govindwar SP (2015) Zinc chloride as a coagulant for textile dyes and treatment of generated dye sludge under the solid state fermentation: Hybrid treatment strategy. Biores Technol 176:38–46. https://doi.org/10.1016/j.biortech.2014.10.137
39. Khelifi E, Gannoun H, Touhami Y, Bouallagui H, Hamdi M (2008) Aerobic decolourization of the indigo dye-containing textile wastewater using continuous combined bioreactors. J Hazard Mater 152:683–689. https://doi.org/10.1016/j.jhazmat.2007.07.059
40. Khosravi A, Karimi M, Ebrahimi H, Fallah N (2020) Sequencing batch reactor/nanofiltration hybrid method for water recovery from textile wastewater contained phthalocyanine dye and anionic surfactant. J Environ Chem Eng 8. https://doi.org/10.1016/j.jece.2020.103701
41. Kim T-H, Lee Y, Yang J, Lee B, Park C, Kim S (2004) Decolorization of dye solutions by a membrane bioreactor (MBR) using white-rot fungi. Desalination 168:287–293. https://doi.org/10.1016/j.desal.2004.07.011
42. Li H, Hou J, Duan L, Ji C, Zhang Y, Chen V (2017) Graphene oxide-enzyme hybrid nanoflowers for efficient water soluble dye removal. J Hazard Mater 338:93–101. https://doi.org/10.1016/j.jhazmat.2017.05.014
43. Li H, Tan L, Ning S, He M (2015) Reactor performance and microbial community dynamics during aerobic degradation and detoxification of Acid Red B with activated sludge bioaugmented by a yeast Candida tropicalis TL-F1 in MBR. Int Biodeterior Biodegradation 104:149–156. https://doi.org/10.1016/j.ibiod.2015.06.006
44. Libra JA, Sosath F (2003) Combination of biological and chemical processes for the treatment of textile wastewater containing reactive dyes. J Chem Technol Biotechnol: Int Res Process, Environ Clean Technol 78:1149–1156. https://doi.org/10.1002/jctb.904
45. Liu Y, Zhang Y, Quan X, Zhang J, Zhao H, Chen S (2011) Effects of an electric field and zero valent iron on anaerobic treatment of azo dye wastewater and microbial community structures. Biores Technol 102:2578–2584. https://doi.org/10.1016/j.biortech.2010.11.109
46. Lucas MS, Dias AA, Sampaio A, Amaral C, Peres JA (2007) Degradation of a textile reactive Azo dye by a combined chemical–biological process: Fenton's reagent-yeast. Water Res 41:1103–1109. https://doi.org/10.1016/j.watres.2006.12.013
47. Mezohegyi G, Kolodkin A, Castro UI, Bengoa C, Stuber F, Font J, Fabregat A, Fortuny A (2007) Effective anaerobic decolorization of azo dye Acid Orange 7 in continuous upflow packed-bed reactor using biological activated carbon system. Ind Eng Chem Res 46:6788–6792
48. Mezohegyi G, van der Zee FP, Font J, Fortuny A, Fabregat A (2012) Towards advanced aqueous dye removal processes: A short review on the versatile role of activated carbon. J Environ Manage 102:148–164. https://doi.org/10.1016/j.jenvman.2012.02.021
49. Moosvi S, Madamwar D (2007) An integrated process for the treatment of CETP wastewater using coagulation, anaerobic and aerobic process. Biores Technol 98:3384–3392. https://doi.org/10.1016/j.biortech.2006.10.043
50. Oller I, Malato S, Sánchez-Pérez JA (2011) Combination of advanced oxidation processes and biological treatments for wastewater decontamination—a review. Sci Total Environ 409:4141–4166. https://doi.org/10.1016/j.scitotenv.2010.08.061
51. Ong S, Toorisaka E, Hirata M, Hano T (2008) Combination of adsorption and biodegradation processes for textile effluent treatment using a granular activated carbon-bioflm confgured packed column system. J Environ Sci 20:952–956. https://doi.org/10.1016/S1001-0742(08)62192-0
52. Parsons S (2004) Advanced oxidation processes for water and wastewater treatment. IWA Publishing
53. Paździor K, Wrębiak J, Klepacz-Smółka A, Gmurek M, Bilińska L, Kos L, Sójka-Ledakowicz J, Ledakowicz S (2017) Influence of ozonation and biodegradation on toxicity of industrial textile wastewater. J Environ Manage 195:166–173. https://doi.org/10.1016/j.jenvman.2016.06.055
54. Qin L, Zhang G, Meng Q, Xu L, Lv B (2012) Enhanced MBR by internal micro-electrolysis for degradation of anthraquinone dye wastewater. Chem Eng J 210:575–584. https://doi.org/10.1016/j.cej.2012.09.006

55. Renault F, Sancey B, Badot PM, Crini G (2009) Chitosan for coagulation/flocculation processes—an eco-friendly approach. Eur Polymer J 45:1337–1348. https://doi.org/10.1016/j.eurpolymj.2008.12.027
56. Rodrigues CSD, Madeira LM, Boaventura RAR (2014) Synthetic textile dyeing wastewater treatment by integration of advanced oxidation and biological processes—performance analysis with costs reduction. J Environ Chem Eng 2:1027–1039. https://doi.org/10.1016/j.jece.2014.03.019
57. Roy U, Das P, Bhowal A (2019) Treatment of azo dye (congo red) solution in fluidized bed bioreactor with simultaneous approach of adsorption coupled with biodegradation: optimization by response surface methodology and toxicity assay. Clean Technol Environ Policy 21:1675–1686. https://doi.org/10.1007/s10098-019-01736-7
58. Salter-Blanc AJ, Bylaska EJ, Lyon MA, Ness SC, Tratnyek PG (2016) Structure–activity relationships for rates of aromatic amine oxidation by manganese dioxide. Environ Sci Technol 50:5094–5102. https://doi.org/10.1021/acs.est.6b00924
59. Sandhya S, Swaminathan K (2006) Kinetic analysis of treatment of textile wastewater in hybrid column upflow anaerobic fixed bed reactor. Chem Eng J 122:87–92. https://doi.org/10.1016/j.cej.2006.04.006
60. Särkkä H, Bhatnagar A, Sillanpää M (2015) Recent developments of electro-oxidation in water treatment—a review. J Electroanal Chem 754:46–56. https://doi.org/10.1016/j.jelechem.2015.06.016
61. Sathya U, Keerthi N, M., Balasubramanian, N. (2019) Evaluation of advanced oxidation processes (AOPs) integrated membrane bioreactor (MBR) for the real textile wastewater treatment. J Environ Manage 246:768–775. https://doi.org/10.1016/j.jenvman.2019.06.039
62. Senthilkumar S, Basha CA, Perumalsamy M, Prabhu HJ (2012) Electrochemical oxidation and aerobic biodegradation with isolated bacterial strains for dye wastewater: combined and integrated approach. Electrochim Acta 77:171–178. https://doi.org/10.1016/j.electacta.2012.05.084
63. Shabbir S, Faheem M, Wu Y (2018) Decolorization of high concentration crystal violet by periphyton bioreactors and potential of effluent reuse for agricultural purposes. J Clean Prod 170:425–436. https://doi.org/10.1016/j.jclepro.2017.09.085
64. Shankar YS, Ankur K, Bhushan P, Mohan D (2019) Utilization of water treatment plant (WTP) sludge for pretreatment of dye wastewater using coagulation/flocculation. In: Advances in waste management. Springer, pp 107–121.
65. Shoukat R, Khan SJ, Jamal Y (2019) Hybrid anaerobic-aerobic biological treatment for real textile wastewater. J Water Process Eng 29. https://doi.org/10.1016/j.jwpe.2019.100804
66. Soares PA, Silva TF, Manenti DR, Souza SM, Boaventura RA, Vilar VJ (2014) Insights into real cotton-textile dyeing wastewater treatment using solar advanced oxidation processes. Environ Sci Pollut Res 21:932–945. https://doi.org/10.1007/s11356-013-1934-0
67. Spagni A, Grilli S, Casu S, Mattioli D (2010) Treatment of a simulated textile wastewater containing the azo-dye reactive orange 16 in an anaerobic-biofilm anoxic–aerobic membrane bioreactor. Int Biodeterior Biodegradation 64:676–681. https://doi.org/10.1016/j.ibiod.2010.08.004
68. Sponza DT, Işik M (2002) Decolorization and azo dye degradation by anaerobic/aerobic sequential process. Enzyme Microb Technol 31:102–110. https://doi.org/10.1016/S0141-0229(02)00081-9
69. Srinivasan A, Viraraghavan T (2010) Decolorization of dye wastewaters by biosorbents: a review. J Environ Manage 91:1915–1929. https://doi.org/10.1016/j.jenvman.2010.05.003
70. Su CX-H, Low LW, Teng TT, Wong YS (2016) Combination and hybridisation of treatments in dye wastewater treatment: a review. J Environ Chem Eng 4:3618–3631. https://doi.org/10.1016/j.jece.2016.07.026
71. Swathi D, P.C., S., Trivedi, A. (2020) Simultaneous decolorization and mineralization of high concentrations of methyl orange in an anoxic up-flow packed bed reactor in denitrifying conditions. J Water Process Eng 101813. https://doi.org/10.1016/j.jwpe.2020.101813

72. Tantak NP, Chaudhari S (2006) Degradation of azo dyes by sequential Fenton's oxidation and aerobic biological treatment. J Hazard Mater 136:698–705. https://doi.org/10.1016/j.jhazmat.2005.12.049
73. Thanavel M, Bankole PO, Selvam R, Govindwar SP, Sadasivam SK (2020) Synergistic effect of biological and advanced oxidation process treatment in the biodegradation of Remazol yellow RR dye. Sci Rep 10:1–9. https://doi.org/10.1038/s41598-020-77376-5
74. Thanavel M, Kadam SK, Biradar SP, Govindwar SP, Jeon B-H, Sadasivam SK (2019) Combined biological and advanced oxidation process for decolorization of textile dyes. SN Appl Sci 1:97. https://doi.org/10.1007/s42452-018-0111-y
75. van Leeuwen J, Sridhar A, Esplugas M, Onuki S, Cai L, Koziel JA (2009) Ozonation within an activated sludge system for azo dye removal by partial oxidation and biodegradation. Ozone: Sci Eng 31:279–286. https://doi.org/10.1080/01919510902907720
76. Verma AK, Dash RR, Bhunia P (2012) A review on chemical coagulation/flocculation technologies for removal of colour from textile wastewaters. J Environ Manage 93:154–168. https://doi.org/10.1016/j.jenvman.2011.09.012
77. Vikrant K, Giri BS, Raza N, Roy K, Kim K-H, Rai BN, Singh RS (2018) Recent advancements in bioremediation of dye: Current status and challenges. Biores Technol 253:355–367. https://doi.org/10.1016/j.biortech.2018.01.029
78. Vikrant K, Kumar V, Kim K-H, Kukkar D (2017) Metal–organic frameworks (MOFs): potential and challenges for capture and abatement of ammonia. J Mater Chem A 5:22877–22896. https://doi.org/10.1039/C7TA07847A
79. Wang H-C, Cheng H-Y, Wang S-S, Cui D, Han J-L, Hu Y-P, Su S-G, Wang A-J (2016) Efficient treatment of azo dye containing wastewater in a hybrid acidogenic bioreactor stimulated by biocatalyzed electrolysis. J Environ Sci 39:198–207. https://doi.org/10.1016/j.jes.2015.10.014
80. Wang J, Yan J, Xu W (2015) Treatment of dyeing wastewater by MIC anaerobic reactor. Biochem Eng J 101:179–184. https://doi.org/10.1016/j.bej.2015.06.001
81. Wang X, Li J, Li X, Du G (2011) Influence of aeration intensity on the performance of A/O-type sequencing batch MBR system treating azo dye wastewater. Front Environ Sci Eng China 5:615–622. https://doi.org/10.1007/s11783-011-0362-y
82. Xu H, Yang B, Liu Y, Li F, Shen C, Ma C, Tian Q, Song X, Sand W (2018) Recent advances in anaerobic biological processes for textile printing and dyeing wastewater treatment: a mini-review. World J Microbiol Biotechnol 34:165. https://doi.org/10.1007/s11274-018-2548-y
83. You S-J, Damodar RA, Hou S-C (2010) Degradation of Reactive Black 5 dye using anaerobic/aerobic membrane bioreactor (MBR) and photochemical membrane reactor. J Hazard Mater 177:1112–1118. https://doi.org/10.1016/j.jhazmat.2010.01.036
84. You S-J, Teng J-Y (2009) Performance and dye-degrading bacteria isolation of a hybrid membrane process. J Hazard Mater 172:172–179. https://doi.org/10.1016/j.jhazmat.2009.06.149
85. Zhang J, Zhang Y, Quan X, Li Y, Chen S, Zhao H, Wang D (2012) An anaerobic reactor packed with a pair of Fe-graphite plate electrodes for bioaugmentation of azo dye wastewater treatment. Biochem Eng J 63:31–37. https://doi.org/10.1016/j.bej.2012.01.008
86. Zheng Y, Chen D, Li N, Xu Q, Li H, He J, Lu J (2017) Highly efficient simultaneous adsorption and biodegradation of a highly-concentrated anionic dye by a high-surface-area carbon-based biocomposite. Chemosphere 179:139–147. https://doi.org/10.1016/j.chemosphere.2017.03.096
87. Zou H, Ma W, Wang Y (2015) A novel process of dye wastewater treatment by linking advanced chemical oxidation with biological oxidation. Arch Environ Prot 41:33–39

Printed in the United States
by Baker & Taylor Publisher Services